AF251336

Springer Series in
MATERIALS SCIENCE   49

Springer

Berlin
Heidelberg
New York
Barcelona
Hong Kong
London
Milan
Paris
Tokyo

Physics and Astronomy   ONLINE LIBRARY

http://www.springer.de/phys/

Springer Series in
MATERIALS SCIENCE

Editors: R. Hull   R. M. Osgood, Jr.   H. Sakaki   A. Zunger

The Springer Series in Materials Science covers the complete spectrum of materials physics, including fundamental principles, physical properties, materials theory and design. Recognizing the increasing importance of materials science in future device technologies, the book titles in this series reflect the state-of-the-art in understanding and controlling the structure and properties of all important classes of materials.

29 **Elements of Rapid Solidification**
Fundamentals and Applications
Editor: M. A. Otooni

30 **Process Technology
for Semiconductor Lasers**
Crystal Growth and Microprocesses
By K. Iga and S. Kinoshita

31 **Nanostructures and Quantum Effects**
By H. Sakaki and H. Noge

32 **Nitride Semiconductors and Devices**
By H. Morkoç

33 **Supercarbon**
Synthesis, Properties and Applications
Editors: S. Yoshimura and R. P. H. Chang

34 **Computational Materials Design**
Editor: T. Saito

35 **Macromolecular Science
and Engineering**
New Aspects
Editor: Y. Tanabe

36 **Ceramics**
Mechanical Properties, Failure
Behaviour, Materials Selection
By D. Munz and T. Fett

37 **Technology and Applications
of Amorphous Silicon**
Editor: R. A. Street

38 **Fullerene Polymers
and Fullerene Polymer Composites**
Editors: P. C. Eklund and A. M. Rao

39 **Semiconducting Silicides**
Editor: V.E. Borisenko

40 **Reference Materials
in Analytical Chemistry**
A Guide for Selection and Use
Editor: A. Zschunke

41 **Organic Electronic Materials**
Conjugated Polymers and Low
Molecular Weight Organic Solids
Editors: R. Farchioni and G. Grosso

42 **Raman Scattering in Materials Science**
Editors: W. H. Weber and R. Merlin

43 **The Atomistic Nature of Crystal Growth**
By B. Mutaftschiev

44 **Thermodynamic Basis of Crystal Growth**
P–T–X Phase Equilibrium
and Nonstoichiometry
By J. Greenberg

45 **Thermoelectrics**
Basic Principles
and New Materials Developments
By G.S. Nolas, J. Sharp, and H. J. Goldsmid

46 **Fundamental Aspects
of Silicon Oxidation**
Editor: Y. J. Chabal

47 **Disorder and Order in Strongly
Nonstoichiometric Compounds**
Transition Metal Carbides, Nitrides
and Oxides
By A.I. Gusev, A.A. Rempel,
and A.J. Magerl

48 **The Glass Transition**
Relaxation Dynamics
in Liquids and Disordered Materials
By E. Donth

49 **Alkali Halides**
A Handbook of Physical Properties
By D. B. Sirdeshmukh, L. Sirdeshmukh,
and K. G. Subhadra

Series homepage – http://www.springer.de/phys/books/ssms/

Volumes 1–28 are listed at the end of the book.

D.B. Sirdeshmukh · L. Sirdeshmukh
K.G. Subhadra

Alkali Halides

A Handbook of Physical Properties

With 60 Figures and 202 Tables

Springer

Professor em. D.B. Sirdeshmukh
Professor L. Sirdeshmukh
Professor K.G. Subhadra
Physics Department, Kakatiya University, Warangal 506 009, India
e-mail: kakatiya@ap.nic.in

Series Editors:

Prof. Alex Zunger
NREL
National Renewable Energy Laboratory
1617 Cole Boulevard
Golden Colorado 80401-3393, USA

Prof. R. M. Osgood, Jr.
Microelectronics Science Laboratory
Department of Electrical Engineering
Columbia University
Seeley W. Mudd Building
New York, NY 10027, USA

Prof. Robert Hull
University of Virginia
Dept. of Materials Science and Engineering
Thornton Hall
Charlottesville, VA 22903-2442, USA

Prof. H. Sakaki
Institute of Industrial Science
University of Tokyo
7-22-1 Roppongi, Minato-ku
Tokyo 106, Japan

ISSN 0933-033x

ISBN 3-540-42180-7 Springer-Verlag Berlin Heidelberg New York

Library of Congress Cataloging-in-Publication Data applied for

Die Deutsche Bibliothek - CHIP-Einheitsaufnahme

Sirdeshmukh, Dinker B.:
Alkali halides : a handbook of physical properties / D. B. Sirdeshmukh ; L. Sirdeshmukh ; K. G. Subhadra. -
Berlin ; Heidelberg ; New York ; Barcelona ; Hong Kong ; London ; Milan ; Paris ; Tokyo ; Springer, 2001
(Springer series in materials science ; 49)
(Physics and astronomy online library)
ISBN 3-540-42180-7

Springer-Verlag Berlin Heidelberg New York
a member of BertelsmannSpringer Science+Business Media GmbH

http://www.springer.de

© Springer-Verlag Berlin Heidelberg 2001
Printed in Germany

Typesetting: Camera-ready copy from the authors
Cover concept: eStudio Calamar Steinen
Cover production: *design & production* GmbH, Heidelberg

Printed on acid-free paper   SPIN: 10785864   57/3020/ma   5 4 3 2 1 0

Preface

The alkali halides constitute a family of versatile crystals. Their simple crystal structure and bonding scheme have made them favourite candidates for testing theories. Their stability and availability in the form of single crystals of meaningful size have made them the subject of numerous experimental investigations. For quite sometime, the alkali halides were looked upon as "model crystals". In recent decades, they have also proved useful in several applications ranging from infrared optical windows to continuous wave tunable lasers. Because of this dual importance – purely scientific as well as technological – a vast amount of information has been generated with regard to all aspects of the alkali halides. This information, expectedly, remains scattered over a large number of journals, books and reference sources. It is therefore considered worthwhile to compile the experimental data on the physical properties of the alkali halides.

This handbook brings together a wide range of information on the experimentally determined physical properties of the alkali halides. In some cases, particularly where experimentally determined values are not available, theoretically calculated values are included. Almost all the important literature between 1935 and 2000 has been scanned. The project was undertaken with the goal of making available all the essential information on the physical properties of alkali halides in one place for researchers in the fields of Solid State Physics and Materials Science.

In the preparation of a compilation of this kind, we necessarily had to depend on co-operation and assistance from many sources. While we express our sincere thanks to those mentioned below, we owe equal gratitude to several others not mentioned here.

We are grateful to Prof. D.E. Schuele (Case Western Reserve University), Prof. S. Haussühl (Köln University), Prof. W.B. Holzapfel and Prof. P.G. Johannsen (both from the Paderborn University, Germany) and Prof. Y.V.G.S. Murti (Indian Institute of Technology, Madras) for encouragement and useful suggestions. Some unpublished results were provided by Dr. Johannsen. We sincerely thank Prof. H.L. Bhat (Indian Institute of Science, Bangalore), Dr. Deepak Sirdeshmukh (Case Western Reserve University, Ohio), Dr. B.J. Rao (Tata Institute of fundamental Research, Bombay), Mrs. Prabhavati Rao (Geological Survey of India, Nagpur), Dr. C. Vijayan (Indian Institute of Technology, Madras), Dr. R. Ramakrishna Reddy (Sri Krishnadevaraya University, Anantapur) and Dr. R. Asokamani (Anna University, Madras) for their willing response whenever we were in need of rare papers. Dr. Ch.V. Purushotham Reddy (Chaitanya Post-graduate College) is thanked for providing computer facilities. Prof. K.G. Prasad (Tata Institute of Fundamental Research, Bombay) and Nandu Sirdeshmukh (C.M.C., Hyderabad) are acknowledged for their assistance in the final stages of the book. Special thanks are due to colleagues in the Physics Department and the Library of the Kakatiya University for their co-operation in various ways. Last, but not least, the authors are grateful to their families for standing by them with patience and for active support.

This work overlapped with the senior author's (DBS) tenure as Professor Emeritus. He thanks the University Grants Commission, New Delhi, for financial support.

Finally, we offer our warm thanks to Dr. Claus Ascheron of Springer-Verlag for constant co-operation and sound advice on all matters related to the publication of this book. We would also like to thank colleagues at the Heidelberg and Berlin offices of Springer-Verlag for their prompt response and for patiently guiding us in preparing the final version of this book.

Warangal, India
Spring 2001

Dinker B. Sirdeshmukh
Lalitha Sirdeshmukh
K.G. Subhadra

Acknowledgements

The following publishers are thanked for permission to reproduce illustrations and material from their publications:

1. Akademie Verlag
2. American Physical Society
3. CRC Press LLC
4. Institute of Physics (UK)
5. International Union of Crystallography
6. John Wiley and Sons Ltd.
7. The Physical Society of Japan
8. The Royal Society (UK)
9. Springer-Verlag

Introduction

Chemical Crystallography

The alkali atoms of the first group and the halogens of the VIIth group combine to form compounds known as the alkali halides. Although the existence of trihalides and pentahalides and some hydrated compounds is also reported, the generic term alkali halide represents the binary anhydrous compounds of the alkalis and halogens. Francium is also an alkali and astatine is a halogen, but their compounds are either rare or unstable and therefore they are not considered here.

The electronic configurations of the alkalis and the halogens favour electron transfer and formation of ions. Thus, the neutral lithium and fluorine atoms have the electronic structure $1s^2$, $2s$ and $1s^2$, $2s^2\,2p^5$ respectively. On the other hand, on charge transfer the electronic structures of the lithium and fluorine ions become $1s^2$ and $1s^2$, $2s^2\,2p^6$ respectively. These correspond to the electronic structures of inert gas atoms He and Ne respectively. This is a simplistic description of the alkali halides as ionic crystals.

The alkali halides crystallise in two structures viz., the NaCl structure and the CsCl structure. In ambient conditions CsCl, CsBr and CsI assume the so-called CsCl structure. The rest crystallise with the NaCl structure. CsCl transforms to the NaCl structure at high temperatures. NaCl, K halides and the Rb halides transform to the CsCl phase at high pressures.

Crystal Growth

With the exception of LiF and NaF, all the alkali halides are soluble in water and can, in principle, be crystallised from solution. Small-sized single crystals can be grown from aqueous solution with or without the addition of impurities. All the alkali halides have congruent melting points and therefore they can be grown from melt. Growth of single crystals has been reported by using a variety of melt growth techniques. Crystals of inch size are commonly grown, but growth of large-sized crystals has also been achieved. In view of their use in many experiments the alkali halide crystals have not only been grown by individual researchers but are also grown and supplied commercially by well-known firms like Harshaw and Optovac in the US and by Korth in Germany. The alkali halides with NaCl structure have a beautiful cleavage along the (100) plane. The CsCl structure alkali halides do not exhibit cleavage.

Alkali Halides as Model Crystals

The alkali halides have a wide range of values for any given property. Thus, for example, the lattice constants vary from 4 Å for LiF to 6 Å for RbI. Again, the melting points vary from 600 °C for Rb halides to 1000 °C for NaF. Thus, they offer themselves as a convenient system for scaling of properties, or for studying empirical trends. The lattice constant, in particular, has been used to scale a large number of properties like the elastic constants, colour centre parameters, Debye temperatures, hardness etc.

The simple structures and knowledge of the chemical bonding in the alkali halides have made them the favourite systems for verifying theories. Thus, the alkali halides were the first system (or one of the first systems) on which Debye's theory of specific heats, Gruneisen's theory of thermal expansion, Born's theory of cohesion, Kellermann's lattice dynamics and Lowdin's very first application of quantum mechanics to crystal elasticity were tested.

It is not as if the alkali halides were used only for testing theories. They were also the materials of first choice in many pioneering experiments. KCl was one of the crystals used by Laue in his famous experiments on X-ray diffraction and NaCl was one of the first crystals which Bragg chose for structure analysis. The potassium halides featured in the innovative experiments of Fizeau on the thermal expansion of crystals in the later half of the nineteenth century. The alkali halides were one of the first systems for measurement of compressibility by Slater in 1924 and by Bridgman in 1936. When Debye and Scherrer developed the method of X-ray powder diffraction LiF was the first material to be studied.

Thus, whether it was testing of theories or trying new experiments, the alkali halides have always played a leading role as model crystals. In 1946 Seitz remarked "In the field of solids, the properties of alkali halides have an enduring interest, since these crystals have continuously yielded to persistent investigation and have gradually provided us with a better and better understanding of the most interesting properties of all solids". This statement remains relevant even after 50 years.

Alkali Halides as Device Materials

In the last few decades the alkali halides emerged as crystals with useful applications. One of the first properties of the alkali halides to be employed for practical application was optical transmission. The alkali halide crystals have useful transmission properties in different ranges of the spectrum from UV to IR. While LiF, NaF and NaCl are useful in UV transmission, KBr, KI, CsBr and CsI are useful in infrared transmission. Several alkali halides, either pure or doped, are employed in energy detection in the X-ray, γ-ray and Cerenkov regions. Alkali halides like LiF and NaCl have been employed as monochromators for X-rays, whereas KCl-KBr mixed crystals have been found useful as neutron monochromators.

The development of lasers revived the interest in alkali halides as materials for optical components. This led to development of alkali halide polycrystalline material for use as optical windows.

The study of thermoluminescence resulted in the use of some of the doped alkali halides as thermoluminescence dosimeter device materials. Tl-doped LiF is the device material in commercial TL dosimeters. Some alkali halides CsI(Na), CsI(Tl) and RbBr(Tl) have proved efficient as X-ray imaging sensors.

The development of the field of colour centres led to the application of alkali halides in two new important fields. The first is information storage. Doped alkali halides have been used for information storage in the visible as well as X-ray regions. The other field is the development of colour-centre-based alkali halide lasers. These alkali halide lasers are tunable and when used in a single wavelength mode they have a laser line of very narrow width. A number of alkali halide-impurity-colour centre combinations have been tried and some of the alkali halide lasers are commercially available. Another application of alkali halides is as pressure markers. The compression of sodium chloride has been studied in great detail both experimentally and theoretically and NaCl is recommended as a pressure marker.

Harmonic generation and superconductivity have also been observed in the alkali halides.

Concept of the Book

Because of the importance of alkali halides as model crystals and their potential as device materials, a very large body of data on physical properties of alkali halides has been generated over the decades. There is considerable literature on all basic properties like elastic constants, thermal expansion, dielectric constants, refractive index, etc., under ambient conditions. In recent years information has also become available on their variation with temperature and pressure. Further, information has become available on new aspects like colour centres, defect properties, electrostriction, positron annihilation, harmonic generation, laser action, superconductivity etc. In spite of the availability of such a large body of information, the alkali halides continue to attract the attention of investigators. This is evidenced by the continuous publication of research papers on alkali halides All this information is naturally dispersed in a large number of journals, reference books and some property-specific monographs spread over a time-span of six to seven decades. It was therefore considered desirable to bring together data on a comprehensive range of physical properties of alkali halides under one cover for the use of researchers in the new millennium.

Selection and Presentation of Data

Data on different aspects of a wide range of physical properties and physical parameters of alkali halides have been compiled in this book. Interrelated properties are put together in nine chapters. A few properties, which were logically not fitting into these nine chapters, have been included in the last "Miscellaneous" chapter. The idea is not just to reproduce data on a physical property from all sources. Rather, the objective is to provide the prospective user with a reliable set of data sifted out of several sources. Where several reports are available for a given property, a representative set of values is chosen based on one or more of the following criteria:

1) Most recent data
2) Most precise data (as claimed by the authors)
3) Data from a source which covers a large number of crystals rather than from a source which reports data for individual crystals.

The uncertainties in the data are mentioned wherever they are reported by authors. As the data have been generated over a considerable period of time, the units differ in older and more recent reports. In many cases, the data are reproduced in the units used by the respective authors and formulae are given for converting the data from one system of units to another. The main data are given in tables, which are numbered section-wise. A different (Roman) numbering scheme is adopted for tables occurring in "Notes and Comments".

This is essentially a handbook. As such, it contains information on physical properties in the form of numerical data. But when this was not possible, data are given in the form of diagrams reproduced from the original sources. Description of experimental techniques and discussion of theories is not possible in a handbook. However in some cases theoretically calculated values have been included, particularly, where experimental values are not available. Unlike most handbooks which contain only compilation of data, this handbook includes some discussion of the properties and their correlations with other properties in the form of "Notes and Comments".

Contents

Preface v
Acknowledgements vii
Introduction ix

1 Structure-Related Parameters 1
   1.1 General Structural Information 1
   1.2 X-ray Powder Diffraction Data 3
   1.3 Lattice Constants 6
       1.3.1 Lattice Constants at Room Temperature 6
       1.3.2 Interionic Distances 7
       1.3.3 Ionic Radii 8
   1.4 Volumes: Molar Volume and Molecular Volume 10
   1.5 Densities 11
   1.6 Polymorphic Transitions – Transition Parameters 12
   References 13

2 Mechanical Properties 15
   2.1 Second Order Elastic Constants 15
       2.1.1 Adiabatic Second Order Elastic Constants at Room
             Temperature 15
       2.1.2 Adiabatic Elastic Compliances at Room Temperature 17
       2.1.3 Elastic Anisotropy 18
       2.1.4 Cauchy Inequality at Room Temperature 19
       2.1.5 Low Temperature Adiabatic Elastic Constants 20
       2.1.6 High Temperature Adiabatic Elastic Constants 21
       2.1.7 Thermoelastic Constants at 0 °C 22
       2.1.8 Temperature Derivatives of Elastic Constants at Constant
             Volume 23
       2.1.9 Pressure Derivatives of Elastic Constants 24
   2.2 Compressibility and Bulk Modulus 25
       2.2.1 Compressibility and Bulk Modulus at RT and 0 K 25
       2.2.2 Pressure and Temperature Derivatives of Compressibility
             (Experimental) 27
   2.3 Polycrystalline Elastic Properties 28
       2.3.1 Elastic Moduli of Polycrystalline Aggregates (RT) 28
       2.3.2 Temperature Derivatives of Polycrystalline Elastic Moduli 30
       2.3.3 Pressure Derivatives of Polycrystalline Elastic Moduli 30
   2.4 Compression Data at High Pressures 31
       2.4.1 Compression Data up to 45 kbars (Experimental) 31
       2.4.2 Equation of State Parameters (Experimental) 34

2.4.3   Equation of State Parameters for CsCl Phase (Experimental)....34
2.4.4   High Temperature Compression Data for NaCl (Experimental) 35
2.4.5   High Temperature Compression Data for NaCl (Theoretical) ...35
2.5   Higher-Order Elastic Constants 38
2.5.1   Third-Order Elastic Constants (Experimental)......................38
2.5.2   Third-Order Elastic Constants (Theoretical) 39
2.5.3   Fourth-Order Elastic Constants (Experimental) 40
2.5.4   Fourth-Order Elastic Constants (Theoretical) at 300 K.........41
2.6   Velocity of Sound 41
2.6.1   Velocities of Longitudinal Waves, Shear Waves and the
        Mean Velocity 41
2.6.2   Second Sound Velocity 43
2.7   Hardness 44
2.7.1   Hardness at Room Temperature 44
2.7.2   Knoop Hardness – Hardness Anisotropy...............................46
2.7.3   Temperature Variation of Hardness 46
2.7.4   Pressure Variation of Hardness 47
2.7.5   Surface Hardness..47
References 49

3   Thermal Properties...51
3.1   Specific Heat..51
3.1.1   Specific Heat at Low Temperatures (Experimental) 51
3.1.2   Specific Heat at Low Temperatures (Theoretical).................54
3.1.3   Specific Heat at High Temperatures (Experimental).............55
3.1.4   Pressure Variation of Specific Heat at RT............................55
3.2   Thermal Expansion 56
3.2.1   Thermal Expansion Coefficient at Room Temperature..............56
3.2.2   Thermal Expansion at Very Low Temperatures (T <12 K) 58
3.2.3   Thermal Expansion at Low Temperatures (10 K $\leq T \leq$ 250 K)..59
3.2.4   Thermal Expansion at High Temperatures
        (300 K $\leq T \leq$ 1000 K)..60
3.2.5   Temperature Variation of Thermal Expansion (Polynomial
        Form)..61
3.2.6   Pressure Variation of Thermal Expansion.............................63
3.3   Thermal Conductivity 64
3.3.1   Thermal Conductivity (Room Temperature)..........................64
3.3.2   Temperature Variation of Thermal Conductivity 65
3.3.3   Pressure Variation of Thermal Conductivity.........................66
3.3.4   Thermal Conductivity of Doped Alkali Halides....................67
3.4   Melting Temperatures..70
3.4.1   Melting Point at Atmospheric Pressure 70
3.4.2   Melting Parameters...71
3.4.3   Pressure Variation of Melting Point (Polynomial Form)...........72
3.4.4   Pressure Variation of Melting Point (Simon Equation)............73
3.4.5   Pressure Coefficient of Melting Point and Entropy Change.......74
3.4.6   Lindemann Parameter..75
3.5   Debye-Waller Factors 76
3.5.1   Debye-Waller Factors at RT (Expt. and Theor.) 76
3.5.2   Debye-Waller Factors at Low Temperatures.........................77

3.5.3   Debye-Waller Factors close to Melting Point 77
3.5.4   Temperature Variation of Debye-Waller factors
         (Experimental).. 77
3.5.5   Temperature Variation of Debye-Waller Factors (Theoretical) . 79
3.6   Debye Temperature 81
3.6.1   Debye Temperatures at Room Temperature............................. 81
3.6.2   Debye Temperatures at ~ 0 K 82
3.6.3   Temperature Variation of Debye Temperatures at Low
         Temperature 83
3.6.4   Temperature Variation of Debye Temperature (High
         Temperatures)... 87
3.6.5   Pressure Variation of Debye Temperature 90
3.7   Gruneisen Parameter.. 91
3.7.1   Gruneisen Parameter (γ) at Room Temperature 91
3.7.2   Temperature Variation of Gruneisen Parameter...................... 92
3.7.3   Low and High Temperature Limits of Gruneisen Parameter 93
3.7.4   Volume Dependence of Gruneisen Parameter.......................... 95
3.7.5   Mode Gruneisen Parameters.. 96
3.8   Anderson-Gruneisen Parameter.. 97
References 99

4   Optical Properties.. 103
4.1   Refractive Index... 103
4.1.1   Refractive Index at RT 103
4.1.2   Dispersion Equations for Refractive Index............................. 105
4.1.3   Temperature Derivative of Refractive Index at Selected
         Wavelengths.. 107
4.1.4   Temperature Derivative of Refractive index – Dispersion
         Equations (Empirical)... 108
4.1.5   Density Derivative of Refractive Index (Experimental)........... 109
4.1.6   Pressure Derivative of Refractive Index................................. 109
4.1.7   Pressure Variation of Refractive Index (Experimental) 110
4.2   Photoelasticity... 111
4.2.1   Strain-Optical Constants.. 111
4.2.2   Stress-Optical Constants... 112
4.2.3   Dispersion of Photoelastic Constants 113
4.2.4   Polycrystalline Photoelastic Constants.................................. 116
4.3   Faraday Effect.. 118
4.3.1   Verdet Constant and its Temperature Coefficient 118
4.3.2   Dispersion of Verdet Constant 120
4.4   Quadratic Electro-optic Effect 121
4.5   Laser-Related Properties... 122
4.5.1   Optical Transmittance 122
4.5.2   Linear Absorption Coefficient... 124
4.5.3   Two-Photon Absorption Coefficient 125
4.5.4   Three-Photon Absorption Coefficient 126
4.5.5   Nonlinear Refractive Index 126
4.5.6   Bulk Damage Threshold Parameters 126
4.6   Harmonic Generation.. 127

4.6.1   Second Harmonic Generation Parameters at Crystal-Glass Boundary 127
4.6.2   Third Harmonic Generation Parameters 127
4.7   Polarisability 128
4.7.1   Electronic Polarisabilities 128
4.7.2   Strain Derivative of Polarisability 129
4.7.3   Wavelength and Temperature Variation of the Strain Polarisability Constant 130
4.7.4   Pressure Variation of Polarisability 131
4.7.5   Quadrupole and Octupole Polarisabilities 133
References 134

5   Dielectric and Electrical Properties 137
5.1   Static Dielectric Constant 137
5.1.1   Static Dielectric Constant at Room Temperature 137
5.1.2   Static Dielectric Constant at Low Temperatures 138
5.1.3   Temperature Coefficient of the Static Dielectric Constant at Low Temperatures 139
5.1.4   First and Second Temperature Derivatives of the Static Dielectric Constant 140
5.1.5   High Temperature Data on Static Dielectric Constant 140
5.1.6   Pressure Coefficient of Static Dielectric Constant 142
5.1.7   Higher-Order Pressure Derivatives of Static Dielectric Constant 143
5.2   Electronic Dielectric Constant 144
5.3   Dielectric Polarisability 145
5.4   Effective Ionic Charge 146
5.4.1   Effective Ionic Charge (Szigeti Charge) 146
5.4.2   Temperature and Volume Derivatives of Effective Ionic Charge 148
5.5   Electrostriction 149
5.6   Electric Breakdown 150
References 152

6   Phonon Spectra 155
6.1   IR Spectra 155
6.1.1   Transverse Optical and Longitudinal Optical Frequencies (RT) 155
6.1.2   Temperature Variation of TO Frequencies (Low Temp.) 157
6.1.3   Temperature Variation of TO and LO Frequencies (High Temp.) 158
6.1.4   Temperature Variation of Damping Constant 159
6.1.5   Pressure Derivative of $k \sim 0$ ν_{TO} Frequency 159
6.1.6   Pressure Variation of $k \sim 0$ ν_{TO}, ν_{LO} Frequencies at High Pressures 160
6.2   Raman spectra 162
6.2.1   Second-Order Raman Spectra 162
6.2.2   Second-Order Laser Raman Spectra 163

6.2.3   Second-Order Raman Spectra of Alkali Fluorides (Expt. and Theor.) 164
6.2.4   Hyper-Raman Spectra (HRS) and Electric Field Induced First-Order Raman Spectra (EFIRS) 165
6.3   Neutron Inelastic Scattering.. 166
6.3.1   Phonon Dispersion Relations from Neutron Inelastic Scattering 166
6.3.2   Phonon Frequencies at Zone Centre and Zone Boundary 171
References 172

7   Chemical-Bond-Related Parameters.. 173
7.1   Lattice Energy.. 173
7.1.1   Interatomic Potentials and Expressions for Theoretical Lattice Energy 173
7.1.2   Structural Parameters in Lattice Energy Formulae for Alkali Halides.. 174
7.1.3   Van der Waal Constants 175
7.1.4   Born Repulsion Parameters 176
7.1.5   Parameters of the Huggins-Mayer form 177
7.1.6   Lattice Energy (Theoretical) 178
7.1.7   Lattice Energy (Experimental) 178
7.2   Electron Affinity.. 180
7.3   Ionicity.. 180
7.4   Electron Density Distribution 182
7.5   Force Constant.. 183
References 185

8   Band-Structure-Related Parameters 187
8.1   Typical Band Structures.. 187
8.2   Band Structure Parameters.. 190
8.2.1   Band Gap Energy 190
8.2.2   Pressure Derivative of Band Gap 191
8.2.3   Valence Band Width 192
8.2.4   Effective Mass.. 193
8.2.5   Interband Transition Energy.. 193
8.3   UV Absorption Spectra.. 194
8.3.1   UV Absorption Spectra (5–12 eV).. 194
8.3.2   Extreme UV Absorption Spectra (50–250 eV)........................ 207
8.4   Exciton Spectra.. 214
8.4.1   Exciton Energy 214
8.4.2   Pressure Derivative of Exciton Energy 216
8.5   UV Photoelectron and X-ray Photon Emission 216
8.6   Characteristic Electron Energy Loss Spectra................................ 217
8.7   Plasma Oscillation Frequency.. 219
8.8   Metallisation and Superconductivity 220
References 221

9   Defect State Parameters 223
   9.1   Schottky Defects...223
      9.1.1   Temperature Variation of Ionic Conductivity (Diagrams) 223
      9.1.2   Temperature Variation of Ionic Conductivity (2-Parameter Equation) 224
      9.1.3   Temperature Variation of Ionic Conductivity (4-Parameter Equation) 225
      9.1.4   Temperature Variation of Ionic Conductivity (6-Parameter Equation) 225
      9.1.5   Diffusion Parameters...226
      9.1.6   Enthalpy of Formation of a Schottky Pair (Different Methods) 227
      9.1.7   Pressure Variation of Ionic Conductivity (σ-P Plots) 228
      9.1.8   Pressure Variation of Ionic Conductivity (Pressure Coefficient and Activation Volume) 231
      9.1.9   Vacancy-Impurity Dipoles 232
      9.1.10 Solution Enthalpy of Divalent Defects....................................233
   9.2   Polarons..233
      9.2.1   Polaron Coupling Constant 233
      9.2.2   Polaron Mass...234
   9.3   Colour Centres 235
      9.3.1   Glossary of Colour Centres 235
      9.3.2   F Centre Parameters 236
      9.3.3   F Centre Formation Energy 238
      9.3.4   EPR and ENDOR Parameters for F Centres 240
      9.3.5   F Centres (Faraday Rotation and Circular Dichroism).............241
      9.3.6   F Centres (Dissociation Energy) 241
      9.3.7   F Centres (Temperature Variation of Peak Position) 242
      9.3.8   F Centres (Temperature Variation of Half-Width)...................244
      9.3.9   F Centres (Pressure Variation) 245
      9.3.10 Deformation Bleaching and Mechanoluminescence Parameters 246
      9.3.11 F Aggregate Centres (R_1, R_2, M, N_1, N_2, V_3 and F_3^+ Centres)..247
      9.3.12 Temperature Variation of M Band Half-Width.......................248
      9.3.13 Pressure Variation of M, R_2 and N Frequencies.....................249
      9.3.14 U Centre (Main UV Bands)..249
      9.3.15 U Centre (Localised IR Bands) 250
      9.3.16 U_2 Centres 251
      9.3.17 F_A Centres...251
      9.3.18 Z Centres 252
      9.3.19 Colour Centre Information Storage 253
      9.3.20 Colour Centre Lasers..254
   9.4   Luminescence 255
      9.4.1   Intrinsic Luminescence...255
      9.4.2   Auger-Free Luminescence...256
      9.4.3   Thermoluminescence...259
   9.5   Dislocations 260
      9.5.1   Slip (Glide) Systems...260
      9.5.2   Stacking Fault Energy 260
      9.5.3   Dislocation Mobility Parameters..261

     9.5.4   Etchants for Observation of Dislocations 261
  References 264

10 Miscellaneous Properties 267
  10.1 Mass-Related Parameters 267
  10.2 Polishing Agents 268
  10.3 Solubility 268
  10.4 Magnetic Susceptibility 270
  10.5 X-ray Monochromator Parameters 271
  10.6 Positron Annihilation 271
      10.6.1 Two-Photon Angular Correlation Cut-off Angle 271
      10.6.2 Momentum Distribution of Annihilating Electron-Positron
             Pairs 272
      10.6.3 Positron Annihilation Lifetimes 273
      10.6.4 Positron Annihilation Rates 274
  10.7 Radiation Detection 276
  10.8 Surface Properties 276
  10.9 Secondary Electron Emission 278
  10.10 Nuclear Quadrupole Relaxation Time 278
  References 280

Index 281

1 Structure-Related Parameters

1.1 General Structural Information

The alkali halides crystallize in either of the two structures viz. the NaCl structure and the CsCl structure. General structural information pertaining to the two structures is given below.

			NaCl Structure	**CsCl Structure**
i)	Space groups		Fm3m	Pm3m
ii)	Point groups		m3m (O_h)	m3m (O_h)
iii)	Coordination numbers		6	8
iv)	Number of molecules per unit cell		4	1
v)	Relation between interionic distance r and lattice parameter a		$r = (1/2)\,a$	$r = (\sqrt{3}/2)\,a$
vi)	Atomic position coordinates		Na: $\quad$ 0 0 0 $\quad\quad$ ½ ½ 0 $\quad\quad$ ½ 0 ½ $\quad\quad$ 0 ½ ½ Cl: $\quad$ ½ ½ ½ $\quad\quad$ 0 ½ 0 $\quad\quad$ ½ 0 0 $\quad\quad$ 0 0 ½	Cs: 0 0 0 Cl: ½ ½ ½
vii)	Basic vectors	a_1 a_2 a_3	$(0, a/2, a/2)$ $(a/2, 0, a/2)$ $(a/2, a/2, 0)$	$(a, 0, 0)$ $(0, a, 0)$ $(0, 0, a)$
viii)	Reciprocal lattice		b.c.c	s.c.
ix)	Reciprocal vectors	b_1 b_2 b_3	$(-1/a, 1/a, 1/a)$ $(1/a, -1/a, 1/a)$ $(1/a, 1/a, -1/a)$	$(1/a, 0, 0)$ $(0, 1/a, 0)$ $(0, 0, 1/a)$

		NaCl Structure	**CsCl Structure**
x)	Madelung constant	α_r 1.74756	1.76267
		α_a 3.49513	2.03536

xi)   Unit cell (Fig. 1.1)

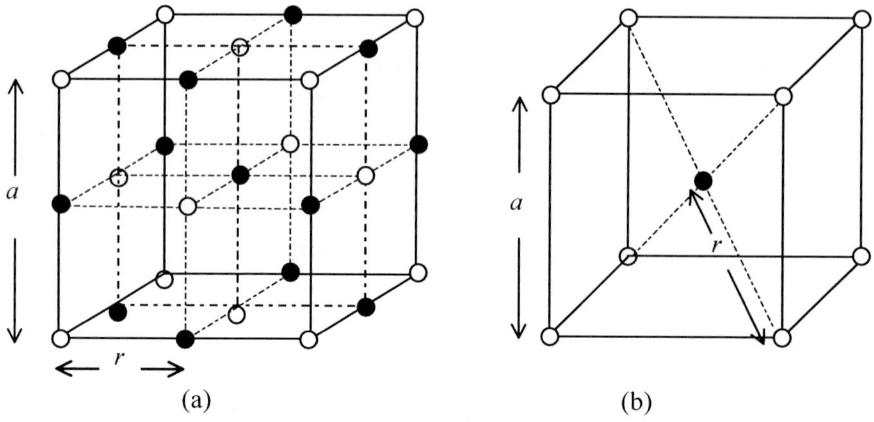

Fig. 1.1 Unit cells of **(a)** NaCl and **(b)** CsCl lattices; ● Na (or Cs) and ○ Cl

xii)   Brillouin zones (Fig. 1.2)

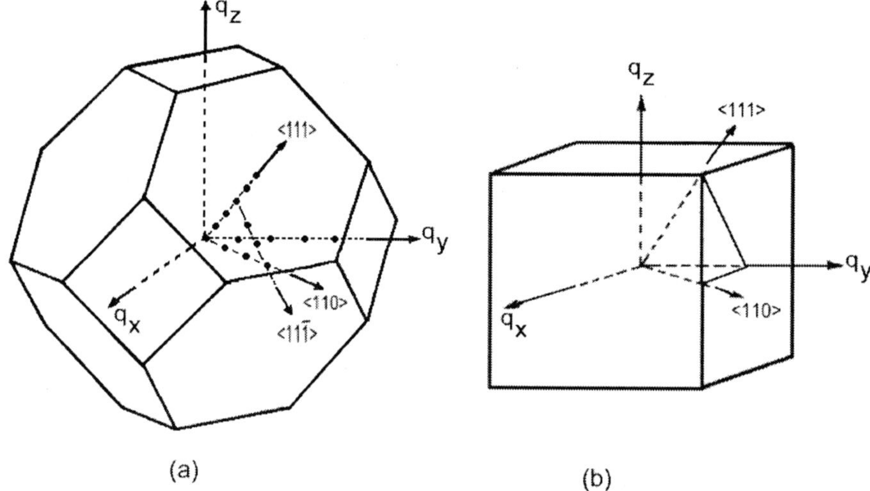

Fig. 1.2 First Brillouin zones for **(a)** NaCl and **(b)** CsCl structures

1.2 X-ray Powder Diffraction Data

Table 1.1 Values of the interplanar spacings d and the relative intensities I for the first few reflections ; the intensity of the most intense reflection is taken as 100. The data are from room temperature X-ray patterns recorded with Cu $K\alpha$ radiation (λ = 1.5405 Å) except in the case of LiI for which Mo $K\alpha$ radiation (λ = 0.7090 Å) was employed. All data are from [1.1]

| Substance | LiF | | LiCl | | LiBr | | LiI | |
Miller indices hkl ↓	d [Å]	I	d [Å]	I	d [Å]	I	d [Å]	I
111	2.325	95	2.967	100	3.180	100	3.47	100
200	2.013	100	2.570	86	2.751	80	3.00	75
220	1.424	48	1.817	58	1.945	60	2.12	40
311	1.214	10	1.550	29	1.659	45	1.81	40
222	1.162	11	1.484	16	1.588	18	1.73	10
400	1.007	3	1.285	4	1.376	8	1.50	6
331	0.924	4	1.179	10	1.262	18	1.38	15
420	0.900	14	1.149	12	1.231	16	1.34	15
422	0.822	13	1.049	8	1.122	12	1.23	8
511	-	-	0.989	9	1.058	10	1.16	8
440	-	-	0.909	2	0.972	4	1.06	1
531	-	-	0.869	10	0.929	8	1.02	6

Table 1.1 (Continued)

| Substance | NaF | | NaCl | | NaBr | | NaI | |
Miller indices hkl ↓	d [Å]	I	d [Å]	I	d [Å]	I	d [Å]	I
111	2.680	3	3.258	13	3.449	64	3.74	80
200	2.319	100	2.821	100	2.988	100	3.23	100
220	1.639	60	1.994	55	2.113	63	2.289	65
311	1.399	2	1.701	2	1.802	21	1.951	40
222	1.338	17	1.628	15	1.725	19	1.868	25
400	1.159	7	1.410	6	1.495	8	1.618	14
331	1.063	< 1	1.294	1	1.371	7	1.485	14
420	1.036	12	1.261	11	1.337	15	1.447	25
422	0.946	8	1.151	7	1.221	9	1.321	20
511	0.892	1	1.086	1	1.151	4	1.246	14
440	0.819	3	0.997	2	1.057	2	1.144	6
531	-	-	0.953	1	1.010	2	1.094	8

Table 1.1 (Continued)

Substance	KF		KBr		KI		RbF	
Miller indices hkl ↓	d [Å]	I	d [Å]	I	d [Å]	I	d [Å]	I
111	3.087	29	3.800	15	4.08	42	3.262	65
200	2.671	100	3.292	100	3.53	100	2.823	100
220	1.890	63	2.333	57	2.498	70	1.998	50
311	1.612	10	1.989	7	2.131	29	1.703	25
222	1.542	17	1.905	16	2.039	27	1.631	14
400	1.337	8	1.649	10	1.767	15	1.413	6
331	1.227	2	1.514	2	1.621	7	1.297	8
420	1.195	14	1.475	17	1.580	24	1.264	12
422	1.091	8	1.346	8	1.442	14	1.154	6
511	1.029	3	1.269	2	1.360	3	1.088	4
440	0.945	3	1.167	3	1.249	2	0.999	4
531	0.904	4	-	-	1.194	3	0.955	2

Table 1.1 (Continued)

Substance	RbCl		RbI		CsF	
Miller indices hkl ↓	d [Å]	I	d [Å]	I	d [Å]	I
111	3.80	30	4.24	8	3.469	80
200	3.29	100	3.67	100	3.003	100
220	2.327	65	2.595	60	2.125	35
311	1.984	14	2.214	4	1.813	25
222	1.900	20	2.119	20	1.737	10
400	1.645	10	1.835	10	1.504	8
331	1.510	4	1.684	2	1.380	6
420	1.472	20	1.642	18	1.345	10
422	1.343	14	1.498	12	1.228	6
511	1.266	4	-	-	1.158	4
440	1.164	4	1.298	4	1.063	2
531	1.112	2	-	-	1.017	4

Table 1.1 (Continued)

Substance	KCl		RbBr	
Miller indices hkl ↓	d [Å]	I	d [Å]	I
200	3.146	100	3.44	100
220	2.224	59	2.436	75
222	1.816	23	1.989	25
400	1.573	8	1.722	12
420	1.407	20	1.541	25
422	1.284	13	1.406	16

Table 1.1 (Continued)

Substance	KCl		RbBr	
Miller indices hkl ↓	d [Å]	I	d [Å]	I
440	1.112	2	1.218	4
600	1.049	6	1.148	8
620	0.995	2	1.089	6
622	0.949	3	1.038	4
444	0.908	1	0.994	2
640	0.873	2	0.955	4
642	0.841	6	0.921	4

Table 1.1 (Continued)

Substance	CsCl		CsBr		CsI	
Miller indices hkl ↓	d [Å]	I	d [Å]	I	d [Å]	I
100	4.12	45	4.29	8	-	-
110	2.917	100	3.039	100	3.23	100
111	2.380	13	2.480	3	-	-
200	2.062	17	2.148	18	2.284	20
210	1.844	14	1.921	6	-	-
211	1.683	25	1.754	43	1.865	35
220	1.457	6	1.519	18	1.615	18
300	1.374	5	1.432	3	-	-
310	1.304	8	1.358	16	1.445	8
311	1.243	2	1.295	< 1	-	-
222	1.190	2	1.240	6	1.319	4
320	1.143	1	1.192	1	-	-
321	1.102	6	1.148	20	1.221	10

Notes and Comments

1. The d values are calculated from the measured Bragg angles θ using Bragg's law $\lambda = 2d \sin\theta$.
2. The X-ray powder diffraction data is useful in X-ray identification of a material, particularly when it occurs mixed with another substance.
3. Note that some of the reflections occurring in NaCl type crystals are absent in KCl and RbBr although they have the same structure. Similarly, some of the reflections in CsCl are absent in CsI though the structures are the same. This is because of the near-equality of the atomic scattering factors of the ions in these crystals.

1.3 Lattice Constants

1.3.1 Lattice Constants at Room Temperature

Table 1.2 Accurate values of the lattice constants at room temperature (t) and atmospheric pressure unless otherwise mentioned (P); uncertainty in last digit indicated in parenthesis wherever reported

Ref. Crystal ↓	[1.2 (i–viii)]			[1.3]	
	t [°C]	a [Å]	Sub-Ref.	t [°C]	a [Å]
NaCl Structure					
LiF	26	4.0270	i	25	4.02620(5)
LiCl	25	5.1396	i	25	5.13988(4)
LiBr	26	5.5013	iv	25	5.501(6)
LiI	25	6.0228	viii	25	6.012(7)
NaF	26	4.6344	i	20	4.6329(5)
NaCl	26	5.6402	ii	25	5.64009(3)
NaBr	26	5.9772	iii	25	5.97299(5)
NaI	26	6.4728	iv	25	6.4728(5)
KF	26	5.347	i	25	5.344(3)
KCl	25	6.2931	i	25	6.29294(8)
KBr	25	6.6000	i	25	6.5982(2)
KI	25	7.0655	i	25	7.06555(15)
RbF	25	5.6516	vi	25	5.6516(1)
RbCl	27	6.5810	iv	20	6.5898(2)
RbBr	25	6.889	v	20	6.8908(2)
RbI	27	7.342	iv	27	7.3466(2)
CsF	25	6.014	vii	20	6.020(6)
High Temperature NaCl Phase					
CsCl				485	7.079(4)
CsCl Structure					
CsCl	25	4.1230	ii	20	4.1200(5)
CsBr	25	4.2953	iii	25	4.2953(5)
CsI	26	4.5679	iv	20	4.5667(5)
High Pressure CsCl Phase					
NaCl (P = 300 kbar)					2.997(4)
KF (P = 35 kbar)					3.06
KCl (P = 20 kbar)					3.674
KBr (P = 18 kbar)					3.867(3)
KI (P = 20.8 kbar)					4.13(2)
RbF (P = 12 kbar)					3.29
RbCl (P = 7.3 kbar)					3.82
RbBr (P = 5 kbar)					4.09
RbI (P = 4.5 kbar)					4.34
CsF (P = 48 kbar)					3.39

Notes and Comments

1. Additional data on lattice constants can be found in Refs. [1.4–1.10].
2. The lattice constant is a basic parameter. In the alkali halides, it is often used as a scaling parameter to indicate trends in various properties like elastic constants, lattice energy, surface energy, Debye temperature, hardness etc. These correlations will be referred to in respective sections.

1.3.2 Interionic Distances

Table 1.3 Values of the interionic distance r at room temperature and 0 K

Crystal	r [Å] at room temp [1.11]	r [Å] at 0 K [1.12]
NaCl Structure		
LiF	2.0131	1.996
LiCl	2.5699	2.539
LiBr	2.7508	2.713
LiI	3.0060	2.951
NaF	2.3164	2.295
NaCl	2.8200	2.789
NaBr	2.9865	2.954
NaI	3.2364	3.194
KF	2.6720	2.648
KCl	3.1464	3.116
KBr	3.2991	3.262
KI	3.5327	3.489
RbF	2.8258	2.789
RbCl	3.2949	3.259
RbBr	3.4454	3.410
RbI	3.6710	3.628
CsF	3.0100	-
CsCl Structure		
CsCl	3.568	3.523
CsBr	3.7198	3.668
CsI	3.9548	3.898

Notes and Comments

1. The interionic distance r is of fundamental importance as it enters most of the theoretical calculations for the alkali halides.
2. Tosi [1.13] points out that the differences between the interionic distances in the corresponding halides of a pair of alkali metals are approximately constant. This applies also to the alkali metal salts of pairs of halogens. This is shown in Table I.

Table I Differences of interionic distances [Å] in the NaCl-type alkali halides*

	F		Cl		Br		I
Li		0.556		0.181		0.249	
	0.303		0.250		0.238		0.237
Na		0.503		0.179		0.248	
	0.357		0.327		0.309		0.296
K		0.473		0.151		0.235	
	0.141		0.144		0.147		0.138
Rb		0.476		0.154		0.226	

*The values in the rows lying between two alkali metals give the differences in the interionic distance in the various halides of the two alkalis, and the values in the columns between two halogens give the differences in the interionic distance in the various alkali metal salts of the two halogens.

3. Shaw [1.14] has shown linear plots between $Z^{-1/3}$ and $r/\bar{n}$ for alkali halides of each cation; $n = (n_A + n_B)/2$ where n_A and n_B are the row numbers of the two constituents in the periodic table and $Z = (Z_A + Z_B)/2$ is the average atomic number. This relationship can be expressed as

$$r(Z^{1/3}/\bar{n}) = C \tag{1.1}$$

Values of C are given in Table II.

Table II Values of C in Eq. (1.1):

	F	Cl	Br	I
Li	6.90	6.96	6.94	6.89
Na	6.27	6.44	6.39	6.47
K	6.09	6.23	6.24	6.30
Rb	6.06	6.22	6.11	6.17

1.3.3 Ionic Radii

Table 1.4 Values of the ionic radii (r_+ and r_-); uncertainties in the last digit indicated in parenthesis, wherever reported

Ion→	r_+ [Å]					r_- [Å]				Ref.
	Li^+	Na^+	K^+	Rb^+	Cs^+	F^-	Cl^-	Br^-	I^-	
Experimental (from X-rays)										
	0.91(8)	-	-	-	-	-	1.66(8)	-	-	[1.15]
	-	-	-	1.71(5)	-	-	1.58(5)	-	-	[1.16]
	0.92(10)	-	-	-	-	1.09(10)	-	-	-	[1.17]
	-	1.15(6)	-	-	-	-	1.67(6)	-	-	[1.18]
	-	-	1.57(7)	-	-	-	-	1.73(7)	-	[1.19]

Table 1.4 (Continued)

Ion→	r_+ [Å]					r_- [Å]				Ref.
	Li^+	Na^+	K^+	Rb^+	Cs^+	F^-	Cl^-	Br^-	I^-	
Experimental (from X-rays)										
	0.78(4)	-	-	-	-	1.23(4)	-	-	-	[1.20]
	-	1.18(10)	-	-	-	-	1.64(10)	-	-	[1.21]
Theoretical										
	-	1.01	1.30	1.50	1.75	1.33	1.72	1.92	2.19	[1.22]
	0.78	0.98	1.33	1.49	1.65	1.33	1.81	1.96	2.20	[1.23]
	0.60	0.95	1.33	1.48	1.69	1.36	1.81	1.95	2.16	[1.24]
	0.68	0.98	1.33	1.48	1.67	1.33	1.81	1.96	2.19	[1.25]
	0.90	1.21	1.51	1.65	1.80	1.19	1.65	1.80	2.01	[1.26]
	0.94	1.17	1.49	1.63	1.86	1.16	1.64	1.80	2.05	[1.27]
	1.038	1.295	1.617	1.759	2.017	1.019	1.534	1.695	1.934	[1.28]
	0.76	1.12	1.38	1.52	1.67	1.33	1.81	1.96	2.20	[1.29]

Notes and Comments

1. The experimental values of ionic radii are determined from electron density maps obtained from X-ray diffraction intensities.
2. Ionic radii are useful in theoretical calculations based on the Born model e.g. in theoretical evaluation of lattice energy, surface energy and energy of formation of Schottky pairs.
3. The radius ratio plays an important role in determining the stability of structures (for a discussion, see [1.13]).
4. Damm and Chovj [1.30] have shown that a number of properties of the alkali halides like atomization energy, valence band separation, exciton absorption band energy, U_2 center absorption energy etc. correlate with the interionic distance r and the radius ratio (r_+/r_-) in the form of a general relation

$$f = A(r)^{-n} (r_+ / r_-)^m \tag{1.2}$$

where f is the particular property and A, n and m are constants. Values of A, n and m for various properties are given in relevant sections.

1.4 Volumes: Molar Volume and Molecular Volume

Table 1.5 Values of the molar volume (V_M) and molecular volume (V_m) at room temperature and at 0 K; for method of calculating the volumes, see Notes and Comments.

Temp.	Room temp.		0 K	
Crystal $\downarrow$	V_M [cm^3]	V_m [Å^3]	V_M [cm^3]	V_m [Å^3]
NaCl Structure				
LiF	9.826	16.316	9.462	15.712
LiCl	20.442	33.945	19.633	32.602
LiBr	25.062	41.616	23.988	39.832
LiI	32.712	54.32	31.234	51.866
NaF	14.970	24.858	14.533	24.132
NaCl	27.012	44.854	26.134	43.397
NaBr	32.083	53.274	31.110	51.641
NaI	40.829	67.798	39.400	65.418
KF	22.977	38.154	22.356	37.123
KCl	37.518	62.300	36.394	60.434
KBr	43.248	71.815	41.986	69.719
KI	53.103	88.180	51.456	85.445
RbF	27.177	45.128	26.478	43.968
RbCl	43.083	71.541	41.784	69.383
RbBr	49.261	81.799	47.883	79.512
RbI	59.584	98.942	57.745	95.887
CsF	32.846	54.542	-	-
CsCl Structure				
CsCl	42.112	69.934	-	-
CsBr	47.852	79.246	-	-
CsI	57.353	95.237	-	-

Notes and Comments

1. The values at room temperature are calculated from the room temperature data on lattice constants given in [1.3] using the relations $V_m = a^3/4$ for the NaCl structure, $V_m = a^3$ for CsCl structure and $V_M = N_0 V_m$, a being the lattice constant and N_0 the Avogadro number ($= 6.02217 \times 10^{23}$ mole^{-1}).
2. The volumes at 0 K are obtained from room temperature values by applying thermodynamic corrections [1.31, 1.32].

1.5 Densities

Table 1.6 Measured density ρ_m [g cm^{-3}] and X-ray density ρ_X [g cm^{-3}] at room temperature and low temperature

	Room temperature		Low temperature
	ρ_m at 22 °C	ρ_X at 25 °C	ρ_m (at 0 K for LiF, LiBr and RbF; at 77 K for KBr and at 4.2 K for other crystals)
Ref.	[1.33]	[1.3]	[1.34]
Crystal ↓			
NaCl Structure			
LiF	2.6402	2.640	2.6757
LiCl	2.0753	2.068	2.1110
LiBr	3.4703	3.464	3.5768
LiI	-	4.061	-
NaF	2.8045	2.809	2.8510
NaCl	2.1641	2.168	2.2170
NaBr	3.1997	3.201	3.299
NaI	3.6714	3.655	3.7620
KF	2.5257	2.48	2.5300
KCl	1.9882	1.989	2.0448
KBr	2.7505	2.750	2.8190
KI	3.1279	3.129	3.2077
RbF	-	3.844	3.9284
RbCl	2.7985	2.76	2.8200
RbBr	3.351	3.349	3.4340
RbI	3.5533	3.564	3.6680
CsF	4.627	4.638	-
CsCl Structure			
CsCl	3.99	3.988	-
CsBr	4.45	4.455	4.6550
CsI	4.52	4.509	4.7120

Notes and Comments

1. The densities are also available in [1.35].
2. The X-ray density $\rho_X = (n\,M / N_0\,V)$ where n is the number of molecules per unit cell, N_0 the Avogadro number, M the molecular weight and V the unit cell volume (a^3 for cubic crystals). ρ_X is also called the ideal or theoretical density since it is not affected by the presence of defects in the crystal.

1.6 Polymorphic Transitions – Transition Parameters

Table 1.7 Values of the transition pressure (P_{tr} in kbar), transition temperature (T_{tr} in K), fractional change in volume ($\Delta V / V_0$) and change in entropy (ΔS in cal mole^{-1} K^{-1}) at the transition; uncertainties in last digit indicated in parenthesis wherever reported.

Crystal	P_{tr}	Ref.	T_{tr}	Ref.	$-(\Delta V / V_0)$	Ref.	ΔS	Ref.
Fm3m→Pm3m								
NaCl	300(10)	[1.36]			0.037	[1.36]	1.5(3)	[1.36]
KF	17.3	[1.37]			0.0696	[1.37]	-	-
KCl	19.3	[1.37]			0.1126	[1.37]	0.025(21)	[1.36]
KBr	17.7	[1.37]			0.1031	[1.37]	-	-
KI	18.0	[1.37]			0.0835	[1.37]	-	-
RbF	9.4	[1.37]			0.0127	[1.37]	-	-
RbCl	5.2	[1.37]			0.1455	[1.37]	−0.044(4)	[1.36]
RbBr	4.3	[1.37]			0.1349	[1.37]	-	-
RbI	3.3	[1.37]			0.1278	[1.37]	-	-
Pm3m→Fm3m								
CsCl	-	-	718	[1.36]	0.165	[1.36]	−1.06(4)	[1.36]

Notes and Comments

1. Some alkali halides with NaCl structure undergo an Fm3m→Pm3m transition at high temperature.
2. Bassett et al. [1.36] found a linear relation between ΔS and $\Delta V/V_0$ for the alkali chlorides for which they proposed the equation

$$\Delta S = 20.0\,(\Delta V / V_0) + 2.25 \tag{1.3}$$

   A thermodynamic explanation has been given for the relation.
3. Values of P_{tr} and $\Delta V/V_0$ have been reported for the potassium and rubidium halides by Dernell and McCollum [1.38] and by Weir and Piermarini [1.39]. The P_{tr} and $\Delta V/V_0$ values by Dernell and McCollum agree with those given in [1.37]. However the P_{tr} values as well as $\Delta V/V_0$ values obtained by Weir and Piermarini are systematically larger.
4. The Pm3m→Fm3m transition in CsCl has been studied by Rao et al. [1.40].
5. A Theoretical estimate for P_{tr} for the Fm3m→Pm3m transition in CsCl has been made by Born and Huang [1.3], and Narayanan and Ramaseshan [1.41].
6. Sinha et al. [1.42] and Krishna et al. [1.43] have confirmed the transition in KI at ~ 19.5 kbar using X-ray and neutron diffraction techniques respectively.

References

1.1   Report S M A–27, 1977; Int. Center for Diffraction. Data, Swarthmore, Penn., USA.

1.2   U.S. National Bureau of Standards Circulars (i) 539, Vol. **I**, 1953; (ii) 539, Vol. **II**, 1953; (iii) 539, Vol. **III**, 1954; (iv) 539, Vol. **IV**, 1955; (v) 539, Vol. **VII**, 1957; (vi) **25**, Sec.8, 1970; (vii) **25**, Sec.3, 1964; (viii) NBS result quoted by K.O. Mclean and C.S. Smith, J. Phys. Chem. Solids, **33**, 279, 1972.

1.3   *Landolt-Bornstein, Numerical data and Functional relationships in Sci. and Tech.*, New Series, Group III, **Vol. 7**, Part a, Ed. K.H. Hellwege and A.M. Hellwege, Springer Verlag, Heidelberg, 1973 and references therein.

1.4   K.K.Srivastava and H.D. Merchant, J. Phys. Chem. Solids, **34**, 2069, 1973.

1.5   V.T. Deshpande and D.B. Sirdeshmukh, Acta Cryst., **14**, 353, 1961.

1.6   V.T. Deshpande, Acta Cryst.,**14**,794,1961.

1.7   D.B. Sirdeshmukh and V.T. Deshpande, Curr. Sci.,**33**, 428, 1964.

1.8   P.D. Pathak and N. G. Vasavada, Acta Cryst., **A26**, 655, 1970.

1.9   P.D. Pathak, J.M. Trivedi and N.G. Vasavada., Acta Cryst.**A29**, 477, 1973.

1.10   P.D. Pathak and N.M. Pandya, Acta Cryst., **A31**, 155, 1975.

1.11   Calculated from values of lattice constants a given in [1.3] using the relation $r = a/2$ for crystals with NaCl structure and $r = (\sqrt{3}/2)a$ for those with CsCl structure.

1.12   Calculated by P.B. Ghate, Phys. Rev., **139A**, 1666, 1965 from room temperature values of γ by applying thermodynamic correction given by Tosi [1.13].

1.13   M.P. Tosi, Solid State Physics, **16**, 1, 1964.

1.14   R.W. Shaw, Phys. Rev. **B5**, 4856, 1972.

1.15   O. Inkinen and M. Jarvinen, Phys. Kondens. Mater., **7**, 372, 1968.

1.16   M.Jarvinen and O. Inkinen, phys. stat. sol., **21**, 127,1967.

1.17   J. Krug, H. Witte and E. Wolfel, Z. Phys. Chem., **4**, 36, 1955.

1.18   K. Kurki-Suonio and L. Fontell, Ann. Acad. Scient. Fenn., **AVI**, 161, 1964.

1.19   V. Meisalo and O. Inkinen, Acta Cryst., **22**, 58, 1967.

1.20   M. Merisalo and O. Inkinen, Ann. Acad. Scient. Fenn., **AVI**, 207, 1966.

1.21   G. Schoknecht, Z. Naturweiss., **A12**, 983, 1957.

1.22   J.A. Wasastjerna, Comment. Phys. Math., **1**, 1, 1923.

1.23   V. M. Goldschmidt, Skr. Norske Vidensk-Akad. Mat-naturv., **1**, k1, 1926.

1.24   L. Pauling, J. Am. Chem. Soc., **49**, 765, 1927.

1.25   W.H. Zachariasen, Z. Kristall. **80**, 137, 1931.

1.26   M. P. Tosi and F.G. Fumi, J. Phys. Chem. Solids, **25**, 45,1964.

1.27   B.S. Gourary and F.J. Adrian, Solid State Physics, **10**, 127, 1960.

1.28   P.A. Sysio, Acta Cryst., **B25**, 2374, 1969.

1.29   R.D. Shannon, Acta Cryst., **A32**, 751, 1976.

1.30   J. Z. Damm and Z. Chovj, phys. stat. sol., **(b)114**, 413, 1982.

1.31   M. Born and K. Huang, *Dynamical Theory of Crystal Lattices*, Oxford University Press, New York, 1954.

1.32   I.P.S. Solanki and J. Shanker, Ind. J. Pure and Appl. Phys., **32**, 863, 1994

1.33   Values of ρ_m: for CsF from S. Haussuhl, Zeit. fur Krist., **138**, 177, 1973; for CsCl, CsBr, CsI from S. Haussuhl, Acta Cryst., **13**, 685, 1960 and the rest from K. Spangenberg and S. Haussuhl, Zeit. fur Krist., **109**, 4, 1957.

1.34   A. Konti and Y.P. Varshni, Can. J. Phys., **49**, 3115, 1971 and references therein.

1.35   *CRC Handbook of Chemistry and Physics*, 76th Ed., CRC Press, Boca Raton, Florida, 1995–96.

1.36 W.A. Basset, T. Takahashi, H.K. Mao and J.S. Weaver, J. Appl. Phys., **39**, 319, 1968.

1.37 S.N. Vaidya and G.C. Kennedy, J. Phys. Chem. Solids, **32**, 951, 1971.

1.38 A.J. Dernell and W.A. McCollum, J. Phys. Chem. Solids, **31**, 805, 1970.

1.39 C.E. Weir and G.I. Piermarini, J. Res. NBS, **68A**, 105, 1964.

1.40 K.J. Rao, G.V.S. Rao and C.N.R. Rao, Trans. Faraday Soc., **63**, 1013, 1967.

1.41 R. Narayanan and S. Ramaseshan, Phys. Rev. Lett., **42**, 992, 1979.

1.42 A. Sinha, A.B. Garg, V. Vijayakumar, B.K. Godwal and S.K. Sikka, Bull. Mat. Sci., **23**, 151, 2000.

1.43 P.S.R. Krishna, A.B. Shinde, S.N. Vaidya and S. K. Paranjape, Sahayoga, **10**, 9, 1999.

2 Mechanical Properties

2.1 Second-Order Elastic Constants

2.1.1 Adiabatic Second-Order Elastic Constants at Room Temperature

Table 2.1 Values of adiabatic second-order elastic constants C_{11}, C_{12} and C_{44} at RT determined from ultrasonic measurements

Crystal	C_{11}	C_{12}	C_{44}	Ref.
	$[10^{11}$ dyne cm^{-2}]; uncertainty $\sim 2\%$			
NaCl Structure				
LiF	11.355	4.76	6.35	[2.1]
LiCl	4.943	2.28	2.46	[2.1]
LiBr	3.937	1.87	1.93	[2.1]
LiI	2.907	1.421	1.407	[2.2]
NaF	9.70	2.43	2.81	[2.1]
NaCl	4.936	1.29	1.265	[2.1]
NaBr	4.012	1.09	0.99	[2.1]
NaI	3.025	0.88	0.74	[2.1]
KF	6.56	1.46	1.25	[2.1]
KCl	4.078	0.69	0.633	[2.1]
KBr	3.476	0.57	0.507	[2.1]
KI	2.76	0.45	0.37	[2.1]
RbF	5.525	1.395	0.925	[2.1]
RbCl	3.634	0.615	0.465	[2.1]
RbBr	3.157	0.495	0.38	[2.1]
RbI	2.583	0.37	0.278	[2.1]
CsF	4.406	1.380	0.703	[2.3]
CsCl Structure				
CsCl	3.64	0.92	0.80	[2.4]
CsBr	3.10	0.84	0.75	[2.4]
CsI	2.45	0.71	0.62	[2.4]

Notes and Comments

1. The internal energy Φ of a deformed crystal may be approximately expressed as

$$\Phi = \Phi_0 + (V/2)C_{ijkl}\,\varepsilon_{ij}\,\varepsilon_{kl} \tag{2.1}$$

where Φ_0 is the internal energy of the undeformed crystal, V the volume of the unit cell, C_{ijkl} are the second-order elastic constants and ε_{ij} the strain components. Applying symmetry conditions, thermodynamic considerations and using Voigt's contracted notation, the number of second-order elastic constants for a cubic crystal reduces to three viz; C_{11}, C_{12} and C_{44}.

2. Subramanyam and Murti [2.5] showed that a linear relationship holds between $\log C_{11}$ and $\log C_{12}$ in the form of equations

$$\log C_{11} = A \log C_{12} + B \tag{2.2}$$

The values of the constants A and B for the alkali halides are given in Table III.

Table III Values of the constants A and B in Eq. (2.2)

	A	B
Li halides	1.25	0.60
Na halides	1.09	1.31
K halides	0.70	1.63
Rb halides	0.65	1.60

3. Haussuhl [2.6] pointed out that the product of the mean elastic constant C and the molecular volume is a constant for a given family. The product is denoted by S. For cubic crystals like the alkali halides, $C = (C_{11}+C_{12} + C_{44})/3$. Taking C in units of 10^{10} N m^{-2}, the values of S (10^{-20} N m) range from 103 for LiI to 125 for CsCl.

4. Motida [2.7] has shown smooth curve plots between C_{44}/C_0 and the effective ionic charge e^* where $C_0 = e^2/r^4$, r being the interionic distance. Further, Motida showed that the plots between $C_{44}{}^a/C_0$ and e^*/e are straight line plots where $C_{44}{}^a = C_{44} - \{(e^*)^2/4r^4\}(\chi - \alpha)$ where α is the Madelung constant and χ a constant occurring in the Krishnan-Roy theory [2.8]. There are four straight lines corresponding to the halides of each alkali ion.

5. Koh and Ng [2.9] observed a linear relationship between C_{44} and the parameter $\beta = r_-/r$. They observed four separate plots for the four alkali ions with equations of the type

$$C_{44} = m\beta - b \tag{2.3}$$

The values of m and b are (-11.53, 8.39), (-2.60, 1.80), (-0.84, 0.55), and (-0.63, 0.09), for the Li, Na, K and Rb halides respectively.

2.1.2 Adiabatic Elastic Compliances at Room Temperature

Table 2.2 Values of adiabatic elastic compliances S_{11}, S_{12} and S_{44} at room temperature determined from ultrasonic measurements [2.10]

Crystal	S_{11}	S_{12}	S_{44}
	[Mbar^{-1}]		
NaCl Structure			
LiF	1.1675	−0.3444	1.5704
LiCl	2.8399	−0.8914	4.0161
LiBr	3.6691	−1.1852	5.2356
LiI	5.1877	−1.7089	7.4074
NaF	1.1588	−0.2357	3.5791
NaCl	2.3181	−0.4917	7.8989
NaBr	2.8036	−0.5646	10.0200
NaI	3.8722	−0.9011	13.6426
KF	1.7105	−0.3387	7.9872
KCl	2.5999	−0.3657	15.8982
KBr	3.0472	−0.4023	19.6850
KI	3.8902	−0.5112	27.1003
RbF	1.9046	−0.3426	10.9890
RbCl	2.8907	−0.4338	20.9205
RbBr	3.3167	−0.4541	26.3158
RbI	4.0706	−0.5293	35.9712
CsCl Structure			
CsCl	3.0105	−0.5841	12.4378
CsBr	3.6477	−0.7537	13.7931
CsI	4.5719	−0.9530	15.8982

Notes and Comments

1. For cubic crystals, the S_{ij}'s are related to the C_{ij}'s by the equations (2.4–2.6)

$$S_{11} = (C_{11} + C_{12}) / [(C_{11} + 2C_{12})(C_{11} - C_{12})] \tag{2.4}$$

$$S_{12} = -C_{12} / [(C_{11} + 2C_{12})(C_{11} - C_{12})] \tag{2.5}$$

$$S_{44} = (1/C_{44}) \tag{2.6}$$

2. The S_{ij} values in Mbar^{-1} may be multiplied by 10^{-12} to convert to cm^2 dyne^{-1}.

2.1.3 Elastic Anisotropy

Table 2.3 Values of the elastic anisotropy A and A^* (for definitions see Notes and Comments)

Crystal	A		A^*
	4.2K	300 K	RT
NaCl Structure			
LiF	1.57	1.81	0.051
LiCl	1.42	1.90	0.045
LiBr	-	1.25	0.046
LiI	-	1.86	0.048
NaF	0.677	0.779	0.008
NaCl	0.577	0.677	0.016
NaBr	0.560	0.673	0.018
NaI	0.536	0.700	0.016
KF	0.430	0.496	0.060
KCl	0.309	0.371	0.111
KBr	0.29	0.355	0.127
KI	0.233	0.326	0.148
RbF	-	0.447	0.076
RbCl	0.270	0.312	0.157
RbBr	0.241	0.290	0.177
RbI	0.205	0.251	0.211
CsF	-	0.464	0.069
CsCl Structure			
CsCl	-	0.588	0.033
CsBr	-	0.663	0.020
CsI	-	0.712	0.014

Notes and Comments

1. A and A^* are defined as:

$$A = 2C_{44} / (C_{11} - C_{12}) \tag{2.7}$$

$$A^* = (G_V - G_R) / G_V \tag{2.8}$$

where G_V and G_R are the Voigt and Reuss average values of the shear modulus.
2. Values of A for cesium halides calculated from data in [2.4] and those for the rest from data in [2.11]. Values of A^* calculated from data in Sec. 2.3.1.
3 A is < 1 in all cases except the lithium halides. Complete isotropy is indicated by $A = 1$ and $A^* = 0$.
4. A is temperature dependent, decreasing at low temperatures.

2.1.4 Cauchy Inequality at Room Temperature

Table 2.4 Values of the Cauchy inequality Δ

Crystal	$\Delta \,[10^{11}\,\text{dyne cm}^{-2}]$
NaCl Structure	
LiF	−1.59
LiCl	−0.18
LiBr	−0.06
LiI	0.014
NaF	−0.38
NaCl	0.025
NaBr	0.10
NaI	0.14
KF	0.21
KCl	0.057
KBr	0.063
KI	0.08
RbF	0.47
RbCl	0.15
RbBr	0.115
RbI	0.092
CsF	0.68
CsCl Structure	
CsCl	0.12
CsBr	0.09
CsI	0.09

Notes and Comments

1. The Cauchy inequality Δ is defined as $\Delta = C_{12}-C_{44}$. The values of Δ given in Table 2.4 have been calculated from the data given in Table 2.1.
2. Being the difference of two parameters of the same order (sometimes very close), the value of Δ is very sensitive to the individual values of C_{12} and C_{44}. Consequently values of Δ calculated from C_{ij} values from different sources show considerable differences.
3. Different views exist regarding the significance of Δ. According to one view, it is a measure of the partial ionicity. On the other hand, Dick [2.12] showed that the difference between C_{12} and C_{44} arises due to exchange forces.
4. Lewis et al. [2.11] pointed out that Δ is highly temperature dependent. Δ is larger at 4.2 K than at RT for Li and Na halides but is smaller for the K and Rb halides. In fact, in some cases (NaCl, NaBr, KCl, KI) the sign of Δ changes at low temperatures.

2.1.5 Low Temperature Adiabatic Elastic Constants

Table 2.5 Values of adiabatic elastic constants C_{ij} at low temperature

Crystal	T [K]	C_{11}	C_{12}	C_{44}
		$[10^{11}$ dyne cm$^{-2}]$; Ref. [2.13]		
NaCl Structure				
LiF	0	12.460	4.240	6.490
LiCl	4.2	6.074	2.270	2.692
LiBr	0	4.721	1.590	2.052
NaF	4.2	10.850	2.290	2.899
NaCl	4.2	5.733	1.123	1.331
NaBr	4.2	4.800	0.986	1.070
NaI	4.2	3.761	0.798	0.781
KF	4.2	7.570	1.350	1.336
KCl	4.2	4.832	0.540	0.663
KBr	77	4.180	0.560	0.520
KI	4.2	3.380	0.220	0.368
RbF	0	6.527	1.255	0.952
RbCl	4.2	4.297	0.649	0.493
RbBr	4.2	3.863	0.474	0.4085
RbI	4.2	3.210	0.360	0.292
CsCl Structure				
CsBr	4.2	3.437	1.035	0.999
CsI	4.2	2.737	0.793	0.825

Notes and Comments

1. The 0 K elastic constants are of importance in theoretical work e.g. in calculation of Debye temperatures, in lattice dynamics etc.
2. Numerical values of C_{ij} of several alkali halides are given as a function of temperature by Lewis et al. [2.11].
3. Varshni [2.14] has proposed the following equation for the temperature variation of the elastic constants at low temperatures:

$$C_{ij} = C_{ij}{}^{0} - s/\{\exp\,(\theta/T) - 1\} \tag{2.9}$$

where $C_{ij}{}^{0}$ is the value of C_{ij} at 0 K and s and θ are constants. The equation has qualitative theoretical support.

2.1.6 High Temperature Adiabatic Elastic Constants

Table 2.6 Values of the elastic constants at selected high temperatures

Crystal	T [°C]	C_{11}	$C_{11} - C_{12}$	C_{44}	Ref.
		$[10^{11}$ dyne cm$^{-2}]$			
NaCl Structure					
LiF	708	4.38	2.69	4.968	[2.10]
NaCl	800	2.27	1.172	0.943	[2.15]
KCl	700	2.04	1.35	0.517	[2.15]
KBr	700	1.54	1.08	0.425	[2.15]

Table 2.7 Regression coefficients A_0, A_1, A_2 of the polynomial
$$C_{ij} \text{ (in } 10^{11} \text{ dyne cm}^{-2}) = A_0 + A_1 T \times 10^{-3} + A_2 T^2 \times 10^{-6} \quad (T \text{ in } \text{C})$$
for high temperature elastic constants

Crystal	Range of T	C_{ij}	A_0	A_1	A_2	Ref.
NaCl Structure						
NaCl	up to T_m	C_{11}	5.039	−3.700	0.436	[2.15]
		$(C_{11}–C_{12})/2$	1.885	−2.021	0.502	
		C_{44}	1.286	−0.308	−0.158	
KCl	up to T_m	C_{11}	4.154	−3.188	0.347	[2.15]
		$(C_{11}–C_{12})/2$	1.729	−1.812	0.385	
		C_{44}	0.634	−0.120	−0.088	
KBr	up to T_m	C_{11}	3.553	−2.872	0.373	[2.15]
		$(C_{11}–C_{12})/2$	1.495	−1.653	0.417	
		C_{44}	0.511	−0.100	−0.077	
CsCl Structure						
CsCl	up to T_m	C_{11}	3.700	−1.445	−0.699	[2.16]
		$(C_{11}–C_{12})/2$	1.400	−0.371	−0.277	
		C_{44}	0.833	−1.018	0.211	
CsBr	up to T_m	C_{11}	3.092	−1.136	−0.651	[2.16]
		$(C_{11}–C_{12})/2$	1.133	−0.238	−0.279	
		C_{44}	0.774	−0.929	0.183	
CsI	up to T_m	C_{11}	2.474	−0.964	−0.505	[2.16]
		$(C_{11}–C_{12})/2$	0.897	−0.232	−0.191	
		C_{44}	0.649	−0.788	0.154	

Notes and Comments

1. There are a few other reports on high temperature elastic constants of NaCl, KCl and KBr, but the work of Slagle and McKinstry [2.15] is by far the most comprehensive.
2. Using the criterion $C_{11}-C_{12} > 0$ for stability, Slagle and McKinstry [2.15, 2.16] estimated the melting temperature (temperature at which $C_{11}-C_{12} = 0$) and obtained values of correct order but always higher than the observed melting point. For the CsCl lattice, the stability condition $C_{44} > 0$ is used.

2.1.7 Thermoelastic Constants at 0 °C

Table 2.8 Values of the thermoelastic constants $T_{ij} = \mathrm{d}\,(\log C_{ij})\,/\,\mathrm{d}T$ at 0 °C

Crystal	T_{11}	T_{12}	T_{44}	Ref.
	$[10^{-3}\,(°C)^{-1}]$			
NaCl Structure				
LiF	−0.66	0.01	−0.28	[2.1]
LiCl	−0.93	−0.15	−0.42	[2.1]
LiBr	−1.02	−0.14	−0.455	[2.1]
LiI	−1.19	−0.26	−0.56	[2.2]
NaF	−0.637	0.18	−0.21	[2.1]
NaCl	−0.80	0.17	−0.266	[2.1]
NaBr	−0.845	0.10	−0.20	[2.1]
NaI	−0.937	0.10	−0.22	[2.1]
KF	−0.721	0.27	−0.207	[2.1]
KCl	−0.835	0.56	−0.212	[2.1]
KBr	−0.852	0.77	−0.234	[2.1]
KI	−0.918	0.89	−0.206	[2.1]
RbF	−0.773	0.145	−0.19	[2.1]
RbCl	−0.867	0.57	−0.234	[2.1]
RbBr	−0.881	0.70	−0.234	[2.1]
RbI	−0.94	0.87	−0.206	[2.1]
CsF	−0.87	0.02	−0.06	[2.3]
CsCl Structure				
CsCl	−0.416	−0.93	−1.29	[2.4]
CsBr	−0.427	−0.82	−1.31	[2.4]
CsI	−0.442	−0.81	−1.30	[2.4]

Notes and Comments

1. These temperature coefficients are measured over a small temperature interval around 0°C and may not be valid at very low or very high temperatures.
2. Note that T_{11} and T_{44} are −ve for all the crystals whereas T_{12} is +ve for all except LiCl, LiBr, LiI, CsCl, CsBr and CsI.

2.1.8 Temperature Derivatives of Elastic Constants at Constant Volume

Table 2.9 Values of the temperature derivatives $(\partial C/\partial T)_V$ where C represents C_{44} or $C' = (C_{11}-C_{12})/2$ or the adiabatic and isothermal bulk moduli B_S and B_T.

Crystal	$(\partial B_S/\partial T)_V$	$(\partial B_T/\partial T)_V$	$(\partial C_{44}/\partial T)_V$	$(\partial C'/\partial T)_V$
	[kbar (°C)$^{-1}$]; Ref. [2.17]			
NaCl Structure				
LiF	0.100	−0.051	−0.088	−0.140
LiCl	0.041	−0.046	−0.038	−0.073
LiBr	0.032	−0.047	−0.025	−0.065
LiI	0.039	−0.038	−0.017	−0.031
NaF	0.055	−0.025	−0.050	−0.125
NaCl	0.033	−0.016	−0.024	−0.071
NaBr	0.022	−0.021	−0.009	−0.059
NaI	0.018	−0.018	0	−0.051
KF	0.029	−0.024	−0.039	−0.098
KCl	0.014	−0.016	−0.021	−0.081
KBr	0.019	−0.007	−0.017	−0.075
KI	0.017	−0.004	−0.011	−0.064
RbF	0.008	−0.028	−0.035	−0.103
RbCl	0.004	−0.018	−0.020	−0.084
RbBr	0.006	−0.015	−0.017	−0.072
RbI	0.009	−0.012	−0.012	−0.059

Notes and Comments

1. For conversion to other units, note that 1 kbar $= 10^9$ dynes cm$^{-2} = 10^8$ N m^{-2} $= 0.1$ GPa.
2. The $(\partial C/\partial T)_V$ values have been derived from experimental data on $(\partial C/\partial T)_P$ and $(\partial C/\partial P)_T$ using the thermodynamic relationship

$$(\partial C/\partial T)_V = (\partial C/\partial T)_P + \beta\, B_T\, (\partial C/\partial P)_T \tag{2.10}$$

   where β is the volume thermal expansion coefficient and B_T is the isothermal bulk modulus.
3. Note that temperature derivative is +ve for B_S and is −ve for C_{44}, C' and B_T.
4. The magnitude of $(\partial C/\partial T)_V$ decreases smoothly in the sequence F–Cl–Br–I for C_{44} and C' for the four alkali halide families.
5. All the $(\partial C/\partial T)_V$ values for each elastic constant and for each alkali halide group form smooth plots when plotted against the interionic distance [2.17].

2.1.9 Pressure Derivatives of Elastic Constants

Table 2.10 Values of the pressure derivatives (dC_{11}/dP), (dC_{12}/dP), (dC_{44}/dP), (dB_S/dP) and (dB_T/dP) at room temperature over a pressure of 1 kbar

Crystal	(dC_{11}/dP)	(dC_{12}/dP)	(dC_{44}/dP)	(dB_S/dP)	(dB_T/dP)	Ref.
NaCl Structure						
LiF	10.01	2.81	1.38	5.21	5.30	[2.2]
LiCl	10.35	2.95	1.70	5.42	5.63	[2.2]
LiBr	10.39	2.89	1.80	5.39	5.68	[2.2]
LiI	11.12	3.12	1.96	5.79	6.15	[2.2]
NaF	11.57	1.99	0.205	5.18	5.25	[2.18]
NaCl	11.66	2.08	0.37	5.27	5.38	[2.18]
NaBr	11.73	2.07	0.46	5.29	5.44	[2.18]
NaI	11.80	2.20	0.61	5.40	5.58	[2.18]
KF	12.27	1.77	−0.43	5.26	5.38	[2.18]
KCl	12.83	1.61	−0.39	5.34	5.46	[2.18]
KBr	12.96	1.60	−0.33	5.38	5.47	[2.18]
KI	13.50	1.44	−0.24	5.47	5.56	[2.18]
RbF	12.14	2.28	−0.70	5.57	5.69	[2.19]
RbCl	13.32	1.56	−0.56	5.48	5.62	[2.19]
RbBr	13.49	1.43	−0.55	5.45	5.59	[2.19]
RbI	13.79	1.27	−0.51	5.44	5.60	[2.19]
CsCl Structure						
CsCl	7.01	5.14	3.69	5.76	-	[2.20]
CsBr	6.71	5.21	3.81	5.71	-	[2.20]
CsI	6.72	5.12	3.84	5.65	-	[2.20]

Notes and Comments

1. The pressure coefficients given are slopes of the C_{ij} versus P curves at $P = 0$ and hence are valid over a small range of pressure. At higher pressures, the C_{ij} versus P curves become nonlinear and hence the values of the coefficients will be different.
2. Plots of (dC_{ij}/dP) versus interionic distance r are monotonic, smooth and with small slopes (i.e. slow variation with r). In the whole series, dC_n/dP (where $C_n = (C_{11}+C_{12}+2C_{44})/2$, is nearly invariant and dB_S/dP is essentially invariant. On the other hand, dC'/dP increases in the sequence Li–Na–K–Rb, where $C' = (C_{11}-C_{12})/2$. Further, dC_{44}/dP is +ve for halides of Li, Na and Cs and −ve for the K and Rb halides.
3. Roberts and Smith [2.18] showed that the plot of dB_T/dP versus n (n being the Born index in the power law) is linear; this conforms with the relation

$$d B_T / d P = (n + 7) / 3 \tag{2.11}$$

4. Motida [2.7] has shown a linear relationship between $(C_{44}{}^a/C_0)$ and $(dC_{44}{}^a/dP)$ where C_{44} and dC_{44}/dP are measured values and

$$C_0 = e^2 / r^4 \tag{2.12}$$

$$(C_{44}{}^a / C_0) = (C_{44} / C_0) - \{(e^*)^2 (\chi - \alpha) / 4r^4\} \tag{2.13}$$

$$(dC_{44}{}^a / dP) = (dC_{44} / dP) + \frac{\psi e^2}{3r^4}\left\{ \frac{\alpha}{6}(n-1) + (\alpha - \chi)\left(\frac{e^*}{e}\right)^2 \right\} \tag{2.14}$$

Here, n is the Born repulsive index, ψ the compressibility, e^* the effective ionic charge, α the Madelung constant and χ a constant parameter from the Krishnan-Roy theory [2.8]. For the NaCl structure, $\chi = 3.14$.

2.2 Compressibility and Bulk Modulus

2.2.1 Compressibility and Bulk Modulus at RT and 0 K

Table 2.11 Values of the adiabatic compressibility ψ at RT and atmospheric pressure, ψ_0 at 0 K and adiabatic bulk modulus B_S at RT

Crystal	ψ [2.21]	$\psi_0{}^*$ [2.22]	$B_S\ (=1/\psi)$
	$[10^{-12}\ \mathrm{cm}^2\ \mathrm{dyne}^{-1}]$		$[10^{11}\ \mathrm{dyne\ cm}^{-2}]$
NaCl Structure			
LiF	1.49	1.27	6.71
LiCl	3.36	2.77	2.97
LiBr	4.20	3.56	2.38
LiI	5.83	4.44	1.71
NaF	2.15	1.88	4.65
NaCl	4.17	3.62	2.40
NaBr	5.02	4.37	1.99
NaI	6.64	5.64	1.51
KF	3.28	2.89	3.05
KCl	5.73	5.01	1.74
KBr	6.75	5.90	1.48
KI	8.55	7.43	1.17
RbF	3.81	3.30	2.62
RbCl	6.40	5.61	1.56
RbBr	7.69	6.57	1.30
RbI	9.48	8.15	1.05
CsF	4.25	-	2.35
CsCl Structure			
CsCl	5.55	-	1.80
CsBr	6.28	-	1.59
CsI	7.83	-	1.28

* ψ_0 calculated from ψ by applying a thermodynamic correction.

Notes and Comments

1. The bulk modulus (or its reciprocal counterpart, the compressibility) is an important parameter as it enters into several theoretical calculations like the cohesive energy, the Gruneisen parameter etc.
2. The bulk modulus and compressibility can be determined directly by the piston displacement method or can be calculated from the elastic constants using Eq. (2.15).

$$B = (1/\psi) = (C_{11} + 2C_{12})/3 \tag{2.15}$$

3. The isothermal bulk modulus (B_T) can be calculated from the relation

$$B_T = B_S / (1 + T\,\beta\,\gamma) \tag{2.16}$$

where β is the volume coefficient of thermal expansion and γ the Gruneisen constant.
4. The values of B (or $1/\psi$) may be converted to other units by the following conversion formulae:
$1\ \mathrm{GPa} = 10^9\ \mathrm{N/m^2} = 0.01\ \mathrm{Mbar} = 1 \times 10^{10}\ \mathrm{dyne\ cm^{-2}}$.
5. The bulk modulus (B) is related to the volume (V) through the relation

$$B = B_0\, V^{-n} \tag{2.17}$$

where B_0 and n are constants. The constant n is equal to 1 although different authors have obtained slightly different values like 1 [2.23], 1.06 [2.24] and 1.13 [2.25] depending on the B and V data used by them. Neumann [2.26] found that the log B versus log V plots resolve into four different plots indicating that the constant B_0 is different for the halides of each alkali ion. He suggested that

$$B_0 = b_0 (1 - b_1 f_i) \tag{2.18}$$

where b_0 and b_1 are constants for the whole family and f_i is the Phillips ionicity. With this modification, he obtained 1.147 for n.
6. The bulk modulus is related to the formation energy of a Schottky pair E_f through the relation

$$E_f = 3.4 \times 10^{-13}\, V_M\, B \tag{2.19}$$

where B is expressed in 10^{12} dyne cm^{-2}. Values of E_f calculated from B are given by Pathak and Vasavada [2.27].
7. Hanneman and Gatos [2.28] pointed out that values of the coefficient of thermal expansion lie on a smooth curve when plotted against the reciprocal of bulk modulus. Sirdeshmukh [2.29] showed that linear plots exist between $\alpha\, V_M$ and B^{-1}, i.e.

$$\alpha\, V_M\, B = \mathrm{constant} \tag{2.20}$$

8. Szigeti [2.30] derived the following relation between the compressibility ψ and the transverse optic mode frequency v_{TO}:

$$\psi = \left[\frac{\varepsilon_\infty + 2}{\varepsilon_0 + 2} \right] \left[\frac{6r}{\mu \, v_{TO}^2} \right] \tag{2.21}$$

where ε_∞ is the high frequency dielectric constant, ε_0 the static dielectric constant, r the interionic distance and μ the reduced mass. Mitra and Marshall [2.23] calculated ψ for several alkali halides from this formula and found fair agreement with experimental values of ψ.

9. Brout [2.31] derived the following relation for ionic crystals with NaCl structure:

$$\sum_{i=1}^{6} 4\pi^2 v_i^2(k) = (18r / \mu \, \psi) \tag{2.22}$$

where v_i are the lattice vibrational frequencies for a given value of the propagation vector. Mitra and Marshall [2.23] calculated ψ for alkali halides using experimental values for v_i at $k = 0$, but found that the agreement with experimental ψ values was not very satisfactory.

2.2.2 Pressure and Temperature Derivatives of Compressibility (Experimental)

Table 2.12 Values of the pressure derivative of compressibility $(1/\psi)(\partial \psi / \partial P)_T$ and its temperature derivative $(1/\psi)(\partial \psi / \partial T)_P$ at RT and at ambient pressure

Crystal	$-(1/\psi)(\partial \psi / \partial P)_T$ $[10^{-12} \, cm^2 \, dyne^{-1}]$ Ref. [2.21]	$(1/\psi)(d\psi / dT)_P$ $[10^{-4} \, K^{-1}]$
NaCl Structure		
LiF	5	2
LiCl	20	7
LiBr	24	8
LiI	37.3	5*
NaF	18	2*
NaCl	21	6.2
NaBr	25	4.1
NaI	28	1.0
KF	20	1.0
KCl	30	3.5
KBr	31	4.8
KI	32	3.4
RbF	17*	2*
RbCl	17	4
RbBr	21	3.8
RbI	21	4.6
CsF	28.4	10.2

Table 2.12 (Continued)

Crystal	$-(1/\psi)(\partial\psi/\partial P)_T$ $[10^{-12}\ \mathrm{cm}^2\ \mathrm{dyne}^{-1}]$	$(1/\psi)(\mathrm{d}\psi/\mathrm{d}T)_P$ $[10^{-4}\ \mathrm{K}^{-1}]$
	Ref. [2.21]	
CsCl Structure		
CsCl	25	5.5
CsBr	27	3.6
CsI	32	5.2

* empirically estimated values

2.3 Polycrystalline Elastic Properties

2.3.1 Elastic Moduli of Polycrystalline Aggregates (RT)

Table 2.13 Bulk modulus B, shear modulus G, Young's modulus E and Poisson's ratio σ, suffixes V, R, and VRH indicate Voigt average, Reuss average and Voigt-Reuss-Hill average; for method of evaluation see Notes and Comments

Crystal	B	G_V	G_R	G_{VRH}	E_V	E_R	E_{VRH}	σ_V	σ_R	σ_{VRH}
	$[10^{11}\ \mathrm{dynes\ cm}^{-2}]$									
NaCl Structure										
LiF	6.958	5.129	4.634	4.882	12.352	11.376	11.864	0.204	0.227	0.216
LiCl	3.168	2.009	1.837	1.923	4.975	4.618	4.797	0.238	0.257	0.248
LiBr	2.559	1.571	1.433	1.502	3.912	3.623	3.767	0.245	0.264	0.255
LiI	1.916	1.141	1.036	1.089	2.856	2.633	2.745	0.252	0.271	0.262
NaF	4.853	3.140	3.091	3.116	7.749	7.649	7.699	0.234	0.237	0.236
NaCl	2.505	1.488	1.441	1.465	3.726	3.627	3.677	0.252	0.258	0.255
NaBr	2.064	1.178	1.137	1.158	2.969	2.882	2.926	0.260	0.267	0.264
NaI	1.595	0.873	0.845	0.859	2.215	2.154	2.185	0.269	0.275	0.272
KF	3.160	1.770	1.570	1.670	4.475	4.041	4.258	0.264	0.287	0.275
KCl	1.819	1.056	0.845	0.951	2.654	2.195	2.425	0.257	0.299	0.278
KBr	1.539	0.885	0.686	0.786	2.228	1.792	2.010	0.259	0.306	0.283
KI	1.220	0.684	0.508	0.596	1.729	1.338	1.534	0.264	0.317	0.290
RbF	2.772	1.381	1.187	1.284	3.553	3.116	3.335	0.286	0.312	0.299
RbCl	1.621	0.883	0.643	0.737	2.242	1.704	1.973	0.270	0.325	0.298
RbBr	1.382	0.760	0.532	0.646	1.927	1.414	1.671	0.268	0.329	0.298
RbI	1.108	0.609	0.397	0.503	1.544	1.064	1.304	0.268	0.340	0.304
CsF	2.389	1.027	0.895	0.961	2.695	2.387	2.541	0.312	0.334	0.323
CsCl Structure										
CsCl	1.827	1.024	0.958	0.991	2.588	2.446	2.517	0.264	0.277	0.270
CsBr	1.593	0.902	0.867	0.885	2.276	2.202	2.239	0.262	0.270	0.266
CsI	1.290	0.720	0.700	0.710	1.821	1.778	1.800	0.265	0.270	0.268

Notes and Comments

1. The Voigt and Reuss values are based on the assumption of uniform strain and uniform stress respectively. These average values can be calculated from the single crystal elastic constants using the following relations

$$B_V = B_R = C_{11} - (2C/3) \tag{2.23}$$

$$G_V = (C/5) + (3C_{44}/5) \tag{2.24}$$

$$G_R = (5CC_{44})/(4C_{44} + 3C) \tag{2.25}$$

$$C = (C_{11} - C_{12}) \tag{2.26}$$

E_V and E_R values can be calculated from G and B values from the relation

$$E = 9BG/(3B + G) \tag{2.27}$$

and values of the Poisson's ratio σ_V and σ_R from

$$(E/G) = 2(1 + \sigma) \tag{2.28}$$

The VRH average values are the mean of the Voigt and Reuss values [2.33].
2. The values in Table 2.13 are calculated from the single crystal elastic constants given in Table 2.1 using Eqs. (2.23–2.28).
3. From the analysis of data on some crystals (other than the alkali halides) Chung and Buessem [2.32] pointed out that the difference between the Voigt and Reuss values increases with anisotropy A^* and that the VRH values are close to the experimental values measured on polycrystalline aggregates. There is only one report (Subramanyam [2.34]) on the experimental determination of the elastic properties of polycrystalline aggregates of alkali halides. His results for nearly fully dense samples (Table IV) agree with the VRH values quoted from Table 2.13.

Table IV Elastic properties of polycrystalline aggregates of alkali halides [2.34]

Crystal	Density [gm cm^{-3}]		G [10^{11} dynes cm^{-2}]		E [10^{11} dynes cm^{-2}]	
	Aggregate	Crystal	Measured	VRH	Measured	VRH
NaCl	2.12	2.168	1.47	1.465	3.65	3.677
KCl	1.96	1.984	1.02	0.951	2.56	2.425
KBr	2.66	2.756	0.80	0.786	2.01	2.010

2.3.2 Temperature Derivatives of Polycrystalline Elastic Moduli

Table 2.14 Values of temperature derivatives of bulk modulus B, Voigt average of shear modulus G_V, Reuss average of shear modulus G_R, Voigt average of Young's modulus E_V and Reuss average of Young's modulus E_R.

Crystal	dB/dT	dG_V/dT	dG_R/dT	dE_V/dT	dE_R/dT
	All temperature derivatives in kbar deg^{-1}; Ref. [2.10]				
NaCl Structure					
NaCl	0.144	0.050	0.018	0.139	0.072
RbCl	−0.080	−0.074	−0.023	−0.176	−0.064
RbBr	−0.070	−0.068	−0.020	−0.162	−0.055
RbI	−0.059	−0.058	−0.014	−0.138	−0.040

2.3.3 Pressure Derivatives of Polycrystalline Elastic Moduli

Table 2.15 Values of the pressure derivatives of bulk modulus B, Voigt average of shear modulus G_V, Reuss average of shear modulus G_R and Voigt-Reuss-Hill average of shear modulus G_{VRH}; these values have been calculated by Kameswara Rao [2.35] from the equations given by Chung [2.36] using data on pressure derivatives of single crystal elastic constants given in Table 2.10

Crystal	dB/dP	dG_V/dP	dG_R/dP	dG_{VRH}/dP
NaCl Structure				
LiF	6.210	2.268	3.299	2.784
LiCl	5.550	2.500	3.385	2.943
LiBr	5.390	2.580	3.529	3.054
LiI	5.787	2.428	3.718	3.073
NaF	5.179	2.039	1.526	1.782
NaCl	5.274	2.138	1.508	1.823
NaBr	5.290	2.208	1.559	1.884
NaI	5.400	2.286	1.632	1.959
KF	5.270	1.842	0.426	1.134
KCl	5.350	2.010	0.144	1.077
KBr	5.390	2.036	0.150	1.093
KI	5.460	2.266	0.216	1.241
RbF	5.570	1.552	0.195	0.874
RbCl	5.480	2.016	−0.200	0.908
RbBr	5.450	2.082	−0.258	0.912
RbI	5.440	2.198	−0.666	0.763
CsCl Structure				
CsCl	5.764	2.580	3.372	2.976
CsBr	5.710	2.586	3.208	2.897
CsI	5.660	2.624	3.168	2.896

Notes and Comments

1. The pressure derivatives are calculated from the following formulae given by Chung [2.36]

$$\mathrm{d}B_V/\mathrm{d}P = \mathrm{d}B_R/\mathrm{d}P = \mathrm{d}B/\mathrm{d}P = (\mathrm{d}C_{11}/\mathrm{d}P) - (2/3)(\mathrm{d}C/\mathrm{d}P) \qquad (2.29)$$

$$\mathrm{d}G_V/\mathrm{d}P = (1/5)(\mathrm{d}C/\mathrm{d}P) + (3/5)(\mathrm{d}C_{44}/\mathrm{d}P) \qquad (2.30)$$

$$\mathrm{d}G_R/\mathrm{d}P = (4/5)(G_R/C)^2(\mathrm{d}C/\mathrm{d}P) + (3/5)(G_R/C_{44})^2(\mathrm{d}C_{44}/\mathrm{d}P) \qquad (2.31)$$

   where $C = (C_{11} - C_{12})$ and $G_R = 5CC_{44}/(3C + 4C_{44})$.

2. The pressure derivatives of Young's moduli can be evaluated from those of the bulk and shear moduli from the relation $E = 9BG/(3B + G)$.
3. Experimental values are not available for comparison.
4. Note that $\mathrm{d}G_R/\mathrm{d}P$ takes very low values for the potassium halides and $-$ve values for the Rb halides.
5. In general the difference $(\mathrm{d}G_V/\mathrm{d}P - \mathrm{d}G_R/\mathrm{d}P)$ increases with the anisotropy A^*.

2.4 Compression Data at High Pressures

2.4.1 Compression Data up to 45 kbar (Experimental)

The compression data obtained by the piston displacement method are given in Table 2.16 (see p. 33).

Notes and Comments

1. Vaidya and Kennedy [2.37] have fitted their compression data to the equation of state

$$\Delta V / V_0 = a_0 + aP + bP^2 + cP^3 \qquad (2.32)$$

The values of the coefficients a_0, a, b and c are given in Table V.

Table V Values of a_0, a, b and c in Eq. (2.32). Phase I and II refer to NaCl and CsCl structures respectively.

Crystal	Phase	Pr. range [kbar]	a_0	$-a \times 10^4$	$+b \times 10^6$	$-c \times 10^5$
LiF	I	0–45	-	15.491	6.5364	
LiCl	I	0–45	-	30.342	13.705	
LiBr	I	0–45	-	38.951	21.239	
LiI	I	0–45	-	57.466	66.674	4.3154
NaF	I	0–45	-	20.648	9.0621	
NaCl	I	0–45	-	42.730	46.578	3.2499
NaBr	I	0–45	-	48.734	51.871	3.4037
NaI	I	0–45	-	66.622	96.809	8.1299

Table V (Continued)

Crystal	Phase	Pr. range [kbar]	a_0	$-a \times 10^4$	$+b \times 10^6$	$-c \times 10^5$
KF	I	0–15	-	61.814	78.118	
KCl	I	0–15	-	54.906	57.969	
KCl	II	20–45	0.13753	34.664	20.600	
KBr	I	0–15	-	60.351	53.606	
KBr	II	20–45	0.12006	45.007	30.882	
KI	I	0–15	-	78.860	112.85	
KI	II	20–45	0.11957	46.086	29.361	
RbF	II	10–30	0.01672	47.177	31.636	
RbCl	II	10–45	0.15178	47.902	68.724	5.5232
RbBr	II	5–45	0.15165	44.797	31.635	
RbI	II	5–45	0.13756	63.083	94.145	7.1946
CsCl		0–45	-	56.401	73.842	5.4194
CsBr		0–45	-	65.652	96.190	7.3963
CsI		0–45	-	78.488	130.42	11.548

2. Vaidya and Kennedy also fitted the data to the Murnaghan equation

$$P = (B / B')\{V_0 / V)^{B'} - 1\} \tag{2.33}$$

and obtained the values of bulk modulus B and pressure derivative B' which are given in Table VI.

Table VI Values of B and B' from Eq. (2.33)

Crystal	B [kbar]	B' [kbar]
NaCl Structure		
LiF	627.9	6.82
LiCl	318.5	3.36
LiBr	242.7	3.50
LiI	168.4	4.32
NaF	467.4	5.18
NaCl	232.3	4.92
NaBr	203.1	4.19
NaI	151.0	4.15
CsCl Structure		
CsCl	171.2	5.09
CsBr	144.0	5.32
CsI	124.7	4.50

Table 2.16 Values of fractional volume V/V_0; pressure P; density ρ; underline indicates commencement of high pressure CsCl phase

Crystal	ρ [g cm^{-3}]	$P\rightarrow$ [kbar] 5	10	15	20	25	30	35	40	45
		V/V_0; Ref. [2.37]								
NaCl Structure										
LiF	2.638	0.9924	0.9852	0.9782	0.9716	0.9654	0.9594	0.9538	0.9485	0.9435
LiCl	2.074	0.9852	0.9710	0.9576	0.9448	0.9327	0.9213	0.9106	0.9006	0.8912
LiBr	3.464	0.9811	0.9632	0.9463	0.9306	0.9159	0.9023	0.8897	0.8782	0.8677
LiI	4.115	0.9728	0.9487	0.9273	0.9083	0.8913	0.8760	0.8620	0.8492	0.8371
NaF	2.802	0.9899	0.9803	0.9711	0.9623	0.9540	0.9462	0.9388	0.9319	0.9254
NaCl	2.164	0.9798	0.9616	0.9452	0.9306	0.9172	0.9050	0.8936	0.8828	0.8724
NaBr	3.200	0.9768	0.9561	0.9374	0.9206	0.9053	0.8913	0.8784	0.8663	0.8547
NaI	3.670	0.9690	0.9422	0.9191	0.8990	0.8812	0.8652	0.8506	0.8364	0.8222
KF	2.524	0.9710	0.9460	0.9249	<u>0.8374</u>	0.8229	0.8105	0.7995	0.7893	0.7794
KCl	1.9865	0.9740	0.9509	0.9307	<u>0.8014</u>	0.7887	0.7770	0.7664	0.7568	0.7482
KBr	2.7533	0.9712	0.9450	0.9215	<u>0.8023</u>	0.7867	0.7727	0.7602	0.7493	0.7399
KI	3.1257	0.9634	0.9324	0.9071	<u>0.8000</u>	0.7836	0.7686	0.7551	0.7431	0.7325
RbF	3.8665	0.9760	<u>0.9393</u>	0.9196	0.9016	0.8851	0.8702	0.8350	0.8200	0.8095
RbCl	2.818	0.9709	<u>0.8066</u>	0.7900	0.7755	0.7628	0.7515	0.7411	0.7312	0.7215
RbBr	3.359	<u>0.8267</u>	0.8067	0.7883	0.7714	0.7561	0.7424	0.7303	0.7198	0.7108
RbI	3.564	<u>0.8322</u>	0.8081	0.7866	0.7682	0.7523	0.7385	0.7261	0.7147	0.7036
CsCl Structure										
CsCl	3.988	0.9736	0.9504	0.9302	0.9124	0.8967	0.8826	0.8698	0.8579	0.8463
CsBr	4.456	0.9695	0.9432	0.9207	0.9012	0.8844	0.8696	0.8563	0.8440	0.8319
CsI	4.525	0.9639	0.9334	0.9078	0.8860	0.8674	0.8509	0.8357	0.8210	0.8059

2.4.2 Equation of State Parameters (Experimental)

Table 2.17 Values of zero pressure bulk modulus B_0 and its pressure derivative B_0' from ultrasonic measurements and the volume V_0

Crystal	$V_0\,[\mathrm{cm^3\ mole^{-1}}]$	$B_0\,[\mathrm{GPa}]$	B_0'
	Ref. [2.38]		
NaCl Structure			
LiF	9.82	66.51	5.30
LiCl	20.42	29.68	5.63
LiBr	25.02	23.52	5.68
LiI	33.17	17.26	6.15
NaF	14.98	46.48	5.25
NaCl	27.01	23.68	5.38
NaBr	32.16	19.47	5.44
NaI	40.81	14.87	5.58
KF	23.00	30.22	5.38
KCl	37.50	17.35	5.46
KBr	43.28	14.64	5.47
KI	53.11	11.51	5.56
RbF	27.18	26.68	5.69
RbCl	43.20	15.58	5.62
RbBr	49.37	13.24	5.59
RbI	59.78	10.49	5.60
CsCl Structure			
CsCl	42.18	16.74	5.98
CsBr	47.73	14.34	5.95
CsI	57.40	11.89	5.93

2.4.3 Equation of State Parameters for CsCl Phase (Experimental)

Table 2.18 Values of V_0 the initial molar volume, B_0 initial bulk modulus and B_0' pressure derivative of bulk modulus at ambient pressure

Crystal	$V_0\,[\mathrm{cm^3\ mole^{-1}}]$	$B_0\,[\mathrm{GPa}]$	B_0'
	Ref. [2.38]		
CsCl Structure			
CsCl	42.185	17.0	5.58
CsBr	47.722	14.8	5.80
CsI	57.397	11.9	5.93
High Pressure CsCl Phase			
KCl	32.4	22.2	4.93
KBr	37.9	17.0	5.38
KI	49.0	10.1	5.77
RbCl	36.7	18.2	5.37
RbBr	42.2	14.8	5.67
RbI	52.8	10.1	5.93

Notes and Comments

1. The values given in Table 2.18 are derived by fitting high pressure lattice parameter data obtained from energy dispersive X-ray diffraction to the following equation of state:

$$P = 3\, B_0 \left(\frac{1-X}{X^5}\right)\, \exp\left[c_0(1-X)\right] \tag{2.34}$$

with $X = (V/V_0)^{1/3}$ and $c_0 = (3/2)(B_0' - 3)$. In the normal CsCl phase, V_0 is the initial molar volume. In the high pressure CsCl phases, V_0 is the molar volume at $P = 0$ obtained by extrapolation of the high pressure volume data.

2.4.4 High Temperature Compression Data for NaCl (Experimental)

Table 2.19 Values of the fractional volume V/V_0 obtained at pressure P and temperature T

T [°C] P [kbar] $\downarrow$	V/V_0; Ref. [2.39]					
	25	100	200	300	400	500
0	1	1.00928	1.02253	1.03682	1.05226	1.06908
5	0.98020	0.98833	0.99989	1.01247	1.02575	1.04014
10	0.96268	0.96999	0.98024	0.99124	1.02290	1.01538
15	0.94690	0.95356	0.96277	0.97248	0.98287	0.99382
20	0.93248	0.93857	0.94699	0.95571	0.96509	0.97478
25	0.91913	0.92471	0.93251	0.94056	0.94913	0.95776
30	0.90667	0.91172	0.91907	0.92678	0.93468	0.94241
35	0.89492	0.89943	0.90647	0.91416	0.92150	0.92845

2.4.5 High Temperature Compression Data for NaCl (Theoretical)

Table 2.20 Theoretically calculated values of the fractional change in volume $\Delta V/V_0$ at pressure P and temperature T (for method of calculation see [2.40, 2.41] and Notes and Comments)

T [°C] $\Delta V/V_0$ $\downarrow$	P [kbar]; Ref. [2.40]						
	0	25	100	200	300	500	800
0.0	-	0.00	2.13	5.00	7.89	13.72	22.48
−0.0060	0.74	1.44	3.57	6.44	9.34	15.16	23.93
−0.0120	2.24	2.94	5.06	7.93	10.83	16.65	25.43
−0.0179	3.77	4.47	6.60	9.47	12.37	18.19	26.97
−0.0238	5.36	6.06	8.18	11.06	13.95	19.78	28.56
−0.0297	7.00	7.70	9.82	12.69	15.59	21.42	30.20
−0.0356	8.68	9.38	11.51	14.38	17.28	23.11	31.89

Table 2.20 (Continued)

T [°C]	P [kbar]; Ref. [2.40]						
$\Delta V/V_0 \downarrow$	0	25	100	200	300	500	800
−0.0414	10.42	11.12	13.24	16.12	19.01	24.85	33.63
−0.0472	12.22	12.91	15.04	17.91	20.81	26.64	35.43
−0.0530	14.07	14.76	16.88	19.76	22.65	28.49	37.28
−0.0588	15.97	16.67	18.79	21.66	24.56	30.40	39.19
−0.0646	17.93	18.63	20.75	23.62	26.52	32.36	41.16
−0.0703	19.96	20.65	22.77	25.65	28.54	34.38	43.19
−0.0760	22.04	22.74	24.86	27.73	30.63	36.47	45.27
−0.0817	24.19	24.88	27.00	29.87	32.77	38.61	47.42
−0.0873	26.40	27.10	29.21	32.08	34.98	40.83	49.64
−0.0930	28.68	29.37	31.49	34.36	37.26	43.11	51.92
−0.0986	31.03	31.72	33.84	36.71	39.61	45.45	54.27
−0.1042	33.44	34.13	36.25	39.12	42.02	47.87	56.69
−0.1097	35.93	36.62	38.74	41.61	44.51	50.35	59.18
−0.1153	38.49	39.18	41.30	44.16	47.07	52.92	61.74
−0.1208	41.13	41.82	43.93	46.80	49.70	55.55	64.38
−0.1263	43.84	44.53	46.64	49.51	52.41	58.26	67.10
−0.1317	46.63	47.32	49.43	52.30	55.20	61.06	69.89
−0.1372	49.51	50.20	52.31	55.18	58.08	63.93	72.77
−0.1426	52.47	53.16	55.26	58.13	61.03	66.89	75.73
−0.1480	55.51	56.20	58.31	61.17	64.08	69.93	78.77
−0.1534	58.64	59.33	61.44	64.30	67.21	73.06	81.91
−0.1588	61.87	62.55	64.66	67.53	70.43	76.29	85.13
−0.1641	65.18	65.87	67.97	70.84	73.74	79.60	88.45
−0.1694	68.59	69.28	71.38	74.25	77.15	83.01	91.86
−0.1747	72.10	72.79	74.89	77.75	80.66	86.52	95.37
−0.1800	75.71	76.40	78.50	81.36	84.26	90.13	98.98
−0.1852	79.42	80.11	82.21	85.07	87.97	93.84	102.70
−0.1904	83.24	83.93	86.02	88.89	91.79	97.65	106.52
−0.1956	87.17	87.85	89.95	92.81	95.71	101.58	110.44
−0.2008	91.21	91.89	93.98	96.84	99.75	105.61	114.48
−0.2060	95.36	96.04	98.13	100.99	103.90	109.76	118.64
−0.2111	99.63	100.31	102.40	105.26	108.16	114.03	122.91
−0.2162	104.01	104.69	106.79	109.64	112.55	118.42	127.30
−0.2213	108.53	109.21	111.30	114.15	117.06	122.93	131.81
−0.2264	113.17	113.84	115.93	118.79	121.69	127.57	136.45
−0.2314	117.94	118.61	120.70	123.56	126.46	132.33	141.22
−0.2364	122.84	123.52	125.60	128.46	131.36	137.23	146.12
−0.2414	127.88	128.55	130.64	133.49	136.40	142.27	151.16
−0.2464	133.06	133.73	135.82	138.67	141.57	147.45	156.34
−0.2514	138.38	139.06	141.14	143.99	146.89	152.77	161.67
−0.2563	143.86	144.53	146.61	149.46	152.36	158.24	167.14
−0.2612	149.48	150.15	152.23	155.08	157.98	163.86	172.76
−0.2661	155.26	155.93	158.01	160.86	163.76	169.64	178.54

Table 2.20 (Continued)

T [°C]	P [kbar]; Ref. [2.40]						
$\Delta V/V_0 \downarrow$	0	25	100	200	300	500	800
−0.2710	161.20	161.87	163.94	166.79	169.70	175.58	184.48
−0.2758	167.30	167.97	170.05	172.90	175.80	181.68	190.59
−0.2807	173.58	174.25	176.32	179.16	182.07	187.95	196.86
−0.2855	180.02	180.69	182.76	185.61	188.51	194.39	203.30
−0.2903	186.65	187.31	189.38	192.23	195.13	201.01	209.93
−0.2950	193.45	194.12	196.18	199.03	201.93	207.81	216.73
−0.2998	200.44	201.11	203.17	206.02	208.92	214.80	223.72
−0.3045	207.63	208.29	210.35	213.20	216.10	221.98	230.91
−0.3092	215.01	215.67	217.73	220.57	223.47	229.36	238.29
−0.3139	222.59	223.26	225.31	228.15	231.05	236.94	245.87
−0.3185	230.38	231.05	233.10	235.94	238.84	244.73	253.66
−0.3232	238.39	239.05	241.10	243.94	246.84	252.73	261.66
−0.3278	246.61	247.27	249.32	252.16	255.06	260.95	269.88
−0.3324	255.06	255.72	257.77	260.60	263.50	269.39	-
−0.3369	263.74	264.39	266.44	269.28	272.17	278.06	-
−0.3415	272.65	273.31	275.35	278.19	281.08	286.97	-
−0.3460	281.81	282.46	284.51	287.34	290.23	296.12	-
−0.3505	291.21	291.87	293.91	296.74	299.63	-	-
−0.3550	300.87	301.53	303.57	306.40	309.29	-	-
−0.3595	310.80	311.45	313.49	316.31	-	-	-

Notes and Comments

1. The fractional change in volume $\Delta V/V_0$ given in Table 2.19 is calculated from
   the Mie-Gruneisen equation of state using for the potential energy ϕ, the Born
   equation

$$\phi = N_0 \left[-(\alpha e^2 / r) - (C / r^6) - (D / r^8) + Q\, b \exp(-r / \rho) \right. \\ \left. + Q' b_- \exp(-r\delta / \rho_-) + Q' b_+ \exp(-r\delta / \rho_+) \right] \tag{2.35}$$

   at pressure P [kbar] and temperature T [°C]. In Table 2.20, the values of pressure P are given at chosen values of T and $\Delta V/V_0$. In the equation the terms in
   serial order represent the Madelung energy, the dipole-dipole Van der Waals
   term, the dipole-quadrupole Van der Waals term, the nearest neighbour repulsion term, second nearest neighbour (SNN) anion-anion repulsion term and the
   SNN cation-cation repulsion terms respectively. Further, $\rho_+ = (2\rho\, r_+)/(r_+ + r_-)$,
   $\rho_- = (2\rho\, r_-)/(r_+ + r_-)$, $b_+ = b\,(\alpha_+ + \alpha_-)/2\alpha_+$ and $b_- = b\,(\alpha_+ + \alpha_-)/2\alpha_-$ where r_+
   and r_- are the radii of the cation and anion and α_+ and α_- are the electronic polarisabilities of the cation and anion. Thus, there are two parameters to be determined viz. b and ρ. The various parameters used in the calculations for NaCl
   are [2.40]:

$$r = 2.8200 \text{ Å} \qquad b = 2.1068 \times 10^{-9} \text{ erg}$$
$$r_+ = 1.20 \text{ Å} \qquad C = 1.40 \times 10^{-58} \text{ erg cm}^6$$
$$r_- = 1.62 \text{ Å} \qquad D = 1.8 \times 10^{-74} \text{ erg cm}^8$$
$$\rho = 0.3121 \text{ Å}$$

2. These calculated values have been found to agree well with several experimental results at various pressure and temperature ranges and hence these results are recommended for pressure calibration and for use of NaCl as marker.

2.5 Higher-Order Elastic Constants

2.5.1 Third-Order Elastic Constants (Experimental)

Table 2.21 Values of the third-order elastic constants C_{ijk} at 25 °C

Crystal	C_{111}	C_{112}	C_{123}	C_{144}	C_{166}	C_{456}	Ref.
	$[10^{11}$ Pa (or 10^{12} dyne cm^{-2})]						
NaCl Structure							
LiF	−14.23	−2.64	1.56	0.85	−2.73	0.94	[2.42]
NaF	−14.8	−2.70	2.80	0.46	−1.14	0.00	[2.43]
NaCl	− 8.43	−0.50	0.46	0.29	−0.60	0.26	[2.42]
KCl	− 7.26	−0.24	0.11	0.23	−0.26	0.16	[2.42]
RbCl	− 6.71	−0.18	0.05	0.11	−0.17	0.14	[2.44]

Notes and Comments

1. The internal energy Φ of a deformed crystal may be expressed as:

$$\Phi = \Phi_0 + (V/2)\, C_{ijkl}\, \varepsilon_{ij}\, \varepsilon_{kl} + (V/6)\, C_{ijklmn}\, \varepsilon_{ij}\, \varepsilon_{kl}\, \varepsilon_{mn} \tag{2.36}$$

   where Φ_0 is the internal energy of the undeformed crystal, V the unit cell volume, C_{ijkl} the second-order elastic constants, C_{ijklmn} the third-order elastic constants (TOEC) and ε_{ij} the components of the strain tensor. By applying symmetry conditions and thermodynamic considerations and by using Voigt's contracted notation, the number of independent TOEC for a cubic crystal reduces to six viz., C_{111}, C_{112}, C_{123}, C_{144}, C_{166} and C_{456}.
2. The absolute value of C_{111} is larger than those of other TOEC. Values of C_{111} are an order of magnitude larger than those of SOEC.
3. Among the alkali halides studied, C_{111}, C_{112} and C_{166} are negative and others are positive.
4. Absolute values of TOEC, particularly C_{111}, show systematic variation in the sequence NaCl–KCl–RbCl.

2.5.2 Third-Order Elastic Constants (Theoretical)

Table 2.22 Values of third-order elastic constants C_{ijk} at room temperature

Crystal	C_{111}	C_{112}	C_{123}	C_{144}	C_{166}	C_{456}
	$[10^{12}$ dyne cm$^{-2}]$; Ref. [2.45]					
NaCl Structure						
LiF	−21.57	−1.38	0.685	0.954	−2.84	1.09
LiCl	−13.87	−1.06	0.319	0.370	−1.34	0.395
LiBr	−11.04	−1.13	0.226	0.281	−1.45	0.309
LiI	−10.27	−1.04	0.202	0.206	−1.06	0.208
NaF	−18.72	−1.18	0.480	0.553	−1.61	0.589
NaCl	−12.30	−0.556	0.230	0.255	−0.706	0.267
NaBr	−10.83	−0.513	0.194	0.204	−0.573	0.209
NaI	− 8.90	−0.590	0.153	0.151	−0.579	0.150
KF	−12.90	−1.181	0.321	0.319	−1.17	0.318
KCl	− 9.61	−0.275	0.149	0.164	−0.366	0.171
KBr	− 9.20	−0.269	0.144	0.139	−0.238	0.137
KI	− 7.49	−0.197	0.085	0.102	−0.307	0.111
RbF	−11.90	−1.155	0.302	0.266	−0.940	0.247
RbCl	− 8.92	−0.321	0.161	0.143	−0.207	0.133
RbBr	− 8.09	−0.235	0.125	0.118	−0.187	0.114
RbI	− 7.13	−0.185	0.100	0.092	−0.135	0.088
CsCl Structure						
CsCl	− 6.42	−0.734	−0.551	−0.592	−0.756	−0.613
CsBr	− 4.83	−0.756	−0.599	−0.638	−0.777	−0.658
CsI	− 3.29	−0.646	−0.520	−0.557	−0.666	−0.575

Notes and Comments

1. While experimental values of TOEC are available for only five alkali halides, theoretical calculations for all the alkali halides have been made by several workers. These should be useful to check any future experimental results on TOEC.
2. Theoretically evaluated values of TOEC are also given in Refs. [2.46–2.48].
3. Note that for crystals with CsCl structure, all the TOEC are negative.

2.5.3 Fourth-Order Elastic Constants (Experimental)

Table 2.23 Values of the linear combinations γ_{11}, γ_{12} and γ_{44} of fourth-order elastic constants C_{ijkl}; for definition of γ_{ij} see Notes and Comments

Crystal	γ_{11}	γ_{12}	γ_{44}	Ref.
	$[10^{11}\ \text{Pa}]$			
NaCl Structure				
NaI	92.4 ± 3.1	13.1 ± 4.1	12.3 ± 0.3	[2.49]
KI	81.0 ± 1.8	0.45 ± 1.4	4.1 ± 0.05	[2.49]
RbCl	112.0 ± 4.0	-2.0 ± 6.0	1.6 ± 0.8	[2.50]
RbBr	91.0 ± 4.0	-3.0 ± 6.0	1.8 ± 0.3	[2.50]
RbI	77.0 ± 3.0	-0.1 ± 4.0	2.1 ± 0.4	[2.50]
CsCl Structure				
CsCl	73.0 ± 13.0	41.0 ± 9.0	46.0 ± 8.0	[2.20]
CsBr	60.0 ± 13.0	36.0 ± 10.0	42.0 ± 8.0	[2.20]
CsI	49.0 ± 10.0	29.0 ± 7.0	35.0 ± 6.0	[2.20]

Notes and Comments

1. The internal energy Φ of a deformed crystal may be expressed as:

$$\Phi = \Phi_0 + (V/2)\, C_{ijkl}\, \varepsilon_{ij}\varepsilon_{kl} + (V/6)\, C_{ijklmn}\, \varepsilon_{ij}\varepsilon_{kl}\varepsilon_{mn}$$
$$+ (V/24)\, C_{ijklmnop}\, \varepsilon_{ij}\varepsilon_{kl}\varepsilon_{mn}\varepsilon_{op} \tag{2.37}$$

   where Φ_0 is the internal energy of the undeformed crystal, V the unit cell volume, C_{ijkl} the second-order elastic constants, C_{ijklmn} the third-order elastic constants, $C_{ijklmnop}$ the fourth-order elastic constants and ε_{ij} components of the strain tensor. By applying symmetry conditions and thermodynamic considerations and by using Voigt's contracted notation, the number of independent FOEC for a cubic crystal reduces to eleven viz., C_{1111}, C_{1112}, C_{1166}, C_{1122}, C_{1266}, C_{4444}, C_{1123}, C_{1144}, C_{1244}, C_{1456} and C_{4466}
2. The linear combinations γ_{ij} are defined as

$$\gamma_{11} = C_{1111} + 4\,C_{1112} + 2C_{1122} + 2\,C_{1123} \tag{2.38}$$

$$\gamma_{12} = 2\,C_{1112} + 2C_{1122} + 5\,C_{1123} \tag{2.39}$$

$$\gamma_{44} = C_{1144} + 2\,C_{1166} + 4C_{1244} + 2\,C_{1266} \tag{2.40}$$

3. Experimental values of the independent FOEC are not available; only linear combinations have been measured.
4. Note that γ_{12} is negative for the rubidium halides.
5. There are large errors in γ_{12}. Thus for RbI the error is ± 4 in 0.1.

2.5.4 Fourth-Order Elastic Constants (Theoretical) at 300 K

The values of the fourth-order elastic constants from theoretical calculations are given in Table 2.24 (see p. 42).

Notes and Comments

1. Experimental values of the independent fourth-order elastic constants are not available. These theoretical values will be useful as a check when experimental values become available.
2. Considering the errors in the experimental values of the linear combinations of FOEC (γ_{ij}'s), there is reasonable agreement between the experimental γ_{ij}'s given in Table 2.22 and the γ_{ij}'s calculated from these theoretical FOEC's.
3. A systematic variation is seen in each FOEC in the sequence F–Cl–Br–I for each alkali halide group.

2.6 Velocity of Sound

2.6.1 Velocities of Longitudinal Waves, Shear Waves and the Mean Velocity

Table 2.25 Values of the velocity of longitudinal waves (V_P), velocity of shear waves (V_S) and mean velocity ($\overline{V}$); Ref. [2.10]

Crystal	V_P [km s^{-1}]	V_S [km s^{-1}]	$\overline{V}$ [km s^{-1}]
NaCl Structure			
LiF	7.149	4.305	4.687
LiCl	5.260	3.058	3.352
LiBr	3.622	2.072	2.283
LiI	2.844	1.609	1.779
NaF	5.663	3.323	3.636
NaCl	4.528	2.591	2.848
NaBr	3.330	1.912	2.108
NaI	2.731	1.518	1.728
KF	4.630	2.548	2.810
KCl	3.915	2.178	2.404
KBr	3.032	1.685	1.865
KI	2.496	1.371	1.522
RbF	3.945	2.133	2.360
RbCl	3.093	1.657	1.839
RbBr	2.587	1.382	1.537
RbI	2.231	1.179	1.314
CsCl Structure			
CsCl	2.839	1.585	1.755
CsBr	2.464	1.393	1.619
CsI	2.212	1.267	1.402

Table 2.24 Theoretically calculated values of fourth order elastic constants C_{ijkl}

Crystal	C_{1111}	C_{1112}	C_{1166}	C_{1122}	C_{1266}	C_{4444}	C_{1123}	C_{1144}	C_{1244}	C_{1456}	C_{4466}
	$[10^{12}$ dynes cm^{-2} or 10^{11} Pa]; Ref. [2.51]										
NaCl Structure											
LiF	262	13.0	13.6	17.0	15.2	14.5	−2.09	−2.65	−2.06	−2.33	−2.41
LiCl	127	7.62	5.47	8.92	6.28	5.79	−1.07	−1.09	−0.780	−0.789	−1.30
LiBr	103	6.45	4.47	7.33	5.11	4.72	−0.857	−0.835	−0.600	−0.589	−1.04
LiI	84	5.71	3.16	6.29	3.71	3.28	−0.727	−0.661	−0.416	−0.384	−0.929
NaF	249	5.09	3.31	6.69	4.39	3.93	−1.45	−1.51	−1.21	−1.24	−1.66
NaCl	132	2.93	1.18	3.37	1.74	1.49	−0.774	−0.703	−0.566	−0.531	−0.887
NaBr	110	2.64	0.791	2.90	1.27	1.04	−0.666	−0.573	−0.452	−0.405	−0.774
NaI	87	2.63	0.646	2.74	1.04	0.804	−0.552	−0.452	−0.329	−0.280	−0.665
KF	184	2.79	−0.564	3.22	0.230	−0.214	−1.09	−0.929	−0.738	−0.623	−1.30
KCl	124	0.978	−0.486	0.789	−0.404	−0.568	−0.530	−0.460	−0.372	−0.338	−0.607
KBr	108	0.490	−0.871	0.528	−0.551	−0.695	−0.453	−0.385	−0.309	−0.276	−0.522
KI	88	0.478	−0.869	0.425	−0.603	−0.738	−0.377	−0.305	−0.238	−0.202	−0.441
RbF	175	5.65	−0.982	5.92	0.020	−0.950	−1.28	−1.02	−0.539	−0.413	−1.73
RbCl	120	0.989	−1.27	0.912	−0.871	−1.13	−0.544	−0.443	−0.312	−0.262	−0.667
RbBr	105	0.498	−1.23	0.406	−0.911	−1.10	−0.437	−0.354	−0.262	−0.221	−0.524
RbI	88	0.133	−1.24	0.010	−0.999	−1.13	−0.341	−0.272	−0.205	−0.171	−0.405
CsCl Structure											
CsCl	16.8	5.69	5.80	6.70	6.77	6.79	5.62	5.70	5.68	5.72	5.73
CsBr	14.5	4.93	5.16	5.72	5.91	5.92	4.72	5.98	4.98	5.16	5.16
CsI	11.9	4.22	4.48	4.79	5.03	5.04	3.96	4.41	4.29	4.52	4.52

Notes and Comments

1. The velocities of sound are calculated from Eqs. (2.41–2.43):

$$V_P = [(B + (4/3)G_{VRH} / \rho]^{1/2} \tag{2.41}$$

$$V_S = [(G_{VRH}) / \rho]^{1/2} \tag{2.42}$$

$$\overline{V} = [(1/3)\{(2/V_S^3) + (1/V_P^3)\}]^{-1/3} \tag{2.43}$$

where ρ is the density, B the bulk modulus and G_{VRH} the VRH shear modulus.

2. Anderson [2.52] pointed out that in general, (V_P/ρ) decreases with increasing molar weight. When (V_P/ρ) is plotted against the molar weight, the data points lie on a smooth curve for each halide group.

3. Anderson [2.53] proposed the following linear relationship connecting the longitudinal velocity V_P, the refractive index n and other molecular properties:

$$V_P /(n^2 - 1) = \text{constant} \ [\sqrt{M/p}] / \rho(\alpha_+ + \alpha_-) \tag{2.44}$$

where M is the molar weight, p the number of atoms (2 for the alkali halides), ρ is the density and α_+ and α_- the electronic polarisabilities of the cation and anion respectively.

2.6.2 Second-Sound Velocity

Table 2.26 Drifting velocity V_{II} and driftless velocity V_{II}' (calculated at 0 K) and the ratio $R = <V^2> /3V_{II}^2$ where V is the sound velocity

	V_{II} [10^5 cm s^{-1}]	V_{II}' [10^5 cm s^{-1}]	$R = <V^2> /3V_{II}^2$
Ref.	[2.54]		
Crystal ↓			
NaCl Structure			
LiF	2.629	2.786	1.529
LiCl	1.962	2.075	1.567
LiBr	1.343	1.415	1.519
LiI	0.937	1.011	1.780
NaF	2.026	2.145	1.581
NaCl	1.586	1.694	1.673
NaBr	1.169	1.251	1.711
NaI	0.940	1.011	1.778
KF	1.527	1.664	1.930
KCl	1.221	1.365	2.245
KBr	0.924	1.041	2.412
KI	0.735	0.841	2.642
RbF	1.045	1.157	2.182
RbCl	0.901	1.022	2.563
RbBr	0.747	0.855	2.715
RbI	0.614	0.714	3.047

Table 2.26 (Continued)

	V_{II} $[10^5$ cm s$^{-1}]$	V_{II}' $[10^5$ cm s$^{-1}]$	$R = <V^2>/3V_{II}^2$
Ref.	[2.54]		
Crystal $\downarrow$			
CsCl Structure			
CsCl	0.918	0.982	1.744
CsBr	0.901	0.950	1.597
CsI	0.812	0.855	1.570

Notes and Comments

1. The second-sound velocities have been calculated by Varshni and Konti [2.13] from the 0 K elastic constants using theoretical formulae given by Hardy [2.54].
2. The percentage difference between the drifting and the driftless sound velocities ranges between 5–16.
3. The ratio $R = <V^2>/3V_{II}^2$ has an expected value of 1. However, in no case is it 1 or <1. The R values lie on a vertical parabola about the ordinate at $A = 1$, when plotted against the elastic anisotropy $A = 2C_{44}/(C_{11}-C_{12})$; strongly anisotropic materials tend to have large R values.
4. "One may anticipate a dependence of second-sound velocities on the nature of the binding" [2.13].

2.7 Hardness

2.7.1 Hardness at Room Temperature

Table 2.27 Values of the Vicker's hardness H_V and Moh hardness H_M

Crystal	H_V [GPa]	Ref.	H_M [see Notes and Comments]
NaCl Structure			
LiF	1.010	[2.55]	3.1
NaF	0.626	[2.55]	2.6
NaCl	0.216	[2.55]	1.9
NaBr	0.129	[2.55]	1.7
NaI	0.101	[2.55]	1.4
KCl	0.128	[2.55]	1.6
KBr	0.098	[2.55]	1.4
KI	0.069	[2.55]	1.3
RbCl	0.093	[2.55]	1.4
RbBr	0.078	[2.55]	1.3
RbI	0.059	[2.55]	1.2

Table 2.27 (Continued)

Crystal	H_V [GPa]	Ref.	H_M [see Notes and Comments]
CsCl Structure			
CsCl	0.21	[2.56]	1.8
CsBr	0.18	[2.56]	1.76
CsI	0.12	[2.56]	1.5

Notes and Comments

1. The older values of H_V are expressed in kg mm^{-2}. These may be multiplied by 9.806×10^{-3} to convert to GPa.
2. H_M values have been obtained from H_V values using the relation

$$H_M = (0.675)\,(H_V)^{1/3} \tag{2.45}$$

   where H_V is in kg mm^{-2} [2.58]; there are no units for H_M.
3. Microhardness is a structure sensitive property which is affected by several factors like impurities, thermal and mechanical history of the sample used for measurement resulting in variation in values from different sources. Thus, for NaCl values of 21.3, 22.1, 27 and 29.2 (in kg mm^{-2}) have been reported [2.57]. This aspect may be kept in view in comparing values given in this and following sections.
4. Thirmal Rao and Sirdeshmukh [2.59] obtained a smooth curve for the plot of H_V versus 'a' (the lattice constant) which resembles a hyperbola.
5. Sirdeshmukh et al. [2.55] obtained a linear plot for log H_V versus log C_{44} with an average value of 0.018 for the ratio (H_V / C_{44}) for NaCl type alkali halides. This ratio (H_V / C_{44}) has been called the Gilman-Chin parameter.
6. Gerk [2.60] obtained a straight line plot for log H versus log G' where G' is

$$G' = (C_{44})\left\{ \frac{(C_{11} - C_{12})/2C_{44}}{1 + (C_{11} - C_{12})/2C_{44}} \right\} \tag{2.46}$$

7. Shukla and Bansigir [2.61] reported the following empirical relation between H_V and E_f the energy of formation of a Schottky pair,

$$H_V = 205.8\, E_f / r^3 \tag{2.47}$$

   where r is the interionic distance.
8. Subramaniam [2.62] showed the existence of the following relation for the alkali halides

$$H_V = 4.30 \times 10^{-12} v_D - 4.26 \tag{2.48}$$

   where H_V is in kg mm^{-2} and v_D is the Debye frequency in s^{-1}.
9. Golovin et al. [2.63] found that the hardness of NaCl subjected to a magnetic field of 20 T for 100 μs decreased from its field-free value by 20 %.

2.7.2 Knoop Hardness – Hardness Anisotropy

Table 2.28 Values of H_{100} (hardness measured on the [100] plane with indentor parallel to <100> direction) and H_{110} (indentor parallel to <110> direction)

Crystal	H_{100} [kg mm^{-2}]	H_{110}	H_{100} / H_{110}	Ref.
NaCl Structure				
LiF	96	103	0.93	[2.64]
NaCl	17.1	15.4	1.11	[2.65]
NaBr	-	-	1.17	[2.65]
KCl	9.55	7.7	1.24	[2.65]
KBr	8.1	6.3	1.28	[2.65]
KI	-	-	1.14	[2.65]
RbCl	9.1	7.4	1.2	[2.59]
RbBr	8.0	7.2	1.1	[2.59]
RbI	7.1	6.3	1.1	[2.59]

Notes and Comments

1. Note that the ratio H_{100} / H_{110} takes values < 1 and also > 1.
2. Brookes et al. [2.64] showed that the hardness anisotropy is related to the slip system. But this does not explain the occurrence of values of H_{100} / H_{110} less than as well as greater than 1 among the alkali halides having the same slip system.
3. Thirmal Rao et al. [2.66] showed that the anisotropy of hardness is related to the point group symmetry. Again, this does not explain the difference in the anisotropy in, say, NaCl and LiF which have the same point group.
4. Sirdeshmukh and Kishan Rao [2.67] pointed out that the hardness anisotropy is closely related to the anisotropy of the Young's modulus. Crystals having $(E_{100} / E_{110}) > 1$ have $(H_{100} / H_{110}) > 1$ and crystals having $(E_{100} / E_{110}) < 1$ have $(H_{100} / H_{110}) < 1$. In fact, they pointed out a linear relationship between (H_{100} / H_{110}) and (E_{100} / E_{110}).

2.7.3 Temperature Variation of Hardness

Table 2.29 Values of hardness at different temperatures

Method	Vicker's hardness H_V [kg mm^{-2}]; [2.68]		Relative hardness H [kg mm^{-2}] from arm length of indentation rosette; [2.69]				
Temp. T [K] Crystal ↓	293	77	303	373	473	573	673
NaCl Structure							
LiF	120	280	-	-	-	-	-
NaF	67	200	-	-	-	-	-
NaCl	20	82	22	21	10	4.6	3.1
KCl	10	35	13.1	13.1	9.8	6.3	5.7

Notes and Comments

1. The hardness decreases with increasing temperature.
2. Kishan Rao and Sirdeshmukh [2.69] showed that the temperature variation of hardness at elevated temperatures is best represented by the equation:

$$H = A \exp(-B\,T) \tag{2.49}$$

where A and B are constants for each crystal.

2.7.4 Pressure Variation of Hardness

Table 2.30 Values of Vicker's hardness H_V; at pressure P

Pressure [GPa] Crystal ↓	H_V [GPa]; Ref. [2.70]	
	0	1.5
NaCl Structure		
LiF	1.170	1.260
NaCl	0.210	0.271
KCl	0.110	0.198
KBr	0.130	0.247

Notes and Comments

1. In all the alkali halides studied, the hardness increases with pressure.
2. The percentage increase in hardness ($\Delta H/H$) for a fixed increase in pressure is more for softer crystals.
3. Barbashov and Tkachenko [2.70] have given plots of H versus P. In KCl and KBr, there is a sudden increase in the slope of the H versus P plot at about 0.7 GPa; this is believed to be related to the structural transitions which these crystals undergo at 1.95 and 1.74 GPa respectively.

2.7.5 Surface Hardness

Table 2.31 Values of H_V^B, bulk Vicker's hardness measured at 200 g and H_V^S, surface Vicker's hardness determined from hardness measurements at different depths h and extrapolated to $h = 0$.

Crystal	H_V^B	H_V^S
	[kg mm^{-2}]; Ref. [2.71]	
NaCl Structure		
LiF	100	370
NaCl	20	60
KBr	9	31
KI	7	28

Notes and Comments

1. Upit and Varchenya [2.71] point out that the depth dependence of hardness is given by

$$H = c\,h^{m} \tag{2.50}$$

   where c is a constant. The value of m for the alkali halides is -0.14.
2. The ratio of surface hardness to bulk hardness for the four alkali halides studied is in the range 3–4.

References

2.1   S. Haussuhl, Zeit. fur Physik, **159**, 223, 1960.

2.2   K.O. McLean and C.S. Smith, J. Phys. Chem. Solids, **33**, 275, 1972; **33**, 279, 1972.

2.3   S. Haussuhl, Zeit. fur Krist., **138**, 177, 1973.

2.4   S. Haussuhl, Acta Cryst., **13**, 685, 1960.

2.5   B. Subramanyam and Y.V.G.S. Murti , phys. stat. sol., **(b)115**, K29, 1983.

2.6   S. Haussuhl, Zeit. fur Physik, **205**, 215, 1993.

2.7   K. Motida, J. Phys. Soc. Japan, **50**, 102, 1981.

2.8   K.S. Krishnan and S.K. Roy, Proc. Roy. Soc. Lond., **210**, 481, 1952.

2.9   A.K. Koh and K.N. Ng, phys. stat. sol., **(b)213**, 11, 1999.

2.10   G. Simmons and H. Wang, *Single Crystal Elastic Constants and Calculated Aggregate Properties – A Handbook*, MIT Press, Cambridge, Mass., USA, 1971 and references therein.

2.11   J.T. Lewis, A. Lehoczky and C.V. Briscoe, Phys. Rev., **161**, 877, 1967 and references therein.

2.12   B.G. Dick, Phys. Rev. **129**, 1583, 1963.

2.13   Y.P. Varshni and A. Konti, Phys. Rev., **B6**, 1532, 1972 and references therein.

2.14   Y.P. Varshni, Phys. Rev. **B2**, 3952, 1970.

2.15   O.D. Slagle and H.A. McKinstry, J. Appl. Phys., **38**, 437, 1967.

2.16   O.D. Slagle and H.A. McKinstry, J. Appl. Phys., **38**, 451, 1967.

2.17   C.S. Smith and L.S. Cain, J. Phys. Chem. Solids, **41**, 199, 1980.

2.18   R.W. Roberts and C.S. Smith, J. Phys. Chem. Solids, **31**, 619, 1970.

2.19   R.W. Roberts and C.S. Smith, J. Phys. Chem. Solids, **31**, 2397, 1970.

2.20   Z.P. Chang and G.R. Barsch, Phys. Rev. Lett., **19**, 1381, 1967.

2.21   M.P. Tosi, Solid State Physics, **16**, 1, 1964.

2.22   I.P.S. Solanki and J. Shanker, Ind. J. Pure and Appl. Phys., **32**, 863, 1994.

2.23   S.S. Mitra and R. Marshall, J. Chem. Phys., **41**, 3158, 1964.

2.24   D.B. Sirdeshmukh and K.G. Subhadra, J. Appl. Phys., **59**, 276,1986.

2.25   A. Jayaraman, B. Batlogg, R.G. Maines and H. Bach, Phys. Rev., **B26**, 3347, 1982.

2.26   H. Neumann, Cryst. Res. Tech., **23**, 531, 1968.

2.27   P.D. Pathak and N.G. Vasavada, J. Phys., **D3**, 1767, 1970.

2.28   R.E. Hanneman and H.C. Gatos, J. Appl. Phys., **36**, 1794, 1965.

2.29   D.B. Sirdeshmukh, J. Appl. Phys., **38**, 4083, 1967.

2.30   B. Szigeti, Trans. Faraday Soc., **45**, 155, 1949; Proc. Roy. Soc. Lond., **A204**, 51, 1950.

2.31   R. Brout, Phys. Rev., **113**, 43, 1959.

2.32   D.H. Chung and W.R. Buessem, J. Appl. Phys., **38**, 2535, 1967.

2.33   R. Hill, Proc. Phys. Soc. Lond., **65**, 349, 1952.

2.34   S.V. Subramanyam, Acustica, **12**, 37, 1962.

2.35   B. Kameswara Rao, Ph.D thesis, Kakatiya University, Warangal, India, 1980.

2.36   D.H. Chung, J. Appl. Phys.,**38**, 5104, 1967.

2.37   S.N.Vaidya and G.C. Kennedy , J. Phys. Chem. Solids, **32**, 951, 1971.

2.38   Values compiled by U. Kohler, P.G. Johannsen and W.B. Holzapfel, J. Phys. Condens. Matter, **9**, 5581, 1997. These values are based on ultrasonic measurements at one kbar [2.2, 2.18, 2.19].

2.39   R. Boehler and G.C. Kennedy, J. Phys. Chem. Solids, **41**, 517, 1980.

2.40   D.L. Decker, J. Appl. Phys., **42**, 3239, 1971. This is a revised version of the earlier papers [2.41].

2.41   D.L. Decker, J. Appl. Phys., **36**, 157, 1965; **37**, 5012, 1966.

2.42   J.R. Drabble and R.E.B. Strathen, Proc. Phys. Soc., **92**, 1090, 1967.

2.43   W.A. Bensch, Phys. Rev., **B6**, 1504, 1972.

2.44   R.J. Wallat and J. Holder, J. Phys. Chem. Solids, **38**, 127, 1977.

2.45   R.P. Singh and J. Shanker, phys. stat. sol., **(b)93**, 373, 1979.

2.46   A.A. Nran'yan, Soviet Phys.– Solid State, **5**, 129, 1963; **5**,1361, 1964.

2.47   P.B. Ghate, Phys. Rev. **139 A**, 1666, 1965.

2.48   R.K. Varshney and J. Shanker, phys. stat. sol., **(b)122**, 65, 1984.

2.49   G.R. Barsch and H.E. Shull, phys. stat. sol., **(b)43**, 637, 1971.

2.50   Z. P. Chang and G.R. Barsch, J. Phys. Chem. Solids, **32**, 27, 1971.

2.51   R.K Varshney and J. Shanker, phys. stat. sol. **(b)129**, 61, 1985.

2.52   O.L. Anderson in *Physical Acoustics*, Vol. **3B**, Academic Press, New York, 1965.

2.53   O.L. Anderson, Amer. Mineralogist, **51**, 1001, 1966.

2.54   R.J. Hardy, Phys. Rev., **B2**, 1193, 1970.

2.55   Compilation by D.B. Sirdeshmukh, K.G. Subhadra, K.Kishan Rao and T. Thirmal Rao, Cryst. Res. Technol., **30**, 861, 1995 and references therein.

2.56   P. Geeta Krishna, Ph.D. Thesis, Kakatiya University, Warangal, India, 1997.

2.57   C. Bhanumathi and K.G. Bansigir, J. Jiwaji University, **2**, 44, 1974.

2.58   B.W. Mott, *Microindentation Hardness Testing*, Butterworths, London, 1956.

2.59   T. Thirmal Rao and D.B. Sirdeshmukh, Cryst. Res. Technol., **26**, K53, 1991.

2.60   A.P. Gerk, J. Mat. Sci., **12**, 735, 1977.

2.61   M. Shukla and K.G. Bansigir, J. Phys., **D9**, L49, 1976.

2.62   B. Subramaniam, Ind. J. Pure and Appl. Phys., **18**, 362, 1980.

2.63   Y.I. Golovin, R.B. Morgunov, D.V. Lopatin and A.A. Baskakov, phys. stat. sol., **(a)160**, R3, 1997

2.64   C.A. Brookes, J.B. O'Neill and B.A.W. Redfern, Proc. Roy. Soc. Lond., **A322**, 73, 1971.

2.65   G.Y. Chin, L.G. Van Uitert, M.L. Green and G. Zydzik, Scripta Met., **6**, 503, 1972.

2.66   T. Thirmal Rao, K. Kishan Rao and D.B. Sirdeshmukh, Cryst. Res. Technol., **26**, K189, 1991.

2.67   D.B. Sirdeshmukh and K. Kishan Rao, J. Mat. Sci. Lett., **7**, 567, 1988.

2.68   Yu. S. Boyarskaya, D.Z. Grabko, D.S. Pishkova and S.S. Shutova, Kristall and Technik, **13**, 975, 1978; data based on Vicker's indentation.

2.69   K. Kishan Rao and D.B. Sirdeshmukh, Pramana, **24**, 887, 1985; data based on arm-length of indentation rosettes.

2.70   V.I. Barbashov and Yu. B. Tkachenko, Sov. Phys. Solid State, **33**, 1564, 1991.

2.71   G.P. Upit and S.A. Varchenya in *The Science of Hardness Testing and its Research Applications,* Amer. Soc. for Metals, Ohio, USA, 1973.

3 Thermal Properties

3.1 Specific Heat

3.1.1 Specific Heat at Low Temperatures (Experimental)

Table 3.1 Values of specific heat C_P [Cal g^{-1} deg^{-1}]; Ref. [3.1]

LiF		LiCl		NaF	
T [K]	C_P	T [K]	C_P	T [K]	C_P
5.14	1.095×10^{-5}	-	-	-	-
10.10	9.101×10^{-5}	-	-	-	-
15.01	3.008×10^{-4}	15.42	1.310×10^{-3}	-	-
20.8	7.710×10^{-4}	20.38	3.406 "	-	-
24.8	1.388×10^{-3}	25.84	7.905 "	-	-
29.9	2.776×10^{-3}	29.27	1.194×10^{-2}	-	-
35.9	5.397 "	36.36	2.325 "	-	-
40.7	8.404 "	40.83	3.430 "	-	-
45.5	1.264×10^{-2}	46.05	4.337 "	-	-
49.2	1.665 "	51.94	5.665 "	54.01	3.951×10^{-2}
60.4	2.984 "	58.29	7.187 "	62.80	5.668 "
70.4	5.181 "	71.80	1.038×10^{-1}	71.81	7.537 "
79.0	6.985 "	80.03	1.208 "	80.98	9.419 "
89.1	9.229 "	88.52	1.370 "	95.40	1.217×10^{-1}
98.3	1.140×10^{-1}	97.03	1.517 "	105.04	1.383 "
149.1	2.256 "	146.06	2.104 "	155.77	2.015 "
200.5	3.029 "	197.56	2.421 "	205.89	2.353 "
248.5	3.498 "	249.94	2.367 "	245.66	2.517 "
271.7	3.676 "	272.31	2.656 "	276.20	2.612 "

Table 3.1 (Continued)

NaCl		NaBr		NaI	
T [K]	C_P	T [K]	C_P	T [K]	C_P
-	-	7.21	3.800×10^{-4}	5	1.961×10^{-4}
10.9	7.53×10^{-4}	10.25	1.118×10^{-3}	10	2.236×10^{-3}
15.3	2.29×10^{-3}	14.45	3.994×10^{-3}	15	7.472×10^{-3}
19.4	4.86 "	20.34	1.092×10^{-2}	20	1.409×10^{-2}
25.1	1.13×10^{-2}	25.70	1.877 "	25	2.060 "
30.7	2.02 "	30.82	2.687 "	30	2.664 "

Table 3.1 (Continued)

NaCl		NaBr		NaI	
T [K]	C_P	T [K]	C_P	T [K]	C_P
35.4	3.04×10^{-2}	35.25	3.393×10^{-2}	-	-
38.4	3.809 "	40.86	4.223 "	40	3.753×10^{-2}
45.1	5.256 "	45.06	4.824 "	-	-
50.7	6.557 "	49.63	5.431 "	50	4.638 "
59.4	8.518 "	60.50	6.558 "	60	5.336 "
69.2	1.028×10^{-1}	71.68	7.675 "	-	-
80.2	1.198 "	81.23	8.366 "	80	6.289 "
89.4	1.308 "	91.92	8.980 "	-	-
98.5	1.416 "	102.00	9.418 "	100	6.861 "
148.0	1.766 "	150.77	1.071×10^{-1}	150	7.605 "
198.1	1.916 "	200.88	1.133 "	200	7.949 "
247.4	1.990 "	250.10	1.165 "	250	8.169 "
267.5	2.007 "	275.63	1.180 "	270	8.246 "

Table 3.1 (Continued)

KF		KCl		KBr	
T [K]	C_P	T [K]	C_P	T [K]	C_P
-	-	5.14	1.35×10^{-4}	5	1.991×10^{-4}
-	-	10.06	1.12×10^{-3}	10	1.948×10^{-3}
16.05	1.992×10^{-3}	15.84	4.83 "	15	6.554 "
20.70	5.195 "	19.9	4.56 "	20	1.347×10^{-2}
24.91	9.402 "	24.8	8.32 "	25	2.138 "
30.33	1.628×10^{-2}	31.5	1.49×10^{-2}	30	2.949 "
33.97	2.233 "	34.9	2.52 "	36.10	4.000 "
38.37	2.969 "	41.2	2.54 "	40	4.467 "
43.08	3.820 "	45.5	2.99 "	45.50	5.344 "
53.43	5.890 "	50.1	3.43 "	50	5.704 "
64.22	7.927 "	61.3	4.39 "	60	6.650 "
70.51	9.006 "	-	-	71.7	7.579 "
80.57	1.057×10^{-1}	81.0	5.53 "	80	7.916 "
87.81	1.156 "	90.4	5.97 "	89.4	8.285 "
102.76	1.326 "	101.7	6.26 "	100	8.659 "
148.20	1.671 "	148.9	7.28 "	150	9.587 "
206.05	1.874 "	201.8	7.71 "	200	1.001×10^{-1}
255.45	1.967 "	251.6	7.97 "	250	1.029 "
277.66	1.998 "	267.6	8.07 "	270	1.039 "

Table 3.1 (Continued)

KI		RbBr		RbI	
T [K]	C_P	T [K]	C_P	T [K]	C_P
5	3.675×10^{-4}				
10	3.479×10^{-3}	10.5	1.88×10^{-3}	11.6	6.0×10^{-3}
15	9.542 "	15.4	4.42 "	14.6	1.03×10^{-2}
20	1.635×10^{-2}	19.8	7.38 "	20.2	1.71 "
25	2.285 "	24.8	1.13×10^{-2}	24.9	2.35 "
30	2.891 "	30.3	1.55 "	29.8	2.97 "
35	3.49 "	36.2	1.94 "	34.8	3.46 "
40	3.830 "	39.4	2.17 "	39.8	3.85 "
46.2	4.48 "	46.2	2.44 "	45.6	4.22 "
50	4.731 "	49.6	2.58 "	51.9	4.53 "
60	5.310 "	60.5	2.89 "	64.3	4.89 "
73.6	5.78 "	72.7	3.09 "	73.5	5.06 "
80	6.048 "	81.5	3.21 "	78.7	5.12 "
92.6	6.10 "	94.5	3.31 "	91.9	5.25 "
100	6.48 "	105.8	3.41 "	102.2	5.37 "
150	7.018 "	148.4	3.62 "	150.3	5.65 "
200	7.271 "	200.1	3.73 "	197.8	6.00 "
250	7.451 "	249.2	3.79 "	254.8	5.85 "
270	7.518 "	272.7	3.78 "	276.9	5.91 "

Table 3.1 (Continued)

CsCl	
T [K]	C_P
9.29	1.556×10^{-3}
15.24	7.115 "
19.18	1.202×10^{-2}
25.36	1.962 "
30.55	2.579 "
35.28	3.112 "
41.43	3.705 "
44.64	3.969 "
51.96	4.514 "
61.29	5.059 "
72.14	5.518 "
79.28	5.772 "
91.69	6.102 "
100.85	6.274 "
150.96	6.845 "
202.22	7.119 "
250.42	7.287 "
279.54	7.394 "

3.1.2 Specific Heat at Low Temperatures (Theoretical)

Table 3.2 Values of theoretically calculated specific heat C_V

Crystal Ref. Temp. [K] $\downarrow$	C_V [Cal mole^{-1} deg^{-1}]				
	LiBr [3.2]	LiI [3.2]	RbF [3.3]	RbCl [3.3]	CsF [3.3]
5	-	0.02	0.01	0.02	0.01
10	0.08	0.29	0.09	0.20	0.19
15	0.37	1.00	0.39	0.72	0.70
20	0.87	1.87	0.91	1.53	1.44
25	1.49	2.66	1.55	2.48	2.21
30	2.10	3.31	2.23	3.46	2.92
35	2.66	3.84	2.90	4.40	3.55
40	3.16	4.28	3.55	5.26	4.13
45	3.60	4.65	4.17	6.03	4.66
50	4.01	4.99	4.74	6.70	5.16
60	4.72	5.61	5.78	7.79	6.08
70	5.35	6.18	6.67	8.61	6.88
80	5.93	6.71	7.41	9.22	7.56
90	6.45	7.19	8.03	9.69	8.15
100	6.93	7.64	8.54	10.05	8.64
120	7.76	8.40	9.34	10.57	9.40
140	8.44	9.02	9.90	10.90	9.94
160	8.99	9.51	10.30	11.12	10.34
180	9.43	9.89	10.60	11.28	10.63
200	9.79	10.20	10.83	11.40	10.85
220	10.09	10.45	11.00	11.49	11.02
240	10.33	10.65	11.07	11.55	11.16
260	10.53	10.81	11.25	11.61	11.26
280	10.69	10.95	11.33	11.65	11.35
300	10.83	11.06	11.41	11.68	11.42

Notes and Comments

1. Karo [3.2, 3.3] has made calculations of specific heat at low temperatures
    using the Born model. His results agree well with experimental results wher-
    ever available. For LiBr, LiI, RbF, RbCl and CsF for which experimental re-
    sults at low temperatures are not available, his calculated data are reproduced.

3.1.3 Specific Heat at High Temperatures (Experimental)

Table 3.3 Values of specific heat C_P [Cal g^{-1} deg^{-1}]; Ref. [3.1]

LiF		KF		KCl		KBr	
T [K]	C_P	T [K]	C_P	T [K]	C_P	T [K]	C_P
300	0.3871	300	0.2014	335	0.1666	270	0.1039
400	0.4295	350	0.2074	377	0.1688	573	0.1126
500	0.4561	404	0.2134	428	0.1724	673	0.1168
600	0.4756	444	0.2174	522	0.1765	773	0.1201
700	0.4923	492	0.2217	608	0.1790	873	0.1230
800	0.5089	530	0.2248	701.5	0.1840	973	0.1258
900	0.5273	-	-	-	-	1003	0.1265
1000	0.5489	-	-	-	-	-	-
1100	0.5752	-	-	-	-	-	-
1200	0.5979	-	-	-	-	-	-

Table 3.3 (Continued)

KI		RbF		CsCl		
T [K]	C_P	T [K]	C_P	T [K]	C_P	
270	0.0757	370	0.1174	279	0.07394	
573	0.07918	400	0.1173	299	0.07447	
673	0.08360	500	0.1215	370	0.07441	
773	0.08657	600	0.1295	400	0.07246	
873	0.08845	700	0.1391	500	0.07304	Pm3m
973	0.08960	800	0.1497	600	0.07934	
-	-	900	0.1607	700	0.08836	
-	-	1000	0.1722	740	0.09242	
-	-	-	-	750	0.03015	
-	-	-	-	800	0.09541	Fm3m
-	-	-	-	900	0.10750	

3.1.4 Pressure Variation of Specific Heat at RT

Table 3.4 Values of the constants C_0 and C_1 in the equation $C_P = C_0 + C_1 P$; specific heat C_P [kJ kg^{-1} deg^{-1}] and pressure P [GPa]; Ref. [3.4]

Crystal	Pressure range	C_0	C_1
NaCl Structure			
KCl	0–1.9	0.680	–0.016
KBr	0–1.7	0.424	–0.014
KI	0–1.8	0.304	–0.017
High Pressure CsCl Phase			
RbCl	0.5–2.5	0.418	–0.006
RbBr	0.4–2.5	0.305	–0.009
RbI	0.4–2.5	0.252	–0.005

Notes and Comments

1. Multiply value in kJ kg^{-1} deg^{-1} by 0.23884 to get value in Cal g^{-1} deg^{-1}.
2. Dzhavadov and Krotov [3.5] observed a reduction in C_P with pressure at the rate of 0.2 % kbar^{-1} in NaCl, KCl and RbCl which was within the limits of their experimental error.
3. Boehler and Kennedy [3.6] observed a decrease of 4 % over a pressure range of 0–30 kbar in NaCl.
4. Gerlich and Andersson [3.7] measured the specific heat of CsCl, CsBr and CsI over a pressure range of 0–2.5 GPa and found that the specific heat was independent of pressure.
5. Hakansson and Andersson [3.8] observed that the specific heat of NaCl and NaI was independent of pressure up to 2 GPa.

3.2 Thermal Expansion

3.2.1 Thermal Expansion Coefficient at Room Temperature

Table 3.5 Values of the linear coefficient of thermal expansion $\alpha\,[10^{-6}\ \mathrm{K}^{-1}]$

Method	X-ray			Dilatometer			Optical interference		
Crystal ↓	Temp. [K]	α	Ref.	Temp. [K]	α	Ref.	Temp. [K]	α	Ref.
NaCl Structure									
LiF	300	32.7	[3.9]	283	32.9	[3.12]	-	-	-
LiCl	288	44.7	[3.10]	300	43.7	[3.13]	-	-	-
LiBr	-	-	-	300	48.9	[3.13]	-	-	-
LiI	-	-	-	RT	59	[3.14]	-	-	-
NaF	300	32.5	[3.9]	-	-		-	-	-
NaCl	300	38.9	[3.9]	283	39.5	[3.12]	290	39.5	[3.16]
NaBr	300	44.8	[3.11]	283	41.7	[3.15]	-	-	-
NaI	-	-	-	283	45.1	[3.12]	290	45.5	[3.16]
KF	-	-	-	300	31.7	[3.13]	-	-	-
KCl	300	35.0	[3.9]	283	36.9	[3.12]	290	36.8	[3.16]
KBr	300	36.8	[3.9]	283	38.5	[3.12]	290	38.7	[3.16]
KI	300	39.4	[3.9]	283	40.0	[3.12]	-	-	-
RbF	-	-	-	RT	34	[3.14]	-	-	-
RbCl	300	39.8	[3.9]	283	35.3	[3.15]	-	-	-
RbBr	300	36.9	[3.9]	283	37.0	[3.15]	-	-	-
RbI	300	38.8	[3.9]	283	39.2	[3.12]	-	-	-
CsF	-	-	-	283	33.8	[3.15]	-	-	-
CsCl Structure									
CsCl	300	46.6	[3.9]	-	-	-	270	45.6	[3.17]
CsBr	300	46.1	[3.9]	283	46.5	[3.15]	270	46.7	[3.17]
CsI	300	46.2	[3.9]	283	48.0	[3.15]	270	48.1	[3.17]

Notes and Comments

1. Data on thermal expansion of alkali halides are included in the compilations by Krishnan [3.18], Touloukian et al. [3.19] and Krishnan et al. [3.20].
2. The product of the coefficient of linear expansion (α) and the melting point ($t_m°C$) is a constant for a family of related crystals. For the alkali halides, the value of αt_m is 0.027 ± 0.003. Van Uitert et al. [3.14] plotted α against t_m and showed that the data points for the alkali halides lie on a smooth hyperbolic curve.
3. Sirdeshmukh [3.21] showed that the product $\alpha e*^2$ is a constant for the alkali halides, where α is the coefficient of linear expansion and $e*$ is the Szigeti effective charge.
4. Touloukian et al. [3.19] plotted the coefficient of linear thermal expansion (α) of the alkali halides against the refractive index (n) and obtained a straight line plot.
5. Khan [3.22] plotted α against the radius ratio r_+/r_- and found that the data points are distributed around a curve with equation

$$\alpha = 113.4 - 290.9(r_+/r_-) + 379.8(r_+/r_-)^2 - 166.6(r_+/r_-)^3 \tag{3.1}$$

Further, Khan [3.22] showed that there is a linear plot between α and $\Delta x = r_{obs} - (r_+ + r_-)$ where r_{obs} is the experimental interionic distance and r_+ and r_- are the ionic radii. This linear plot can be represented by the equation

$$\alpha = 42.6 + 139.5\Delta x \tag{3.2}$$

with α in 10^{-6} deg^{-1} and Δx in Å. From this equation, Khan predicted a value of 42.6×10^{-6} deg^{-1} for α of RbF.
6. Hanneman and Gatos [3.23] proposed the relation

$$\alpha = c\psi \tag{3.3}$$

for the alkali halides, where c is a constant and ψ the compressibility. Sirdeshmukh [3.24] suggested a modified relation

$$\alpha V_M = c'\psi \tag{3.4}$$

where V_M is the molar volume and c' a constant.
7. Watanabe et al. [3.25] determined the surface linear expansion ($\alpha_{surface}$) of LiF using a He-atom diffraction technique and found that $\alpha_{surface} = 12.4 \times 10^{-6}$ K^{-1} (compared with the bulk value of 34.2×10^{-6} K^{-1}).

3.2.2 Thermal Expansion at Very Low Temperatures (T < 12 K)

Table 3.6 Values of the linear thermal expansion coefficient $\alpha\,[10^{-8}\,\mathrm{K}^{-1}]$ at temperature T. Data for K-halides from Ref. [3.12] and for the other alkali halides from Ref. [3.15]

NaF		NaCl		NaBr		NaI	
T [K]	α	T [K]	α	T [K]	α	T [K]	α
5.463	0.24	2.735	0.23	2.877	0.33	2.315	0.61
6.219	0.19	3.775	0.33	3.498	0.68	3.019	1.17
7.374	0.53	4.489	0.54	4.008	0.91	4.043	3.19
8.071	0.71	5.194	0.80	4.455	1.45	4.860	5.67
8.755	0.90	5.909	1.21	5.341	2.76	5.666	9.47
9.518	1.17	7.211	2.20	6.072	4.12	6.408	14.10
10.538	1.61	7.884	2.86	6.862	6.19	7.311	23.10
-	-	9.458	5.04	7.934	9.96	7.880	30.41
-	-	10.073	6.14	9.484	18.46	9.221	53.3
-	-	10.627	7.28	10.310	24.65	10.749	90.1

Table 3.6 (Continued)

KCl		KBr		KI	
T [K]	α	T [K]	α	T [K]	α
6	0.9	5	1.2	4	1.8
7	1.6	6	2.5	5	3.6
8	2.5	7	4.4	6	6.6
10	5.5	8	7.2	7	12.0
12	10.0	10	16.0	8	21.0
-	-	12	35.0	10	53.0
-	-	-	-	12	100.0

Table 3.6 (Continued)

RbCl		RbBr		RbI		CsF	
T [K]	α	T [K]	α	T [K]	α	T [K]	α
2.382	−0.02	3.493	−0.13	2.462	−0.19	3	0.18
3.015	−0.07	4.004	−0.16	3.096	−0.50	4	0.45
3.775	−0.01	4.410	−0.26	3.611	−0.89	5	0.95
4.429	−0.02	4.854	−0.42	4.038	−1.37	6	2.0
4.845	0.01	4.907	−0.54	4.434	−1.87	7	4.2
5.269	0.01	5.465	−0.67	5.486	−3.64	8	8.3
6.001	0.05	6.024	−0.84	6.640	−3.89	9	14.8
6.809	0.22	6.592	−1.15	7.143	−2.91	10	25
7.833	0.80	7.656	−0.58	7.677	−0.55	-	-
8.492	1.41	8.373	0.69	8.261	3.87	-	-
9.930	5.31	9.043	2.88	8.908	10.81	-	-
10.695	9.07	9.682	6.21	9.668	22.63	-	-
-	-	10.578	13.49	10.558	42.51	-	-

Table 3.6 (Continued)

CsBr		CsI	
T [K]	α	T [K]	α
1.846	0.53	2.332	2.01
2.422	1.27	2.881	3.75
3.045	2.71	3.575	7.25
3.586	4.42	4.007	10.7
4.028	6.35	4.440	14.5
4.809	11.38	4.875	19.8
5.612	19.30	5.581	31.5
6.337	30.5	6.292	48.2
7.213	48.5	7.176	77.4
8.401	83.3	8.354	132.9
9.085	110.4	9.032	172.6
9.800	143.2	9.744	222.3
10.618	186.3	10.605	293.9

3.2.3 Thermal expansion at Low Temperatures (10 K $\leq T \leq$ 250 K)

Table 3.7 Values of the linear coefficient of thermal expansion α at low temperatures
10 K $\leq T \leq$ 250 K

T [K] Crystal $\downarrow$	$\alpha\,[10^{-6}\,\mathrm{K}^{-1}]$									Ref.
	10	20	30	40	50	100	150	200	250	
NaCl Structure										
LiF	-	0.06	0.24	0.4	1.7	11.5	18.1	24.2	29.0	[3.12, 3.13]
LiCl	-	-	-	-	-	24.4	32.6	38.4	41.5	[3.13]
LiBr	-	-	-	-	-	29.7	37.6	41.4	46.5	[3.13]
NaF	0.01	0.12	0.5	1.7	3.1	15.4	22.1	27.1	30.7	[3.13, 3.15]
NaCl	0.06	0.6	2.5	5.7	9.7	25.2	32.1	36.4	38.0	[3.16]
NaBr	0.22	2.3	6.4	11.7	16.1	25.5	33.9	37.4	39.5	[3.13, 3.15]
NaI	0.65	4.7	10.3	16.0	21.3	34.8	39.5	41.9	43.9	[3.16]
KF	-	-	-	-	-	19.1	24.7	27.8	30.1	[3.13]
KCl	0.05	0.7	3.1	7.1	11.4	25.4	30.8	33.5	35.4	[3.16]
KBr	0.16	2.1	6.8	11.6	16.7	29.3	33.1	35.3	37.1	[3.16]
KI	0.53	4.5	10.3	16.8	21.3	32.1	34.6	36.2	38.1	[3.12, 3.13]
RbCl	0.05	1.8	6.2	10.8	15.1	27.2	30.8	32.8	34.3	[3.15, 3.26]
RbBr	0.07	4.5	10.2	14.9	18.9	30.0	33.4	34.9	35.9	[3.15, 3.27]
RbI	0.3	6.2	13.5	19.1	23.4	32.5	34.2	36.1	38.5	[3.15, 3.26]
CsF	0.25	2.45	5.55	9.0	12.9	23.6	29.1	32.2	33.6	[3.15]
CsCl Structure										
CsCl	-	-	13.6	19.6	24.5	35.6	40.2	43.0	45.0	[3.17]
CsBr	1.56	10.15	19.5	26.0	30.2	38.8	42.1	44.3	46.0	[3.15, 3.17]
CsI	2.50	13.9	22.9	29.8	33.7	40.7	43.4	45.4	47.3	[3.15, 3.17]

Notes and Comments

1. Low temperature thermal expansion data are very useful for comparison with lattice dynamical results.
2. White and Collins [3.15] and Bailey and Yates [3.17] have calculated the Gruneisen constant as a function of temperature (at low temperatures) and have estimated the limiting values (γ_0 and γ_∞). These are given separately in Sec. 3.7.3.

3.2.4 Thermal Expansion at High Temperatures ($300\ \mathrm{K} \le T \le 1000\ \mathrm{K}$)

Table 3.8 Values of the linear coefficient of thermal expansion α

T [K] Crystal ↓	$\alpha\,[10^{-6}\,\mathrm{K}^{-1}]$								Ref.
	300	400	500	600	700	800	900	1000	
NaCl Structure									
LiF	35.5	38.0	40.7	43.6	46.5	49.4	53.7	59.2	[3.28]
LiCl	43.7	47.0	52.2	55.8	-	-	-	-	[3.13]
LiBr	48.9	52.5	57.8	70.8	-	-	-	-	[3.13]
NaF	34.2	36.4	38.7	41.2	43.8	46.4	49.4	52.8	[3.29]
NaCl	40.4	43.3	46.3	49.4	52.6	56.2	61.1	68.7	[3.30]
NaBr	41.9	44.3	48.1	54.9	-	-	-	-	[3.13]
NaI	45.0	46.8	51.8	62.6	-	-	-	-	[3.13]
KF	31.7	34.3	38.7	48.9	-	-	-	-	[3.13]
KCl	37.3	40.1	43.4	46.4	49.4	52.8	57.8	67.3	[3.30]
KBr	39.2	41.4	43.9	46.4	50.0	54.8	61.6	74.0	[3.29]
KI	37.9	40.6	43.1	45.8	48.0	53.4	60.4	74.5	[3.31]
RbCl	40.2	42.3	44.4	46.5	48.6	50.7	52.8	-	[3.9]
RbBr	39.0	41.8	44.6	48.0	52.1	57.7	66.2	-	[3.29]
RbI	38.0	40.6	43.3	46.3	50.3	56.4	69.8	-	[3.31]
CsCl Structure									
CsCl	47.0	52.0	57.0	62.0	67.0	-	-	-	[3.9]
CsBr	46.7	50.9	54.9	58.6	63.1	68.6	74.3	-	[3.30]
CsI	46.6	51.3	55.9	60.5	65.1	69.7	74.4	-	[3.9]

Notes and Comments

1. The data for LiCl, LiBr, NaBr, NaI and KF (Table 3.8) have been determined with a dilatometer; the rest of the data are from X-ray diffractometry.
2. Rapp and Merchant [3.13] used the high temperature thermal expansion data to calculate the thermal Gruneisen constant (γ) for 14 alkali halides and concluded that γ increases very slightly with temperature (see Sect. 3.7.2).
3. The temperature variation of the coefficient of thermal expansion follows Gruneisen's formula:

$$\alpha = Q\,C_v/3\,(Q - K\,E_T)^2 \tag{3.5}$$

where, $Q = (V_M/\gamma\psi)$, V_M the molar volume, γ the Gruneisen constant and ψ the compressibility, $K = \gamma + (2/3)$ and E_T the thermal energy is given by

$$E_T = \int_0^T C_V\,dT \cdot$$

4. The coefficient of thermal expansion increases linearly (or slightly nonlinearly) upto a certain temperature ($\approx 500\ °C$) and thereafter it shows a rapid increase. This increase $\Delta\alpha$ (difference between observed α and the value of α extrapolated from linear /slightly nonlinear trend) is attributed to the thermally generated defects. From Arrhenius plots of log $\Delta\alpha$ and T^{-1}, Pathak and coworkers [3.28–3.31] have estimated values of the energy of formation of Schottky pairs; these values are given in Sec. 9.1.6.

3.2.5 Temperature Variation of Thermal Expansion (Polynomial Form)

Table 3.9 The linear coefficient of thermal expansion α expressed as a polynomial in the temperature T [K] for different temperature ranges

Crystal	Temp. range	Polynomial	Ref.
Very low temperatures			
NaCl Structure			
LiF	<35 K	$\alpha = [0.83\,T^3\,]\times 10^{-11}$	[3.12]
NaF	<20 K	$\alpha = [1.34\,T^3 + (5.5\times10^{-4})T^5\,]\times 10^{-11}$	[3.15]
NaCl	<10 K	$\alpha = [5.73\,T^3 + (4\times10^{-3})T^5\,]\times 10^{-11}$	[3.15]
NaBr	<12 K	$\alpha = [16\,T^3 + (6\times10^{-2})T^5\,]\times 10^{-11}$	[3.15]
NaI	< 7 K	$\alpha = [43.5\,T^3 + 0.25\,T^5\,]\times 10^{-11}$	[3.15]
KCl	<10 K	$\alpha = [4.8\,T^3 + (7\times10^{-3})T^5\,]\times 10^{-11}$	[3.15]
KBr	< 8 K	$\alpha = [9.5\,T^3 + (6\times10^{-2})T^5\,]\times 10^{-11}$	[3.12]
KI	< 7 K	$\alpha = [24\,T^3 + (20\times10^{-2})T^5\,]\times 10^{-11}$	[3.12]
RbCl	< 5 K	$\alpha = [(0\pm1)\,T^3\,]\times 10^{-11}$	[3.15]
RbBr	< 6 K	$\alpha = [(-2\pm2)T^3\,]\times 10^{-11}$	[3.15]
RbI	< 5 K	$\alpha = [(-16\pm4)\,T^3 - 0.02\,T^5\,]\times 10^{-11}$	[3.15]
CsF	< 6 K	$\alpha = [(6\pm1)T^3 + 0.1\,T^5\,]\times 10^{-11}$	[3.15]
CsCl Structure			
CsBr	< 6 K	$\alpha = [(91\pm4)\,T^3 + 0.6\,T^5\,]\times 10^{-11}$	[3.15]
CsI	< 5 K	$\alpha = [(151\pm5)\,T^3 + 0.8\,T^5\,]\times 10^{-11}$	[3.15]

Table 3.9 (Continued)

		Intermediate and High Temperatures
Crystal	Temp. range [K]	Polynomial; Ref. [3.19]

NaCl Structure

LiF	293–1100	$\alpha = [3.424 \times 10^{-3} + (3.466 \times 10^{-7})T + (20.235 \times 10^{-10})T^2] \times 10^{-2}$
LiCl	75–293	$\alpha = [0.427 \times 10^{-3} + (2.464 \times 10^{-5})T - (3.918 \times 10^{-8})T^2] \times 10^{-2}$
	293–600	$\alpha = [3.458 \times 10^{-3} + (2.588 \times 10^{-6})T + (19.224 \times 10^{-10})T^2] \times 10^{-2}$
LiBr	293–550	$\alpha = [8.033 \times 10^{-3} - (18.804 \times 10^{-6})T + (29.01 \times 10^{-9})T^2] \times 10^{-2}$
NaF	10–293	$\alpha = [-7.813 \times 10^{-4} + (2.626 \times 10^{-5})T - (4.290 \times 10^{-8})T^2] \times 10^{-2}$
	293–1200	$\alpha = [3.163 \times 10^{-3} + (6.950 \times 10^{-7})T + (14.928 \times 10^{-10})T^2] \times 10^{-2}$
NaCl	293–600	$\alpha = [3.205 \times 10^{-3} + (2.942 \times 10^{-6})T - (25.677 \times 10^{-11})T^2] \times 10^{-2}$
	600–1000	$\alpha = [4.030 \times 10^{-3} - (9.108 \times 10^{-7})T + (3.855 \times 10^{-9})T^2] \times 10^{-2}$
NaBr	50–300	$\alpha = [1.243 \times 10^{-3} + (19.918 \times 10^{-6})T - (3.546 \times 10^{-8})T^2] \times 10^{-2}$
	293–850	$\alpha = [3.589 \times 10^{-3} + (12.094 \times 10^{-7})T + (24.990 \times 10^{-10})T^2] \times 10^{-2}$
NaI	150–550	$\alpha = [4.347 \times 10^{-3} - (19.444 \times 10^{-7})T + (7.218 \times 10^{-9})T^2] \times 10^{-2}$
KF	150–600	$\alpha = [2.330 \times 10^{-3} + (19.620 \times 10^{-7})T + (24.072 \times 10^{-10})T^2] \times 10^{-2}$
KCl	150–1000	$\alpha = [2.889 \times 10^{-3} + (2.212 \times 10^{-6})T + (10.764 \times 10^{-10})T^2] \times 10^{-2}$
KBr	293–1000	$\alpha = [4.314 \times 10^{-3} - (2.72 \times 10^{-6})T + (5.310 \times 10^{-9})T^2] \times 10^{-2}$
KI	150–700	$\alpha = [3.511 \times 10^{-3} + (4.360 \times 10^{-7})T + (4.284 \times 10^{-9})T^2] \times 10^{-2}$
RbCl	150–900	$\alpha = [2.292 \times 10^{-3} + (5.418 \times 10^{-6})T - (27.303 \times 10^{-10})T^2] \times 10^{-2}$
RbBr	50–293	$\alpha = [1.921 \times 10^{-3} + (11.406 \times 10^{-6})T - (18.492 \times 10^{-9})T^2] \times 10^{-2}$
	293–900	$\alpha = [3.785 \times 10^{-3} - (4.352 \times 10^{-7})T + (24.540 \times 10^{-10})T^2] \times 10^{-2}$
RbI	150–900	$\alpha = [3.190 \times 10^{-3} + (2.276 \times 10^{-6})T + (4.659 \times 10^{-10})T^2] \times 10^{-2}$

CsCl Structure

CsCl	150–700	$\alpha = [3.475 \times 10^{-3} + (3.494 \times 10^{-6})T + (14.754 \times 10^{-10})T^2] \times 10^{-2}$
CsBr	100–875	$\alpha = [3.501 \times 10^{-3} + (3.858 \times 10^{-6})T + (6.540 \times 10^{-10})T^2] \times 10^{-2}$
CsI	150–850	$\alpha = [3.760 \times 10^{-3} + (3.272 \times 10^{-6})T + (13.152 \times 10^{-10})T^2] \times 10^{-2}$

Notes and Comments

1. The linear coefficient of thermal expansion varies with temperature in a complex manner. The increase with temperature is slow at very low temperatures. This is followed by a steep increase. Finally, at elevated temperatures, the increase is, again slow. It is, therefore, not possible to represent the thermal expansion coefficient by a single polynomial. Different authors have proposed polynomials for different temperature ranges.
2. Polynomials to describe the temperature variation of thermal expansion have also been given in [3.13, 3.18, 3.20].

3.2.6 Pressure Variation of Thermal Expansion

Table 3.10 Values of the volume coefficient of expansion $\beta = (1/V)\,(dV/dT)$ at different pressures

Crystal	$\beta\,[10^{-4}\,(°C)^{-1}]$					Ref.
	Experimental Results (up to 30 kbar)					
P [Kbar]→	0	10	20	30		
NaCl Structure						
LiF	0.96	0.93	0.87	0.79		[3.32]
NaCl	1.169	0.968	0.841	0.756		[3.6]
	Experimental Results (up to 80 kbar)					
P [Kbar]→	0	20	40	60	80	
LiF	0.96	0.89	0.86	0.83	0.80	[3.33]
NaF	0.94	0.84	0.79	0.75	0.73	[3.33]
KF	0.97	0.84	0.78	-	-	[3.33]
CsCl Structure						
CsCl	1.38	0.85	0.54	-	-	[3.33]
High Pressure CsCl Phase						
KF	(1.78)		1.04	0.85	0.70	[3.33]
Extrapolated from high pressure values						
	Theoretical Results (up to 40 kbar)					
P [Kbar]→	0	10	20	30	40	
NaCl Structure						
LiCl	1.32	1.07	0.90	0.78	0.69	[3.34]
LiBr	1.50	1.15	0.94	0.79	0.68	[3.34]
LiI	1.80	1.25	0.96	0.78	0.66	[3.34]
NaBr	1.26	0.96	0.77	0.65	0.56	[3.34]
NaI	1.37	0.96	0.74	0.60	0.51	[3.34]

Notes and Comments

1. In all cases, the coefficient of expansion decreases with increasing pressure.
2. In the experimental results, the decrease in β for a given pressure change is nearly the same for crystals with NaCl structure; the corresponding change for crystals in CsCl phase is much larger.
3. Kumar [3.34] calculated the coefficient of thermal expansion at various pressures from the thermodynamically derived equation

$$\beta(P) = \beta(0)[1 + (\delta\psi) + (P - P_0)]^{-1} \tag{3.6}$$

where δ is the Anderson-Gruneisen parameter and ψ the compressibility and obtained good agreement with experimental values for LiF and NaCl. Kumar's calculated values for LiCl, LiBr, LiI, NaBr and NaI are quoted in Table 3.10, as there is no experimental data for these crystals.

4. Assuming that the parameter $(\gamma C_V / V)$ is independent of pressure (γ being the Gruneisen parameter), we get the result that the product (βB) is independent of pressure, B being the bulk modulus. From the experimental data on LiF, NaF, KF and CsCl, Yagi [3.33] observed that the product (βB) is indeed independent of pressure up to moderately high pressures. It is therefore suggested that the pressure variation of the oefficient of thermal expansion may be estimated from experimental data on pressure variation of the bulk modulus.
5. KF is the only crystal for which pressure variation of thermal expansion has been studied beyond the Fm3m→Pm3m transition. However, the pressure ($\approx$38 kbar) at which Yagi [3.33] observed the transition is much higher than the values of 14–18 kbar obtained by other methods.

3.3 Thermal Conductivity

3.3.1 Thermal Conductivity (Room Temperature)

Table 3.11 Values of the coefficient of thermal conductivity λ

	λ [W m^{-1} K^{-1}]			
Ref. Crystal ↓	[3.35]	[3.36]	[3.8]	[3.7]
NaCl Structure				
LiF	14.2	-	-	-
NaF	-	10.5	-	-
NaCl	6.32	-	6.02	-
NaBr	-	2.5	-	-
NaI	-	-	1.33	-
KF	-	7.1	-	-
KCl	6.7	-	-	-
KBr	-	3.8	-	-
KI	2.1	2.9	-	-
RbCl	-	2.1	-	-
RbI	-	3.3	-	-
CsCl Structure				
CsCl	-	-	-	0.95
CsBr	-	-	-	0.86
CsI	-	-	-	0.97

Notes and Comments

1. In earlier literature, the coefficient of thermal conductivity is expressed in units of Cal deg^{-1} cm^{-1} s^{-1}; to convert to W m^{-1} K^{-1}, multiply by 418.68.
2. Slack [3.37] pointed out that the magnitude of λ depends on the mass ratio. As the ratio M (alkali atom)/ M (halogen atom) decreases, λ also decreases.

3. Assuming that only acoustic phonons contribute to λ, the theoretical expression for λ is

$$\lambda_{\text{Theor}} = B\, n^{1/3}\, \overline{M}\, (V_m\, /\, n)\, (\theta_\infty)^2\, (\gamma_\infty)^{-2} \tag{3.7}$$

where, B is a constant $= 3.04 \times 10^{-8}$, $n = 2$ for the alkali halides, $\overline{M}$ is the average atomic mass in g, (V_m/n) is the average atomic volume ($a^3/4n$ for NaCl lattice and a^3/n for CsCl lattice, a being the lattice constant) and θ_∞ and γ_∞ the high temperature limiting values of the Debye temperature and Gruneisen parameter respectively. Here λ pertains to $T = \theta_\infty$. The theoretical values of λ agree with the experimental values only to the extent of 40 % [3.37].

3.3.2 Temperature Variation of Thermal Conductivity

Table 3.12 Values of λ [W cm^{-1} K^{-1}], the coefficient of thermal conductivity at selected temperatures; T [K]; Ref.. [3.38]

LiF		NaCl		KCl		KBr	
T	λ	T	λ	T	λ	T	λ
2.6	3.10	2.08	0.414	2.0	0.58	2.04	0.556
2.8	3.05	3.72	1.10	2.3	1.0	2.63	0.794
3.5	4.60	5.68	1.68	2.8	1.6	2.93	0.855
5.6	10.0	8.04	2.03	3.4	2.2	3.46	1.05
7.7	16.0	10.4	2.00	4.0	2.8	4.18	1.21
9.0	17.0	12.9	1.78	4.3	3.2	4.65	1.24
18.0	18.5	14.7	1.54	5.4	3.9	15.0	0.693
20.0	15.5	20.7	0.958	7.5	3.9	15.9	0.615
22.0	15.7	29.7	0.514	8.2	3.6	18.0	0.53
27.0	13.5	44.1	0.317	13.0	2.3	20.0	0.49
30.0	11.0	65.5	0.237	37.0	0.75	78.2	0.142
35.0	8.7	87.9	0.193	77.0	0.325	80.3	0.140
45.0	6.7	106.2	0.166	85.0	0.335	89.5	0.129
52.0	3.8	134.6	0.137	140.0	0.172	241.0	0.055
58.0	2.65	175.0	0.110	200.0	0.111	275.0	0.050
74.0	1.60	314.0	0.069	260.0	0.080	318.2	0.048
83.0	0.93	-	-	315.0	0.062	-	-
378.0	0.025	-	-	-	-	-	-

3.3.3 Pressure Variation of Thermal Conductivity

Table 3.13 Values of $(1/\lambda_0)(\partial\lambda/\partial P)_T$, the pressure coefficient of thermal conductivity and the density derivative $g = (\partial \log \lambda /\partial \log \rho)_T$; in those cases where a Fm3m→Pm3m transition takes place, the fractional change in thermal conductivity $(\Delta\lambda/\lambda)$ at the transition is also given; Ref. [3.39

Crystal	Pressure range [GPa]	$(1/\lambda_0)(\partial\lambda/\partial P)_T$ $[(GPa)^{-1}]$	g	$(\Delta\lambda/\lambda)_{P=P(\text{transition})}$
NaCl Structure				
LiF	0–0.5	0.18	11.3	-
NaF	0–0.5	0.12	5.7	-
NaCl	0–0.5	0.32	7.3	-
KCl	0–0.5	0.37	6.6	−0.60
KBr	0–0.5	0.58	8.8	−0.64
KI	0–0.5	1.15	12.0	−0.77
RbCl	0–0.5	0.63	9.2	−0.54
RbBr	0–0.5	0.61	7.5	−0.56
RbI	0–0.5	0.72	7.1	−0.57
CsCl Structure				
CsCl	0–0.5	0.41	7.1	-
CsBr	0–0.5	0.64	8.8	-
CsI	0–0.5	0.80	9.5	-
KCl	2–2.5	0.30	5.0	-
KBr	2–2.5	0.35	5.0	-
KI	2–2.5	0.23	3.6	-
RbCl	0.5–1.5	0.30	5.3	-
RbBr	0.5–1.5	0.53	7.2	-
RbI	0.5–1.5	0.70	7.3	-

Notes and Comments

1. The parameter g is obtained from the pressure derivative of λ using the relation

$$g = (\partial \log \lambda / \partial \log \rho)_T = (\psi / \lambda)(\partial\lambda/\partial P)_T \qquad (3.8)$$

   where ψ is the compressibility.
2. In the definition of the pressure derivative $(1/\lambda_0)(\partial\lambda/\partial P)_T$, λ_0 is the thermal conductivity at $P = 0$. In the case of the CsCl phase of K and Rb halides λ_0 is obtained by extrapolation of high pressure data.
3. The plots of g and the mass ratio (MR) are smooth plots. In the case of alkali halides with NaCl structure, the g versus MR plot has a small +ve slope indicating that g increases with MR; in the case of the CsCl phase crystals, the plot has a small −ve slope indicating a decrease in g with increase in MR [3.39].
4. The fractional change in thermal conductivity $(\Delta\lambda/\lambda)$ at the Fm3m→Pm3m transition correlates with the fractional change in density $(\Delta\rho/\rho)$ at the transition; the plot is linear with $(\Delta\lambda/\lambda)$ decreasing with $(\Delta\rho/\rho)$ (Ross et al. [3.39]).

3.3.4 Thermal Conductivity of Doped Alkali Halides

The temperature variation of thermal conductivity of alkali halides doped with divalent impurities is shown in Figures 3.1–3.3.

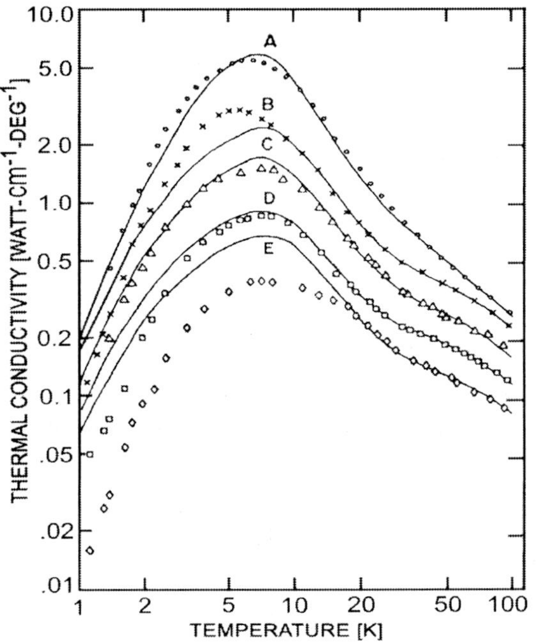

Fig. 3.1 Thermal conductivity versus temperature for KCl containing various divalent impurities: (A) pure KCl; (B) KCl: $EuCl_2$; (C) KCl: $CaCl_2$; (D) KCl: $BaCl_2$; (E) KCl: $SrCl_2$. Solid lines are theoretical curves whose calculation is discussed in Notes and Comments below. Curves A and B are as measured. The data were multiplied by 0.70, 0.50 and 0.40 to give curves C, D and E respectively. Concentrations of divalent ions (in units of 10^{18} cm^{-3}) are: Eu^{++}, 3.3; Ca^{++}, 1.8; Ba^{++}, 5.8; Sr^{++}, 5.4 (after Schwartz and Walker [3.40])

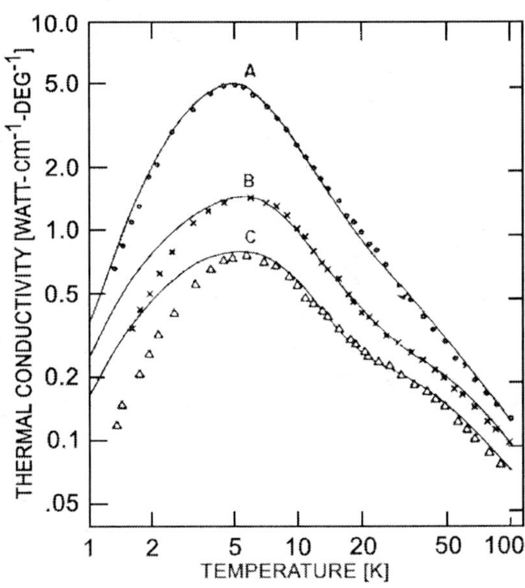

Fig. 3.2 Thermal conductivity versus temperature for KBr containing divalent impurities: (A) Pure KBr; (B) KBr: $SrBr_2$; (C) KBr: $BaBr_2$. Solid lines are theoretical curves whose calculation is discussed in Notes and Comments below. Curves A and B are as measured. The data were multiplied by 0.70 to give curve C. Concentration of divalent ions: Sr^{++}, 4.4; Ba^{++}, 3.5 (in units of 10^{18} cm^{-3}) (after Schwartz and Walker [3.40])

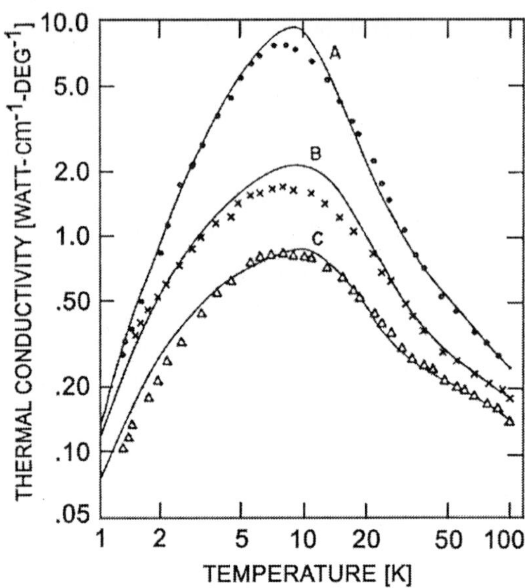

Fig. 3.3 Thermal conductivity versus temperature for NaCl containing divalent impurities: (A) Pure NaCl; (B) NaCl: CaCl$_2$; (C) NaCl: SrCl$_2$. Solid lines are theoretical curves whose calculation is discussed in Notes and Comments below. Curves A and B are as measured. The data were multiplied by 0.70 to give curve C. Concentrations of divalent ions (in units of 10^{18} cm^{-3}): Ca^{++}, 5.8; Sr^{++}, 11.0; (after Schwartz and Walker [3.40])

Notes and Comments

1. Data on thermal conductivity of NaCl: Ag^{+}; KCl: Ag^{+}; KCl: Li^{+} ; KBr: Li^{+}; NaCl: Cu^{+}; KBr: F^{-}; KCl: I^{-}; KCl: Tl^{+} are quoted by Klein [3.41].
2. Typically, thermal conductivity versus temperature curves at low temperature for pure alkali halides consist of two regions. In the first region the thermal conductivity increases with increasing temperature while in the second region, it decreases with increasing temperature. The peak of the curve occurs in the region 5–10 K. Within the framework of the relaxation time and Debye approximations, the lattice thermal conductivity λ is given by

$$\lambda = (1/2\pi^2\, \bar{\upsilon}) \int_0^{\omega_\mathrm{D}} \tau(\omega,T)\,(\hbar^2\omega^4/k_\mathrm{B}T^2) \left\{ \frac{\exp(\hbar\omega/k_\mathrm{B}T)}{[\exp(\hbar\omega/k_\mathrm{B}T)-1]^2} \right\} d\omega \qquad (3.9)$$

where ω_D is the Debye frequency, $\bar{\upsilon}$ is the mean sound velocity, τ the total relaxation time, T the temperature and $\hbar$ and k_B are the Planck and Boltzmann constants respectively. The inverse relaxation time τ^{-1} has three contributions viz., boundary scattering, isotope scattering and intrinsic scattering. By a proper choice of these three terms, the observed data can be fitted to the theoretical expression. The Debye temperature and inverse relaxation time needed to account for the observed thermal conductivity data for pure KCl, KBr and NaCl shown in Figs. 3.1–3.3 are given in Table VII.

Table VII Debye temperature and inverse relaxation time needed to fit pure-crystal data

Crystal	Debye temperature [K]	Inverse relaxation time [3.40]		
		Boundary scattering term $[10^5\,\mathrm{s}^{-1}]$	Isotope term $[10^{44}\,\mathrm{s}^{-1}]$	Intrinsic term $[10^{18}\,\mathrm{s}^{-1}]$
KCl	230	6.0	$10.0\,\omega^4$	$2.7\,\omega^2 T\exp(-34\,°\mathrm{K}/T)$
KBr	165	5.5	$8.0\,\omega^4$	$4.8\,\omega^2 T\exp(-30°\mathrm{K}/T)+0.022\,\omega^2 T$
NaCl	310	10.0	$5.0\,\omega^4$	$5.0\,\omega^2 T\exp(-55\,°\mathrm{K}/T)$

3. The effect of addition of small concentrations of divalent impurities is seen in the thermal conductivity versus temperature curves in the form of a 'dip' typically in the region 25–50 K. In a model due to Krumhansl [3.42] this effect is attributed to resonant scattering due to the change in force constant caused by the presence of the impurity. This results in an additional contribution to the inverse relaxation time:

$$\tau^{-1} = \frac{(R\,S^2)\,(\omega/\omega_\mathrm{D}^{\,4})}{[4(1+\tfrac{2}{3}S+2\,S\,\omega^2/\omega_\mathrm{D}^{\,2})^2 + \pi^2\,S^2\,\omega^6/\omega_\mathrm{D}^{\,6}]} \qquad (3.10)$$

where ω_D is the Debye frequency, S is the fractional change in force constant and R is given by

$$R = [(N\,\bar{v}\pi^2 a^2)/(4K_0^{\,4}a^4)]\left|\sin^2(\boldsymbol{K}_0\cdot\boldsymbol{a})\sin^2(\boldsymbol{K}_\mathrm{S}\cdot\boldsymbol{a})\right| \qquad (3.11)$$

where N is the concentration of defects, $\bar{v}$ the sound velocity, $\boldsymbol{a}$ the lattice vector (whose magnitude a is the lattice constant), $\boldsymbol{K_0}$ the incident phonon wave vector and K_0 its magnitude and $\boldsymbol{K}_\mathrm{S}$ the scattered phonon wave vector. Eq. (3.10) is equivalent to

$$\tau^{-1} = A\,\omega^4 / [(\omega_0^{\,2}-\omega^2)^2 + B\,\omega^6] \qquad (3.12)$$

where ω_0 is a resonance frequency. The parameters in these equations with which the observed thermal conductivity of doped KCl, KBr and NaCl shown in Figs. (3.1–3.3) is accounted for are given in Table VIII:

Table VIII Parameters to fit doped crystal data; Ref. [3.40]

System	$\omega_0\,[\mathrm{rad\,s}^{-1}]$	S	$R\,[\mathrm{s}^{-1}]$	$N\,[\mathrm{cm}^{-3}]$
KCl: $EuCl_2$	0.99×10^{13}	-1.1	2.1×10^{10}	3.3×10^{18}
KCl: $CaCl_2$	0.99×10^{13}	-1.1	2.1×10^{10}	1.8×10^{18}
KCl: $BaCl_2$	0.75×10^{13}	-1.25	1.1×10^{10}	5.8×10^{18}
KCl: $SrCl_2$	0.99×10^{13}	-1.1	4.1×10^{10}	5.4×10^{18}
KBr: $SrBr_2$	1.04×10^{13}	-0.80	1.4×10^{11}	4.4×10^{18}
KBr: $BaBr_2$	0.75×10^{13}	-1.05	5.0×10^{10}	3.5×10^{18}
NaCl: $CaCl_2$	1.72×10^{13}	-0.90	3.2×10^{11}	5.8×10^{18}
NaCl: $SrCl_2$	0.88×10^{13}	-1.30	4.0×10^{10}	11.0×10^{18}

3.4 Melting Temperatures

3.4.1 Melting Point at Atmospheric Pressure

Table 3.14 Values of the melting point (t_m, T_m) at atmospheric pressure

Ref. Crystal ↓	t_m [°C] [3.36a]	t_m [°C] [3.36b]	T_m [K] [3.43]	[3.44]	[3.45]	[3.46]
NaCl Structure						
LiF	845	848	-	-	-	-
LiCl	605	610	878 ± 5	-	-	-
LiBr	550	552	-	-	-	-
LiI	449	469	-	-	-	-
NaF	993	996	1265 ± 5	1265 ± 6	-	-
NaCl	801	801	1073 ± 5	1073 ± 6	-	-
NaBr	747	747	1014 ± 10	1014 ± 6	-	-
NaI	661	660	928 ± 10	928 ± 6	-	-
KF	858	858	-	-	1121 ± 15	-
KCl	770	771	1043 ± 5	-	1043 ± 15	-
KBr	734	734	-	-	1006 ± 15	-
KI	681	681	-	-	957 ± 15	-
RbF	795	833	-	-	-	1071 ± 8
RbCl	718	715	990 ± 5	-	-	991 ± 8
RbBr	693	682	-	-	-	950 ± 8
RbI	647	642	-	-	-	913 ± 8
CsF	682	703	-	-	-	-
CsCl Structure						
CsCl	645	645	917 ± 5	-	-	-
CsBr	636	636	-	-	-	-
CsI	626	621	-	-	-	-

Notes and Comments

1. For all practical purposes, the value given in °C is useful. However, in all theoretical expressions, the melting point in K is used. Hence values of t_m [°C] as well as T_m [K] are given for ready reference.
2. Considerable differences exist in different listings of the melting point. Thus for RbF, the values quoted in [3.36a] and [3.36b] are 795 and 833 °C respectively. Errors ranging from ± 3 to ± 15 are quoted in experimental work.
3. The product of the average coefficient of linear expansion (α) in units of 10^{-6} (°C)$^{-1}$ and melting point t_m [°C] is a constant with a value of 0.027 [3.14].
4. The plots between T_m and the Phillip's ionicity parameter f_i are straight lines, the halides of each alkali ion falling on a different straight line [3.47].
5. Tateno [3.48] proposed the relation

$$T_m = C\,n\,U\,/\,(\varepsilon_\infty - 1) \tag{3.13}$$

where C is a constant, n the index of the repulsion term in lattice energy, U the lattice energy and ε_∞ the dielectric constant at high frequency. A straight line plot is obtained between T_m and $\{n\,U\,/\,(\varepsilon_\infty-1)\}$.

6. Pietronero [3.49] derived the relation

$$r_0\,/\,r_1 = [\sqrt{2}\,/(2-\sqrt{2}\,)] - [1\,/\,r_1(2-\sqrt{2}\,)](k_B T_m\,/\,n)^{1/2} \tag{3.14}$$

where r_2 and r_1 are radii of the larger and smaller ions, k_B is Boltzmann's constant and n is a constant. Pietronero reported straight line plots between $r_2\,/r_1$ and T_m for the alkali halides with NaCl structure with a separate plot for halides of each alkali ion.

7. Ubbelohde [3.50] proposed the relation

$$T_m\,\bar{r} = \text{constant} \tag{3.15}$$

where $\bar{r} = (r_+ + r_-)$, r_+ and r_- being the ionic radii. For the alkali halides (with both NaCl and CsCl structures), the constant is $(3 \pm 0.5) \times 10^{-5}$ with T in [K] and $\bar{r}$ in cm.

8. Shanker and Kumar [3.51] obtained straight line plots between T_m and the effective ionic charge $e*/e$.

9. Barr and Dawson [3.52] showed the existence of the following empirical relation between T_m and E_f, the formation energy of Schottky defect pairs in alkali halides:

$$E_f\ (\text{in eV}) = 2.14 \times 10^{-3}\,T_m\ (\text{in K}) \tag{3.16}$$

10. According to the Lindemann theory the melting point T_m is related to the Debye temperature θ_D through the relation

$$\theta_D = C\,[T_m\,/\,M\,V_M^{\,2/3}]^{1/2} \tag{3.17}$$

where C is a constant, M the molar mass and V_M the molar volume.

3.4.2 Melting Parameters

Table 3.15 Values of the change in entropy ΔS [Cal mole^{-1} K^{-1}]; fractional volume change $\Delta V\,/V_S$ (where V_S is the volume of the solid at melting); and enthalpy of melting ΔH_m [kJ mole^{-1}]

Parameter	$\Delta V\,/V_S$	ΔS	ΔH_m
Ref.	[3.50]	[3.50]	[3.53]
Crystal ↓			
NaCl Structure			
LiF	0.294	5.78	26.4
LiCl	0.262	5.6	13.4
LiBr	0.243	4.9	13.0
NaF	0.274	5.5	33.6
NaCl	0.250	6.7	28.8

Table 3.15 (Continued)

Parameter	$\Delta V/V_S$	ΔS	ΔH_m
Ref.	[3.50]	[3.50]	[3.53]
Crystal $\downarrow$			
NaCl Structure			
NaBr	0.224	6.0	23.1
KI	0.159	4.3	26.4
RbCl	0.143	4.4	-
RbBr	0.135	3.9	-
CsCl Structure			
CsCl	0.105	3.5	-

Notes and Comments

1. Shanker et al. [3.47] obtained straight line plots between $(\Delta V/V_S)$ and the Phillips ionicity f_i, with a separate plot for the halides of each alkali ion.
2. Furukawa [3.54] showed that values of $(\Delta V/V_S)[(\alpha_+ + \alpha_-)/2]^{1/3}$ lie on a smooth curve when plotted against the radius ratio (r_+/r_-); α_+ and α_- are the electronic polarisabilities of the cation and anion respectively and r_+ and r_- are the radii of the corresponding ions.
3. M. Kumar [3.55] obtained linear plots between ΔS and (i) Phillips ionicity f_i and (ii) radius ratio (r_+/r_-) with a separate plot for halides of each alkali ion.
4. Sangwal [3.53] pointed out the following empirical relation between ΔH_m and the melting point T_m:

$$(\Delta H_m / R T_m) = 3 \tag{3.18}$$

where R is the gas constant.

3.4.3 Pressure Variation of Melting Point (Polynomial Form)

Table 3.16 Values of the constants in the polynomial $t_m = a_0 + a_1 P + a_2 P^2 + a_3 P^3$ (t_m in °C and P in kbar)

Crystal	a_0	a_1	a_2	a_3	Ref.
NaCl Structure					
NaF	992	15.47	−0.219	1.429×10^{-3}	[3.44]
NaCl	802.5	21.61	−0.201	0.576×10^{-3}	[3.44]
NaBr	742	26.41	−0.529	6.095×10^{-3}	[3.44]
NaI	655	27.66	−0.388	2.064×10^{-3}	[3.44]
KF	858	14.91	−0.1587	-	[3.45]
KCl	775	22.75	−0.4458	-	[3.45]
KBr	742	23.66	−0.5059	-	[3.45]
KI	684	29.40	−0.8658	-	[3.45]

Table 3.16 (Continued)

Crystal		a_0	a_1	a_2	a_3	Ref.
CsCl Structure						
KCl	P > 18.7 kbar	625	24.92	−0.1156	-	[3.45]
KBr	P > 17.4 kbar	478	36.66	−0.3191	-	[3.45]
KI	P > 17.3 kbar	332	41.95	−0.3748	-	[3.45]

3.4.4 Pressure Variation of Melting Point (Simon Equation)

Table 3.17 Values of the parameters of the Simon equation: $(P-P_0) = A\,[(T/T_0)^C - 1]$; T_0 is the melting point at pressure P_0

Crystal	T_0 [K]	P_0 [kbar]	A [kbar]	C	Ref.
NaCl Structure					
LiCl	878	0	14.5	2.5	[3.43]
NaF	1265	0	12.2	5.762	[3.44]
NaCl	1073	0	15.0	2.969	[3.44]
NaBr	1014	0	11.1	3.356	[3.44]
NaI	928	0	7.13	3.649	[3.44]
KF	1121	0	7.38	6.743	[3.45]
KCl	1043	0	5.98	5.990	[3.45]
KBr	1006	0	3.57	7.442	[3.45]
KI	957	0	1.39	10.73	[3.45]
RbF	1071	0	13.8	5.13	[3.46]
RbCl	991	0	5.6	6.91	[3.46]
RbBr	950	0	4.4	6.71	[3.46]
RbI	913	0	5.6	5.03	[3.46]
CsCl Structure					
CsCl	933	0.950	8.4	2.3	[3.43]
KCl	1323	18.7	29.83	2.097	[3.45]
KBr	1270	16.8	12.49	3.499	[3.45]
KI	1207	16.9	9.72	3.656	[3.45]
RbCl	1125	7.80	15.7	2.085	[3.46]
RbBr	1081	6.1	8.20	2.853	[3.46]
RbI	1033	5.0	7.14	2.798	[3.46]

3.4.5 Pressure Coefficient of Melting Point and Entropy Change

Table 3.18 Values of $(dT_m/dP)_{P=0}$ the pressure coefficient at zero pressure; ΔV, the change in volume at melting and ΔS, the change in entropy at melting

Crystal	$(dT_m/dP)_{P=0}$ [deg bar^{-1}]	ΔV [cm^3 mole^{-1}]	ΔS [Cal mole^{-1} deg^{-1}]	$(\Delta V/\Delta S)_{T=T_m}$ [deg bar^{-1}]	Ref.
NaCl Structure					
LiCl	0.0242	5.88	5.6	0.025	[3.43]
NaF	0.0151	4.15–4.64	5.5–6.2	0.016–0.020	[3.44]
NaCl	0.0241	7.55	6.3–6.7	0.027–0.029	[3.44]
NaBr	0.0272	8.07	5.9–6.0	0.032–0.033	[3.44]
NaI	0.0357	8.58	5.6	0.037	[3.44]
KF	0.0226	4.45	5.8	0.019	[3.45]
KCl	0.0291	7.20	6.2	0.028	[3.45]
KBr	0.0379	7.98	4.9	0.039	[3.45]
KI	0.0646	9.30	4.3	0.052	[3.45]
RbF	0.015	2.5	3.9	-	[3.46]
RbCl	0.026	6.72	4.4	0.037	[3.46]
RbBr	0.032	7.26	3.9	0.045	[3.46]
RbI	0.033	-	-	-	[3.46]
CsCl Structure					
CsCl	0.017	5.69	3.9	0.035	[3.43]

Notes and Comments

1. According to the Clausius-Clapeyron equation

$$(\Delta V / \Delta S)_{T=T_m} = (dT_m/dP)_{P=0} \tag{3.19}$$

There is reasonable agreement between the experimental values for the two quantities.

2. The pressure coefficient of the melting point is related to the Gruneisen parameter (γ) through the equation

$$\gamma = \frac{1}{2}\left[\frac{1}{\psi\, T_m}\left(\frac{dT_m}{dP} \right)_{P=0} + \frac{2}{3} \right] \tag{3.20}$$

where ψ is the compressibility. Vaidya and Gopal [3.56] calculated γ of some alkali halides using this relation and found good agreement with γ calculated from thermal expansion particularly when the high temperature value of the compressibility is used.

3.4.6 Lindemann Parameter

Table 3.19 Values of the Lindemann parameter δ (for definition see Notes and Comments)

Crystal ↓	δ		
	Experimental		Theoretical
Ref.	[3.57]	[3.58]	[3.59]
NaCl Structure			
LiF	0.114	-	0.161
NaCl	0.112	0.166	0.158
NaBr	-	-	0.158
NaI	-	-	0.162
KCl	0.110	0.184	0.154
KBr	0.114	-	0.156
KI	-	-	0.162
RbCl	-	-	0.152
RbBr	-	-	0.156
RbI	-	-	0.159

Notes and Comments

1. The Lindemann parameter δ is defined as the ratio (rms amplitude of vibration / interionic distance) at the melting point.
2. The experimental values of δ have been obtained from experiments on Mossbauer scattering [3.57] and X-ray diffraction [3.58].
3. The theoretical values of δ [3.59] have been obtained from lattice dynamical calculations based on a 7-parameter bond-bending model.
4. According to Lindemann theory, the parameter δ has a constant value for each family of crystals. The experimental values of δ are in the range 0.11–0.18. The theoretical values are constant with a value of 0.16.

3.5 Debye-Waller Factors

3.5.1 Debye-Waller Factors at RT (Expt. and Theor.)

Table 3.20 Values of the Debye-Waller factors B_A and B_B for crystal AB and the mean Debye-Waller factor $\overline{B}$ at room temperature; B's in Å^2; uncertainties in last digit given given in parenthesis

Crystal	Experimental				Theoretical [3.63]	
	B_A	B_B	$\overline{B}$	Ref.	B_A	B_B
NaCl Structure						
LiF	1.05(1)	0.65(1)	0.76(1)	[3.60]	0.895	0.620
LiCl	2.33(10)	1.18(10)	1.37(10)	[3.61]	1.845	1.360
LiBr	-	-	-	-	2.247	1.592
LiI	-	-	-	-	3.429	1.881
NaF	0.91(1)	0.91(1)	0.91(1)	[3.60]	0.850	0.850
NaCl	1.72(2)	1.41(1)	1.53(2)	[3.60]	1.539	1.292
NaBr	1.55(15)	1.14(10)	1.23(15)	[3.62]	1.894	1.555
NaI	2.63(20)	1.81(15)	1.94(25)	[3.62]	2.413	1.958
KF	-	-	1.21(2)	[3.60]	1.179	1.308
KCl	2.17(1)	2.16(1)	2.17(1)	[3.60]	1.771	1.813
KBr	2.36(4)	2.38(4)	2.37(6)	[3.60]	2.105	2.158
KI	3.52(22)	2.80(15)	2.97(27)	[3.60]	2.692	2.550
RbF	-	-	1.40(25)	[3.60]	1.352	1.565
RbCl	2.14(7)	2.27(7)	2.18(10)	[3.60]	1.976	2.029
RbBr	-	-	2.24(16)	[3.60]	2.287	2.439
RbI	-	-	3.36(40)	[3.60]	3.097	2.781
CsF	-	-	-	-	1.663	1.892
CsCl Structure						
CsCl	1.83(2)	1.89(5)	1.84(5)	[3.60]	1.842	1.805
CsBr	2.00(9)	2.24(12)	2.09(15)	[3.60]	2.071	1.877
CsI	2.27(11)	2.19(12)	2.24(16)	[3.60]	2.244	2.184

Notes and Comments

1. Details of experimental determination of Debye-Waller factors can be found in [3.61, 3.62, 3.69, 3.70]. Details of theoretical evaluation of Debye-Waller factors from lattice dynamics can be found in [3.63, 3.70, 3.77].
2. Linkaoho [3.64] pointed out that the ratio B/a^2 for a given ion (where a is the lattice constant) increases linearly with the number of electrons of the companion ion. This was clearly shown in the case of the Cl ions in LiCl, NaCl, KCl and RbCl using room temperature data.
3. McIntyre et al. [3.65] pointed out that in the alkali halides at room temperature, the ratio $(B_A + B_B)/a^2$ is nearly constant with a value of $(10.5 \pm 0.5) \times 10^{-2}$.

3.5.2 Debye-Waller Factors at Low Temperatures

Table 3.21 Experimental values of Debye-Waller factors B_A and B_B for crystal AB at low temperatures; uncertainties in last digit given in parenthesis wherever reported

Crystal	T [K]	B_A [Å^2]	B_B [Å^2]	Ref.
NaCl Structure				
LiF	78	0.93(10)	0.41(10)	[3.61]
NaF	80	0.43(7)	0.40(7)	[3.66]
NaCl	80	0.62(6)	0.41(6)	[3.67]
KF	91	0.69	0.94	[3.68]
KCl	86	0.59	0.59	[3.69]
KBr	91	1.10	0.70	[3.68]
KI	91	1.40	0.94	[3.68]
CsCl Structure				
CsCl	90	0.40(2)	0.50(7)	[3.70]
CsBr	78	0.56(4)	0.63(7)	[3.71]

3.5.3 Debye-Waller Factors close to Melting Point

Table 3.22 Values of mean Debye-Waller factor $\overline{B}$ close to melting point ($\sim$ 20 K below melting point)

Crystal	$\overline{B}$ [Å^2]	Method	Ref.
NaCl Structure			
LiF	2.97	Mossbauer	[3.57]
NaCl	8.44	Mossbauer	[3.57]
	6.31(20)	X-ray	[3.58]
KCl	9.57	Mossbauer	[3.57]
	9.50	X-ray	[3.58]
KBr	11.40	Mossbauer	[3.57]

Notes and Comments

1. Martin and O'Connor [3.57] found that the ratio of the mean amplitude of vibration to the interionic distance at melting point has a value of 0.11 for the alkali halides; Viswamitra and Jayalaxmi [3.58] found larger values in the range 0.16–0.20 for the same ratio.

3.5.4 Temperature Variation of Debye-Waller factors (Experimental)

The temperature variation of Debye-Waller factors obtained from X-ray diffraction is shown in Fig. 3.4 as a plot between the Debye-Waller factors and $(T/\theta_D)^{3/2}$, θ_D being the Debye temperature [K] and T the temperature [K].

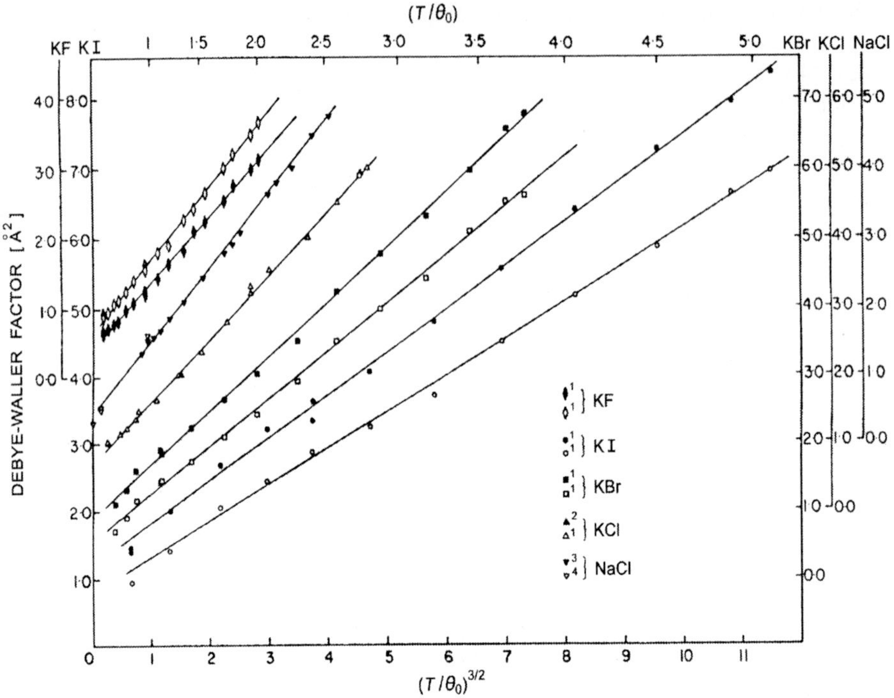

Fig. 3.4 After Bastow et al. [3.72]; θ_D values: KF, 316 K; KBr, 170 K; KCl, 227 K; KI, 121 K and NaCl, 307 K. Superscripts on the symbols in the diagram refer to data sources: (1) Tompson [3.73], individual Debye-Waller factors given for KF, KBr and KI; open symbols for anions and filled symbols for cations; for KCl only the mean Debye-Waller factor is given (open symbol); (2) Patomaki and Linkoaho [3.74], only the mean Debye-Waller factor for KCl given (filled symbol); (3) Vadets and Giller [3.75] only the mean Debye-Waller factor for NaCl given (filled symbol); (4) Butt and Cheetham [3.76] and Linkoaho [3.64]; only the mean Debye-Waller factor for NaCl given (open symbol)

Notes and Comments

1. Data on temperature variation of the mean Debye-Waller factors for some alkali halides obtained from Mossbauer scattering are given by Martin and O'Connor [3.57].
2. Values of Debye-Waller factors of alkali halides at different temperatures have been calculated from a 7-parameter bond-bending model by Kushwaha [3.59].
3. Bastow et al. [3.72] showed that the temperature variation of Debye-Waller factors can be represented by the equation

$$B(T) = a + bT^{3/2} \tag{3.21}$$

where a and b are constants for a given material and T the temperature [K].

3.5.5 Temperature Variation of Debye-Waller Factors (Theoretical)

Experimental values of Debye-Waller factors as a function of temperature are available for only a few alkali halides. Gao et al. [3.77] calculated the Debye-Waller factors for twelve alkali halides with NaCl structure and the three cesium halides at different temperatures from 1 K to 1000 K using a 11-parameter breathing shell model of lattice dynamics. The mean Debye-Waller factors calculated from those of the individual ions agreed with experimental values within 10 %. The calculated values were fitted to a polynomial of the form:

$$B(T) = a_0 + a_1 T + a_2 T^2 + a_3 T^3 + a_4 T^4 \tag{3.22}$$

where B is in Å^2 and T is in K. The polynomial parameters are given in Table 3.23.

Table 3.23 Polynomial regression form fitting parameters of the Debye-Waller factors using the 11-parameter breathing shell model; ME is the maximum error (% difference between actual calculated value and the value from the polynomial)

(0-80K)

Crystal	Atom	a_0	a_1	a_2	a_3	a_4	ME (%)
NaCl Structure							
KF	K	0.27569	−3.97001E-5	1.90082E-5	1.45132E-7	−1.36815E-9	0.07
	F	0.41449	−1.32321E-5	1.90294E-5	3.95364E-8	−4.43675E-10	0.01
KCl	K	0.34407	−2.71951E-4	5.20042E-5	−7.64989E-9	−1.40092E-9	0.11
	Cl	0.33970	−1.59544E-4	4.57814E-5	−5.53940E-8	−7.72741E-10	0.05
KBr	K	0.36082	−3.06273E-4	9.25366E-5	−5.74224E-7	1.42586E-9	0.07
	Br	0.25412	−6.33280E-4	1.18614E-4	−9.10515E-7	2.66978E-9	0.27
KI	K	0.38640	−2.04061E-4	1.26784E-4	−1.00081E-6	3.44973E-9	0.07
	I	0.22061	−8.82914E-4	1.92263E-4	−2.01719E-6	8.20798E-9	0.38
NaF	Na	0.31824	3.36863E-5	3.28677E-6	1.30257E-7	−6.58199E-10	0.01
	F	0.32146	5.17396E-6	6.33234E-6	3.29345E-8	−1.57649E-10	0.01
NaCl	Na	0.41303	−5.80361E-6	1.52557E-5	2.51515E-7	−1.81479E-9	0.03
	Cl	0.30318	−3.03046E-5	1.73283E-5	1.99679E-7	−1.66904E-9	0.06
NaBr	Na	0.46778	−5.23749E-5	4.03194E-5	5.83251E-8	−1.16063E-9	0.02
	Br	0.23260	−4.85717E-4	7.24684E-5	−3.24320E-7	−4.86362E-11	0.27
NaI	Na	0.51388	−9.52548E-5	7.31870E-5	−2.61936E-7	1.43879E-10	0.08
	I	0.19861	−7.92322E-4	1.32936E-4	−1.21423E-6	4.31378E-9	0.39
RbF	Rb	0.20574	−3.25100E-4	6.34232E-5	−3.20302E-7	2.71684E-10	0.19
	F	0.44368	3.77853E-6	4.20515E-5	−2.49335E-7	9.64524E-10	0.04
RbCl	Rb	0.24532	−6.61109E-4	1.22166E-4	−9.70945E-7	2.96674E-9	0.27
	Cl	0.36621	−1.50104E-4	8.05590E-5	−4.90654E-7	1.29254E-9	0.02
RbBr	Rb	0.25697	−6.16459E-4	1.52916E-4	−1.39827E-6	5.08502E-9	0.22
	Br	0.25998	−5.05979E-4	1.41578E-4	−1.26004E-6	4.50533E-9	0.18
RbI	Rb	0.27878	−5.25372E-4	2.07079E-4	−2.13499E-6	8.69368E-9	0.12
	I	0.22175	−5.39958E-4	2.10075E-4	−2.31512E-6	9.83668E-9	0.17
LiF	Li	0.55882	−4.51204E-6	4.05664E-6	−1.12501E-8	1.42384E-10	0.01
	F	0.28529	4.69857E-6	2.88026E-6	2.42906E-8	−4.19277E-11	0.02

Table 3.23 (Continued)

(0-80K)

Crystal	Atom	a_0	a_1	a_2	a_3	a_4	ME (%)
CsCl Structure							
CsCl	Cs	0.19200	−9.32891E-4	1.24655E-4	−1.09899E-6	3.72778E-9	0.52
	Cl	0.35291	−7.61578E-5	3.20808E-5	1.43255E-7	−1.64323E-9	0.04
CsBr	Cs	0.20343	−9.21443E-4	1.54253E-4	−1.50007E-6	5.67996E-9	0.45
	Br	0.24988	−4.28096E-4	9.70985E-5	−6.16537E-7	1.26817E-9	0.19
CsI	Cs	0.21257	−8.97273E-4	1.76420E-4	−1.79416E-6	7.08823E-9	0.39
	I	0.21701	−8.53885E-4	1.65725E-4	−1.60810E-6	6.06843E-9	0.37

(80-1000K)

Crystal	Atom	a_0	a_1	a_2	a_3	a_4	ME (%)
NaCl Structure							
KF	K	0.14194	0.00318	2.02989E-6	−2.19253E-9	8.53360E-13	0.59
	F	0.27178	0.00297	3.69801E-6	−3.95570E-9	1.53053E-12	0.96
KCl	K	0.14761	0.00537	2.16668E-6	−2.35307E-9	9.19088E-13	0.58
	Cl	0.16026	0.00477	2.32750E-6	−2.52224E-9	9.83875E-13	0.61
KBr	K	0.14868	0.00659	2.19080E-6	−2.38050E-9	9.30057E-13	0.48
	Br	0.07524	0.00652	1.13348E-6	−1.23774E-9	4.85304E-13	0.27
KI	K	0.15024	0.00803	2.23466E-6	−2.43511E-9	9.53368E-13	0.39
	I	0.04829	0.00787	7.37853E-7	−8.08101E-10	3.17417E-13	0.16
NaF	Na	0.21945	0.00175	2.93528E-6	−3.12712E-9	1.20651E-12	1.00
	F	0.24767	0.00128	3.14520E-6	−3.31513E-9	1.27028E-12	1.20
NaCl	Na	0.23672	0.00378	3.34096E-6	−3.60005E-9	1.39937E-12	0.89
	Cl	0.15733	0.00338	2.25586E-6	−2.43726E-9	9.48641E-13	0.64
NaBr	Na	0.24414	0.00540	3.51803E-6	−3.80606E-9	1.48289E-12	0.64
	Br	0.07509	0.00507	1.13152E-6	−1.23607E-9	4.84749E-13	0.33
NaI	Na	0.24823	0.00712	3.61419E-6	−3.91746E-9	1.52803E-12	0.59
	I	0.04800	0.00609	7.30330E-7	−7.99154E-10	3.13594E-13	0.18
RbF	Rb	0.06966	0.00444	1.04785E-6	−1.14546E-9	4.49484E-13	0.41
	F	0.27795	0.00397	3.83827E-6	−4.11514E-9	1.59392E-12	0.83
RbCl	Rb	0.07044	0.00650	1.06161E-6	−1.15833E-9	4.53694E-13	0.27
	Cl	0.16240	0.00613	2.37878E-6	−2.58312E-9	1.00912E-12	0.51
RbBr	Rb	0.07064	0.00749	1.06457E-6	−1.16088E-9	4.54347E-13	0.17
	Br	0.07548	0.00725	1.13991E-6	−1.24437E-9	4.87347E-13	0.18
RbI	Rb	0.07118	0.00931	1.08335E-6	−1.18566E-9	4.65529E-13	0.15
	I	0.04818	0.00858	7.36715E-7	−8.07694E-10	3.17577E-13	0.16
LiF	Li	0.52670	2.69660E-4	5.38908E-6	−5.41633E-9	2.01237E-12	0.96
	F	0.23424	7.51248E-4	2.85120E-6	−2.97887E-9	1.13506E-12	1.20
CsCl Structure							
CsCl	Cs	0.04602	0.00567	7.00518E-7	−7.67034E-10	3.01533E-13	0.31
	Cl	0.16333	0.00485	2.40028E-6	−2.60727E-9	1.01869E-12	0.88
CsBr	Cs	0.04348	0.00665	6.22486E-7	−6.68495E-10	2.58682E-13	0.43
	Br	0.07168	0.00615	1.02629E-6	−1.10544E-9	4.29313E-13	0.66
CsI	Cs	0.04607	0.00737	6.97347E-7	−7.60489E-10	2.97656E-13	0.29
	I	0.04834	0.00725	7.34509E-7	−8.02746E-10	3.14886E-13	0.31

3.6 Debye Temperature

3.6.1 Debye Temperatures at Room Temperature

Table 3.24 Values of Debye temperature θ_M from X-ray /neutron diffraction or Mossbauer scattering; θ_{el} from room temperature elastic constants and θ_ψ from compressibility; uncertainty in last digit given in parenthesis

Crystal	θ_M [K]	Ref.	θ_{el} [K]	Ref.	θ_ψ [K] Ref. [3.81]
NaCl Structure					
LiF	594(4)	[3.60]	739	[3.79]	685
LiCl	352(13)	[3.61]	394	[3.79]	484
LiBr	-	-	249	[3.79]	425
LiI	-	-	177	[3.79]	375
NaF	426(2)	[3.60]	492	[3.79]	429
NaCl	278(2)	[3.60]	322	[3.79]	292
NaBr	202(6)	[3.62]	224	[3.79]	241
NaI	144(6)	[3.62]	167	[3.79]	210
KF	316(3)	[3.60]	328	[3.79]	335
KCl	206(1)	[3.60]	236	[3.79]	229
KBr	155(2)	[3.60]	172	[3.79]	181
KI	117(5)	[3.60]	131	[3.79]	156
RbF	216(19)	[3.60]	212	[3.79]	294
RbCl	161(4)	[3.60]	169	[3.79]	191
RbBr	135(5)	[3.60]	136	[3.79]	140
RbI	95(5)	[3.60]	108	[3.79]	116
CsF	109(1)	[3.78]	158	[3.80]	273
CsCl Structure					
CsCl	148(2)	[3.60]	159	[3.79]	175
CsBr	123(4)	[3.60]	149	[3.79]	125
CsI	108(4)	[3.60]	126	[3.79]	102

Notes and Comments

1. The θ_M values in the compilation by Butt et al. [3.60] are based on Debye-Waller factors which are the mean of several values in the literature. Therefore it is not possible to attribute these values to a particular source or a particular diffraction technique.

2. The θ_ψ values given above are calculated from the formulae

$$\theta_\psi = (h / 2\pi \, k_B)\,(5r / \mu\psi)^{1/2} \qquad \text{for NaCl structure and}$$

$$\theta_\psi = (h / 2\pi \, k_B)\,(20r / 3\sqrt{3}\mu\psi)^{1/2} \qquad \text{for CsCl structure} \qquad (3.23)$$

Here r is the interionic distance, μ the reduced mass and ψ the compressibility.

3. The Debye temperature is related to the melting temperature (T_m) through the Lindemann law

$$\theta = \text{constant } (T_m / MV_M^{2/3})^{1/2} \tag{3.24}$$

where M is the molecular mass and V_M the molar volume.

4. The Debye temperature is related to the formation energy of a Schottky pair (E_f) through the relation

$$\theta = \text{constant } (E_f / MV_M^{2/3})^{1/2} \tag{3.25}$$

Values of E_f calculated from this formula for the alkali halides are given by Pathak and Trivedi [3.82] with a value of 4183 for the constant.

5. Bansigir [3.83] has shown that the plot of log θ and log V for the alkali halides is linear, and the slope of this plot is the mean Gruneisen constant γ for the family of crystals. For the alkali halides, the slope and, hence, the Gruneisen constant is 1.67.

6. Hoinks et al. [3.84] found from LEED measurements that the surface Debye temperature of LiF is 415 K (compared with the bulk value of 732 K).

3.6.2 Debye Temperatures at ~ 0 K

Table 3.25 Values of the Debye temperature (θ_D^0) from specific heats and (θ_{el}^0) from elastic constants at ~ 0 K; uncertainty in last digit given in parenthesis; Ref. CsCl [3.85], rest [3.86]

Crystal	θ_{el}^0 [K]	θ_D^0 [K]
NaCl Structure		
LiF	733	737(9)
LiCl	429	422(6)
LiBr	274	-
LiI	210	-
NaF	492	-
NaCl	320.8	321(1)
NaBr	224	-
NaI	167.5	164(1)
KF	332.8	-
KCl	236.1	233(3)
KBr	172	174(1)
KI	130.8	132(1)
RbCl	168.8	165
RbBr	136.3	131
RbI	107.8	103
CsCl Structure		
CsCl	168	174(2)
CsBr	149.5	-
CsI	126.2	128

Notes and Comments

1. θ_D^0 values are obtained from specific heat data at low temperatures using the Debye-T^3 law.
2. There is close agreement between θ_D^0 and θ_{el}^0.
3. Konti and Varshni [3.86] obtained linear plots when θ_{el} was plotted against the reduced mass. But instead of a single plot for all the alkali halides, they obtained separate plots for halides of a common alkali ion. In fact, the θ_{el} value for LiI has been interpolated from such a plot.

3.6.3 Temperature Variation of Debye Temperatures at Low Temperature

The temperature variation of Debye temperature from specific heats at low temperature is shown in Figs. 3.5–3.12.

Notes and Comments

1. From the variation of Debye temperature at low temperatures, the following parameters have been derived by Barron et al. [3.88]:
   i) θ_0, the low temperature limit of Debye temperature
   ii) θ_∞, the high temperature limit of Debye temperature
   iii)   Moments μ_2, μ_4 and μ_6 in the Thirring expansion of specific heats
   iv)   Zero point energy E_Z
   v)   Anharmonicity parameter λ.
   The values of these parameters are given in Table IX.

Table IX Values of θ_0, θ_∞, μ_2, μ_4, μ_6, E_Z and λ

Crystal	θ_0 [K]	θ_∞ [K]	$\mu_2 \times 10^{-24}$ [s^{-2}]	$\mu_4 \times 10^{-49}$ [s^{-4}]	$\mu_6 \times 10^{-74}$ [s^{-6}]	E_Z [Cal mole^{-1}]	$\lambda \times 10^2$
KCl	235.1	235.1	14.45	27.5	59	1037	4.35
KBr	174.3	187.5	9.17	12.4	21	811	3.92
KI	132.3	162.5	6.90	7.6	10.2	684	3.69
NaCl	320.6	290	21.9	63	220	1280	5.05
NaI	164.2	195	9.90	18.6	53	811	4.54

Barron et al. [3.88] also pointed out that the product $(\theta_0/\theta_\infty)(1-\eta^2)^{1/2}$ is nearly constant for the halides of a given alkali ion, the values of this product being ≈ 1 for potassium halides, ≈ 0.85 for sodium halides, 0.75 for LiF and ≈ 0.9 for rubidium halides [here $\eta = (m_1 - m_2)/(m_1 + m_2)$]. For the cesium halides, the product is 0.83 [3.93]

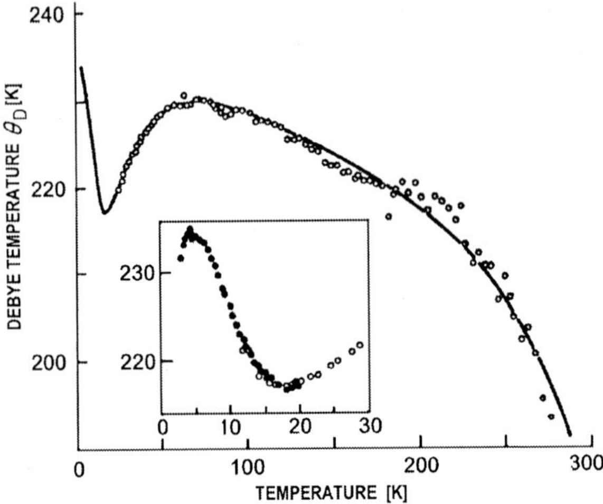

Fig. 3.5 The Debye characteristic temperature of potassium chloride. In this and in Figs. 3.6–3.8, the open and filled circles denote the results from two different calorimeter assemblies (after [3.87])

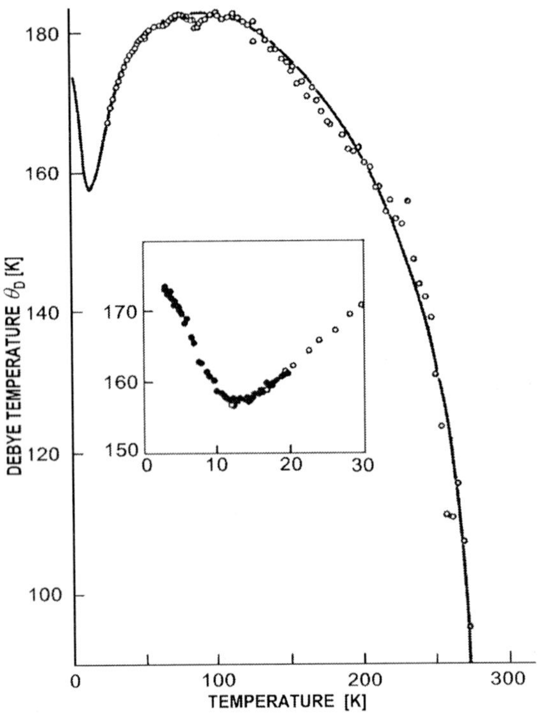

Fig. 3.6 The Debye characteristic temperature of potassium bromide (after [3.87])

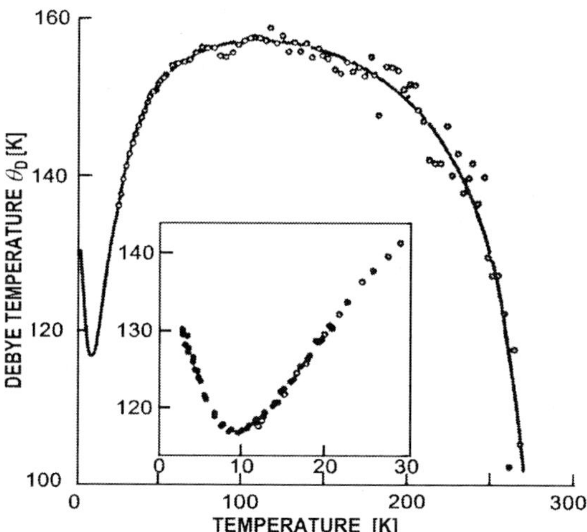

Fig. 3.7 The Debye characteristic temperature of potassium iodide (after [3.87])

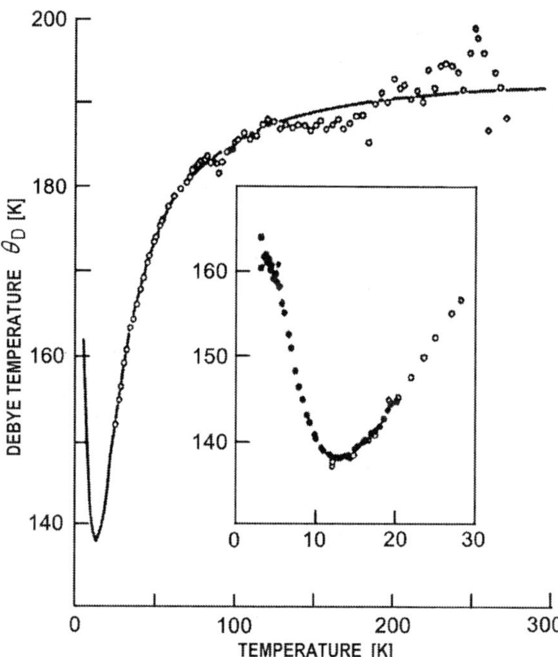

Fig. 3.8 The Debye characteristic temperature of sodium iodide (after [3.87])

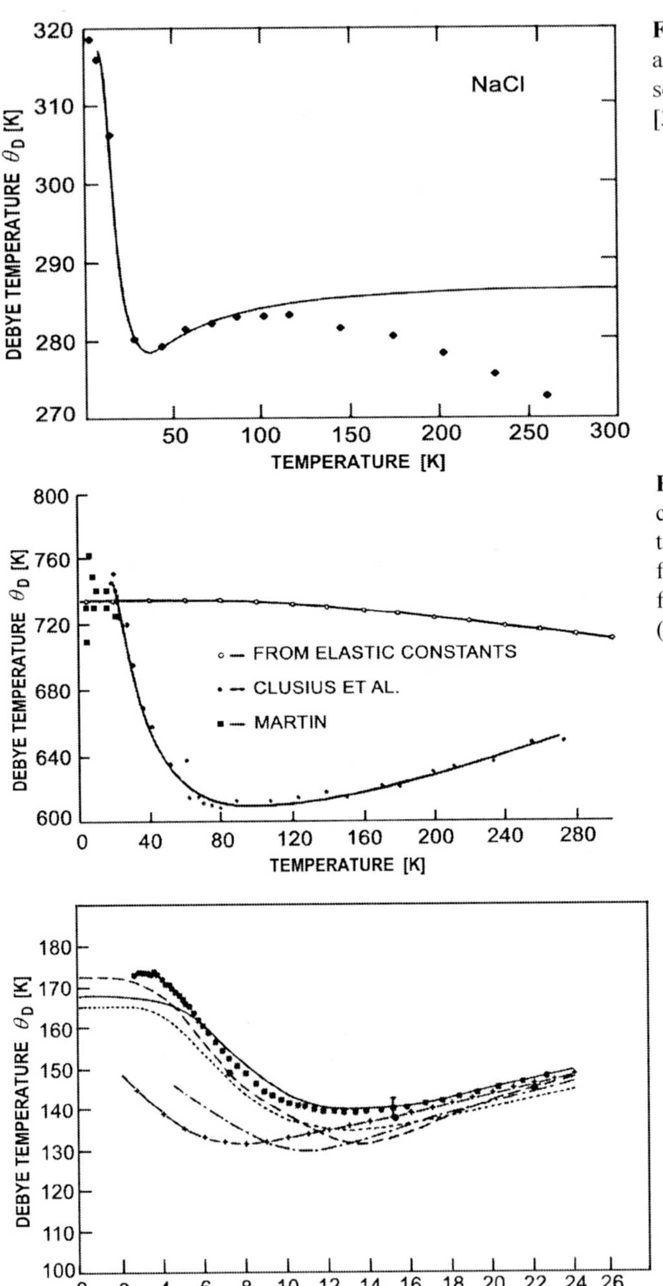

Fig. 3.9 The Debye characteristic temperature of sodium chloride (after [3.88])

Fig. 3.10 The Debye characteristic temperature of lithium fluoride from various sources as a function of temperature (after [3.89])

Fig. 3.11 Debye characteristic temperature of CsCl as a function of temperature: ● experimental from [3.90]; ■ experimental from [3.85]; —— 78 K, ---- 298 K, calculated from phonon spectra measured at these temperatures [3.91]; other curves from lattice dynamical calculations

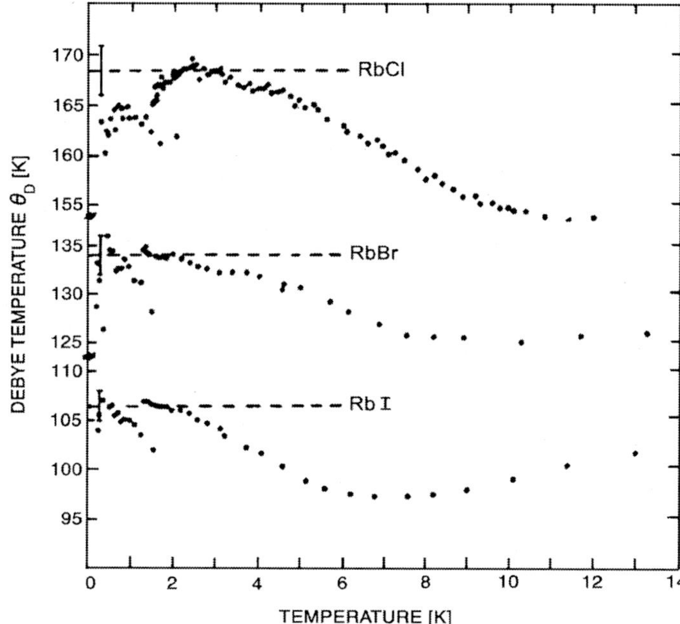

Fig. 3.12 Debye temperature of RbCl, RbBr and RbI. The apparent decrease in the Debye temperature at low temperatures is probably due to very small concentrations of tunnelling impurities (after [3.92])

3.6.4 Temperature Variation of Debye Temperature (High Temperatures)

The temperature variation of Debye temperature θ_M determined from X-ray diffraction at high temperatures is shown in Figs. [3.13-3.19].

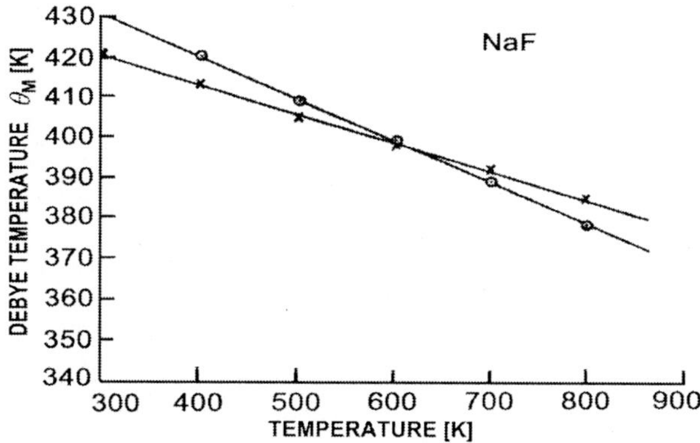

Fig. 3.13 The Debye characteristic temperature θ_M versus temperature for NaF: × Paskin's method, o Chipman's method (after [3.83])

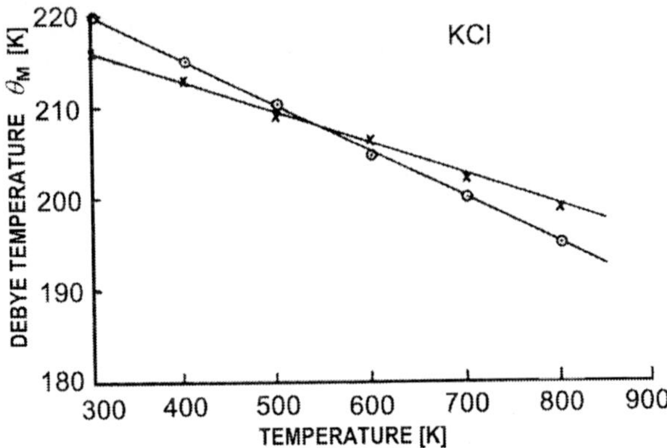

Fig. 3.14 The Debye characteristic temperature θ_M versus temperature for KCl: × Paskin's method, o Chipman's method (after [3.94])

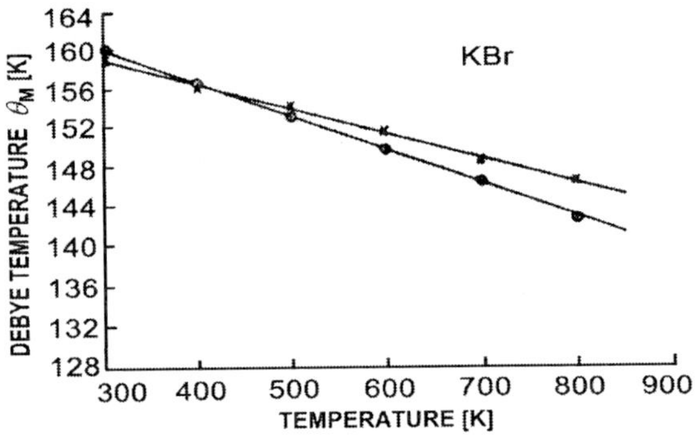

Fig. 3.15 The Debye characteristic temperature θ_M versus temperature for KBr: × Paskin's method, o Chipman's method (after [3.94])

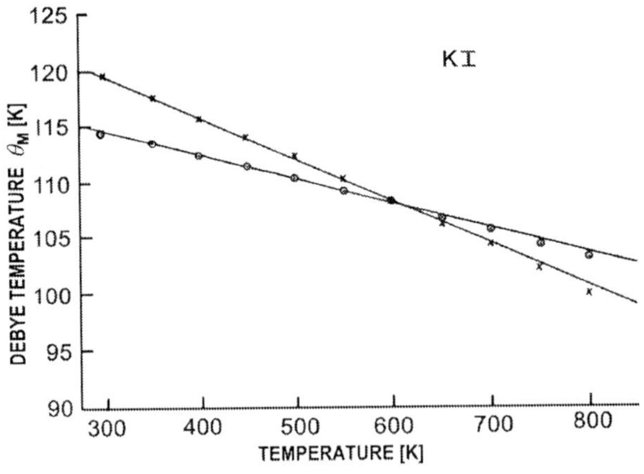

Fig. 3.16 The Debye characteristic temperature θ_M versus temperature for KI: × Paskin's method, o Chipman's method (after [3.95])

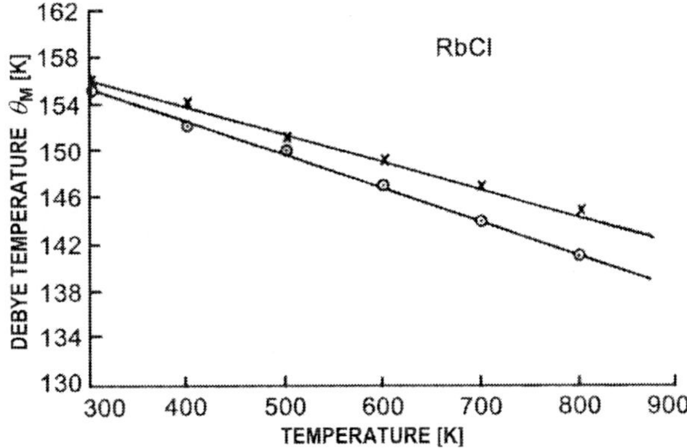

Fig. 3.17 The Debye characteristic temperature θ_M versus temperature for RbCl: × Paskin's method, o Chipman's method (after [3.94])

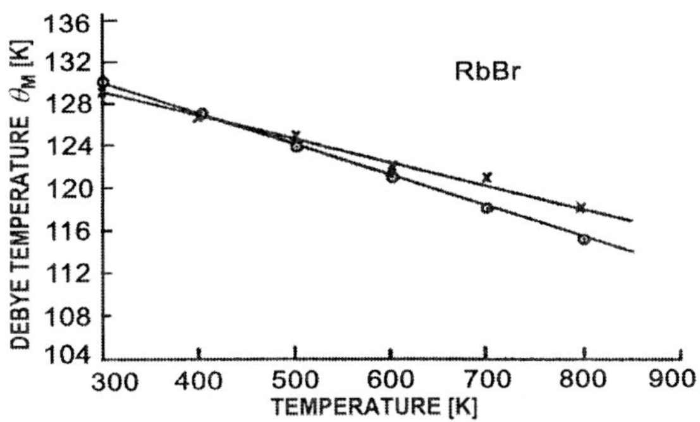

Fig. 3.18 The Debye characteristic temperature θ_M versus temperature for RbBr: × Paskin's method, o Chipman's method [after [3.82]]

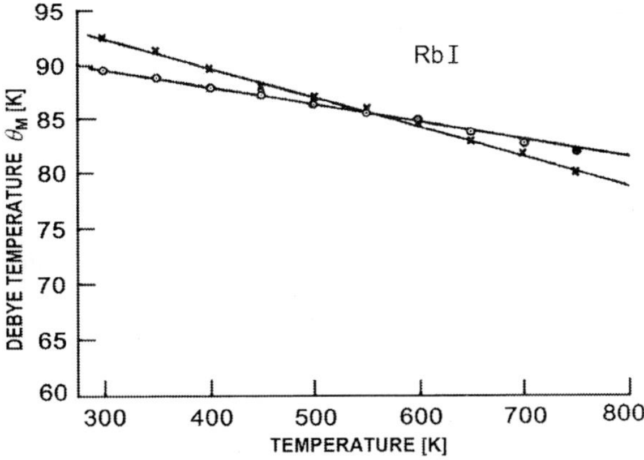

Fig. 3.19 The Debye characteristic temperature θ_M versus temperature for RbI: × Paskin's method, o Chipman's method (after[3.95])

Notes and Comments

1. Pathak and Trivedi [3.94] have shown that there is a law of corresponding states for the temperature variation of thermal expansion and that of Debye temperature. This law is expressed in the form

$$(\alpha / \alpha_{m/2}) = 0.70 + 0.36 \times 10^{-11} (T / Aa^2 \theta^2)$$
(3.26)

where α, a and θ are the coefficient of thermal expansion, lattice constant and Debye temperature respectively at temperature T [K], $\alpha_{m/2}$ is the coefficient of thermal expansion at half the melting point and A is the mean atomic weight. Values of $(\alpha / \alpha_{m/2})$ and $(T/a^2\theta^2)$ are shown to lie on a single linear plot for several alkali halides.

3.6.5 Pressure Variation of Debye Temperature

Table 3.26 Values of Debye temperature θ_0 at zero pressure; θ_P at pressure P calculated from data on elastic constants at high pressures [3.96]

Crystal	θ_0 [K]	θ_P [K]	θ_P / θ_0	
		$P = 3$ kbar	from columns 2, 3	calculated
NaCl Structure				
LiF	697.4	703.3	1.0085	1.0074
LiCl	390.5	399.3	1.0225	1.0183
LiBr	246.8	254.0	1.0292	1.0248
NaF	474.5	478.0	1.0074	1.0099
NaCl	306.1	310.4	1.0145	1.0198
NaI	156.1	160.5	1.0282	1.0340
KCl	224.0	225.0	1.0044	1.0244
KI	125.4	126.7	1.0104	1.0446
RbCl	161.5	161.4	0.9994	1.0267
RbBr	128.5	128.3	0.9984	1.0328
RbI	100.8	100.4	0.9960	1.0446

Notes and Comments

1. The pressure variation of Debye temperature is shown in Fig. 3.20.
2. The pressure variation of Debye temperature is a small effect.
3. The pressure variation is ion-size dependent, increasing as the size of the halogen ion increases and decreasing for increasing cation size.
4. The calculated values of (θ_P / θ_0) are from the relation

$$\theta_P = [1 + (\gamma \psi)P] \theta_0$$
(3.27)

where γ is the Gruneisen constant and ψ the compressibility [3.97]. There is qualitative agreement in the case of the halides of Li and Na. In the case of the rubidium halides (θ_P/θ_0) is < 1 for values calculated from Eq. (3.27).

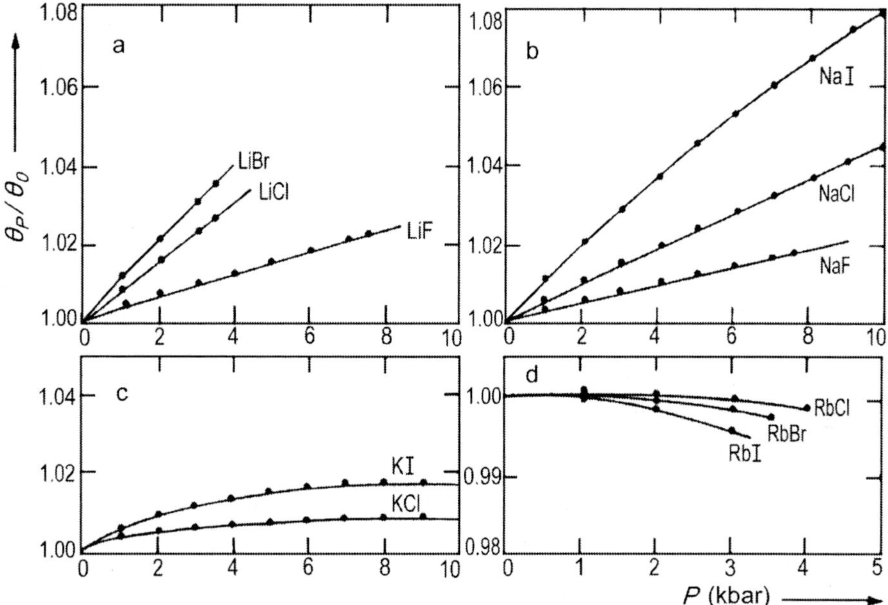

Fig. 3.20 (θ_P/θ_0) obtained from elastic constants as a function of pressure (P): (a) lithium halides, (b) sodium halides, (c) potassium halides, (d) rubidium halides (after [3.96])

3.7 Gruneisen Parameter

3.7.1 Gruneisen Parameter (γ) at Room Temperature

Table 3.27 Values of the Gruneisen parameter (γ) calculated from thermal expansion data and elastic properties at room temperature

Crystal	γ from thermal expansion data [3.98]	γ from elastic properties [3.99]
NaCl Structure		
LiF	1.64	2.94
LiCl	1.69	2.61
LiBr	1.88	2.53
LiI	2.03	2.73
NaF	1.83	2.42
NaCl	1.64	2.47
NaBr	1.72	2.48

Table 3.27 (Continued)

Crystal	γ from thermal expansion data [3.98]	γ from elastic properties [3.99]
NaCl Structure		
NaI	1.66	2.53
KF	1.58	2.47
KCl	1.49	2.51
KBr	1.46	2.53
KI	1.45	2.56
RbF	1.37	2.62
RbCl	1.57	2.57
RbBr	1.43	2.56
RbI	1.51	2.55
CsF	1.49	-
CsCl Structure		
CsCl	2.24	2.72
CsBr	1.93	2.69
CsI	2.00	2.66

Notes and Comments

1. The γ values from thermal expansion data are calculated from the definitive formula

$$\gamma = 3\alpha V / \psi C_V \tag{3.28}$$

   where α is the coefficient of linear expansion, V_M the molar volume, ψ the compressibility and C_V the molar specific heat.

2. The γ values from elastic properties are calculated using the Slater formula

$$\gamma = -(1/6) - (1/2)(\psi^{-2} d\psi / dP) \tag{3.29}$$

   where ψ is the compressibility.

3. The γ(thermal) vs r plots are smooth showing either no dependence (halides of Na, K and Cs) or slight positive dependence (halides of Li and Rb) of γ(thermal) on the interionic distance r.

4. The γ(elastic) values are systematically larger than the γ(thermal) values.

3.7.2 Temperature Variation of Gruneisen Parameter

Low Temperatures

The temperature variation of the thermal Gruneisen parameter calculated from Eq. (3.28) at low temperatures is shown in Fig. 3.21.

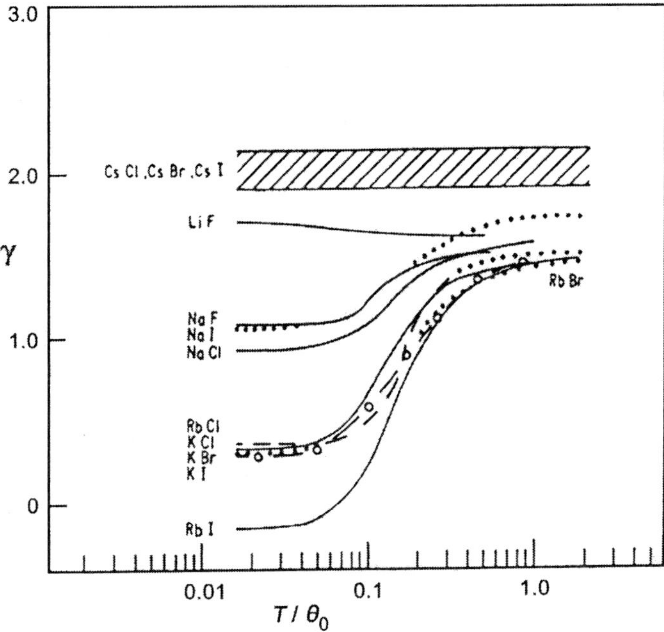

Fig. 3.21 The Gruneisen parameter of alkali halides displayed as functions of reduced temperature (after [3.93])

High Temperatures

The temperature variation of the thermal Gruneisen parameter calculated from Eq. (3.28) at high temperatures is shown in Fig. 3.22 (see p. 95).

3.7.3 Low and High Temperature Limits of Gruneisen Parameter

Table 3.28 Values of the low temperature limit (γ_0) and the high temperature limit (γ_∞) of the Gruneisen parameter

Crystal	γ_0	γ_∞	Ref.
NaCl Structure			
LiF	1.70	1.58	[3.12]
NaF	0.92	1.5	[3.15]
NaCl	0.90	1.57	[3.12]
NaBr	0.99	1.55	[3.15]
NaI	1.04	1.71	[3.12]
KCl	0.32	1.45	[3.12]
KBr	0.29	1.49	[3.12]
KI	0.28	1.47	[3.12]
RbCl	0.00	1.48	[3.15]
RbBr	−0.03	1.46	[3.15]
RbI	−0.18	1.42	[3.12]

Table 3.28 (Continued)

Crystal	γ_0	γ_∞	Ref.
CsCl Structure			
CsCl	1.98	2.06	[3.17]
CsBr	1.9	2.0	[3.17]
CsI	2.01	2.01	[3.17]

Notes and Comments

1. In all cases, $\gamma_0 \le \gamma_\infty$.
2. Krishnan et al. [3.20] point out that γ_0 shows a strong dependence on cation size, increasing as the cation size decreases (Rb–K–Na–Li).
3. The low and high temperature limiting Gruneisen parameters can also be calculated from elastic constant data using Eqs. (3.30–3.33) as follows:

The ith mode Gruneisen parameter is given by

$$\gamma_i = -d \log v_i \, / \, d \log V \tag{3.30}$$

The mode gammas can be obtained from data on the pressure variation of elastic constants:

$$\gamma_i = -\frac{1}{6} + \frac{B_T}{C_i}\left(\frac{\partial C_i}{\partial P}\right)_T \tag{3.31}$$

where C_i is the appropriate elastic constant. Finally,

$$\gamma_0 = \frac{E(v_i)\gamma_i}{E(v_i)} \tag{3.32}$$

where $E(v_i)$ is the Einstein function for v_i, and

$$\gamma_\infty = \frac{\gamma_i}{3N} \tag{3.33}$$

where $3N$ is the total number of modes of vibration. Values of γ_0 and γ_∞ for LiF, NaCl KCl and RbI calculated by D.E. Schuele from Eq.s (3.31–3.33) are quoted in [3.12].
4. From Sec. 3.2.2 it is seen that the thermal expansion coefficient for RbBr and RbI has small negative values at very low temperatures (< 6 K) resulting in a value of –0.18 for γ_0 for RbI (given in [3.12]) and a value of $\gamma_0 = -0.03$ for RbBr given by White and Collins [3.15]. Negative γ_0 values for RbBr and RbI have also been obtained by Roberts and Smith [3.100] and Fontanella and Schuele [3.101] from the pressure variation of elastic constants.

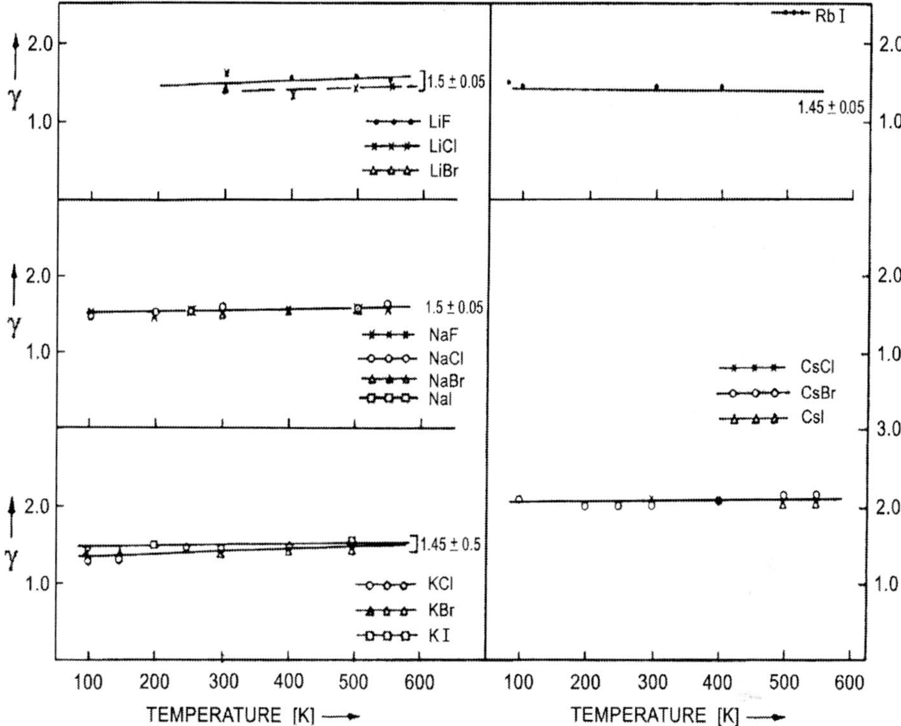

Fig. 3.22 Temperature dependence of Gruneisen parameter of alkali halides at high temperatures (after [3.13])

3.7.4 Volume Dependence of Gruneisen Parameter

Table 3.29 Values of the volume derivative $q = (\mathrm{d} \log \gamma / \mathrm{d} \log V)_{P, T = 0}$

	q			
Method	Lattice dynamics	Pr. variation of thermal expansion	Pr. variation of thermal conductivity	Higher order elastic constants
Ref. Crystal ↓	[3.102]	[3.103]	[3.103]	[3.103]
NaCl Structure				
LiF	-	−1.8	3.57	2.5
LiBr	3.5	-	-	-
LiI	3.4	-	-	-
NaF	2.3	-	0.77	−0.46
NaCl	2.0	-	1.48	2.2
NaBr	1.9	-	2.8	3.6
NaI	1.9	-	-	-
KF	2.4	-	-	−0.03
KCl	2.1	-	1.29	0.51

Table 3.29 (Continued)

Method	q Lattice dynamics	Pr. variation of thermal expansion	Pr. variation of thermal conductivity	Higher order elastic constants
Ref. Crystal↓	[3.102]	[3.103]	[3.103]	[3.103]
NaCl Structure				
KBr	2.0	-	2.39	1.2
KI	2.0	-	2.99	1.9
RbF	2.9	-	-	-
RbCl	2.2	-	-	-
RbBr	2.2	-	-	-
RbI	2.4	-	-	-
CsCl Structure				
CsCl	-	-	0.57	−5.23
CsBr	-	-	1.63	−4.40
CsI	-	-	2.01	−2.24

Notes and Comments

1. q is called the second Gruneisen parameter.
2. q results from higher order effects. Hence, small errors in experimental inputs create large errors in values of q. Note sign difference in values from different methods.
3. Nava [3.104] derived an expression for the pressure derivative $(d\gamma/dP)$ in terms of the second order elastic constants and their temperature derivatives (but without using data on pressure derivatives of elastic constants). The values thus obtained by Nava for $d\gamma/dP$ at room temperature are given in Table X.

Table X Values of $d\gamma/dP$ [GPa^{-1}]; Ref. [3.104]

Crystal	NaCl	KCl	KBr
$d\gamma/dP$	−0.10	−0.16	−0.18

3.7.5 Mode Gruneisen Parameters

Table 3.30 Values of the mode Gruneisen parameters γ_{TO} and γ_{LO}; uncertainty in last digit given in parenthesis

	γ_{TO} at 290 K					γ_{LO} at 290K
Ref. Crystal ↓	[3.105]	[3.106]	[3.107]	[3.108]	[3.109]	[3.106]
NaCl Structure						
LiF	2.35(16)	2.59	2.6	-	-	-

Table 3.30 (Continued)

	γ_{TO} at 290 K					γ_{LO} at 290K
Ref. Crystal ↓	[3.105]	[3.106]	[3.107]	[3.108]	[3.109]	[3.106]
NaCl Structure						
NaF	2.08(18)	2.95	3.0	-	-	0.64
NaCl	2.35(16)	-	2.4	-	-	-
NaBr	2.37(20)	-	3.0	-	-	-
KCl	2.28(18)	-	2.9	2.83(10)	-	-
KBr	2.06(13)	-	2.6	2.46(10)	-	-
KI	2.20(6)	-	-	-	3.1	-
RbCl	2.16(10)	-	-	-	-	-
RbBr	2.39(16)	-	-	-	-	-
RbI	2.09(15)	-	-	-	2.5	-
CsCl Structure						
CsCl	3.14(10)	-	-	-	-	-
CsBr	2.74(5)	-	-	-	-	-
CsI	2.42(8)	-	-	-	-	-

Notes and Comments

1. The mode Gruneisen parameters are calculated from data on pressure variation
   of IR spectral frequencies using the relation

$$\gamma_i = -\psi_i^{-1} \omega_i^{-1} (d\,\omega_i\,/\,d\,P) \tag{3.34}$$

   where ω_i is the $k \sim 0$ TO or LO frequency and ψ is the compressibility.
2. Mitra [3.110] showed a linear correlation between γ_{TO} and the effective ionic
   charge e^* for the alkali halides and also offered an explanation based on the
   rigid ion model.
2. There are several reports on lattice dynamical evaluation of the mode γ's
   [3.111–3.113].

3.8 Anderson-Gruneisen Parameter

Table 3.31 Values of the adiabatic and isothermal Anderson-Gruneisen parameters δ_S
and δ_T; [3.114]

Crystal	$\delta_S = -(1/\beta)(\partial \log B_S\,/\,\partial T)_P$	$\delta_T = -(1/\beta)(\partial \log B_T\,/\,\partial T)_P$
NaCl Structure		
LiF	3.56	6.00
LiCl	4.09	6.77
LiBr	4.12	7.01
LiI	4.06	7.32

Table 3.31 (Continued)

Crystal	$\delta_S = -(1/\beta)(\partial \log B_S / \partial T)_P$	$\delta_T = -(1/\beta)(\partial \log B_T / \partial T)_P$
NaCl Structure		
NaF	3.75	5.77
NaCl	3.80	5.85
NaBr	4.11	6.23
NaI	4.13	6.43
KF	4.08	6.20
KCl	4.38	6.22
KBr	4.02	5.88
KI	3.93	5.76
RbF	4.97	6.80
RbCl	4.93	6.76
RbBr	4.72	6.60
RbI	4.47	6.52

Notes and Comments

1. The Anderson-Gruneisen parameter was originally defined as

$$\delta = -(1/\beta)(\partial \log B / \partial T)_P \tag{3.35}$$

   where B is the bulk modulus and β the volume coefficient of expansion. However, $(1/\beta)(\partial \log B_S / \partial T)_P \neq (1/\beta)(\partial \log B_T / \partial T)_P$ where B_S and B_T are adiabatic and isothermal bulk moduli. Several workers have reported values without making this distinction. Shanker and Singh [3.114] define two Anderson-Gruneisen parameters δ_S and δ_T pertaining to the two temperature derivatives of B and calculate values for δ_S and δ_T.
2. Values of the Anderson-Gruneisen parameter are also given in [3.115–3.117].
3. Pandey [3.118] derived the following formula for the interionic distance $r(T)$ by assuming that the Anderson-Gruneisen parameter δ is independent of temperature:

$$r(T) = r_0[1 - A^{-1}\{\log (1 - A\alpha(T - T_0))\}] \tag{3.36}$$

   where $A = (\delta + 1)$ and α is the linear expansion coefficient. Using this formula, Pandey evaluated $r(T)$ at various temperatures up to the melting point.

References

3.1   Y.S. Touloukian and E.H. Buyco in *Thermophysical Properties of Matter*, Vol.5, *Specific heat–Nonmetallic Solids* (Plenum Press, New York, 1970) and references therein.

3.2   A.M. Karo, J. Chem. Phys., **31**, 1489, 1959.

3.3   A.M. Karo, J. Chem. Phys., **33**, 7, 1960.

3.4   P. Andersson, J. Phys. C: Solid State Physics, **18**, 3943, 1985.

3.5   L.N. Dzhavadov and Y.J. Krotov, Sov. Phys. Solid State, **20**, 379, 1978.

3.6   R. Boehler and G.C. Kennedy, J. Phys. Chem. Solids, **41**, 517, 1980.

3.7   D. Gerlich and P. Andersson, J. Phys. C: Solid State Phys., **15**, 5211, 1982.

3.8   B. Hakansson and P. Andersson, J. Phys. Chem. Solids, **47**, 355, 1986.

3.9   K.K. Srivastava and H.D. Merchant, J. Phys. Chem. Solids, **34**, 2069, 1973.

3.10   M.E. Straumanis and A. Ievins, Z. Anorg. Allg. Chem., **238**, 175, 1938.

3.11   V.T. Deshpande, Acta Cryst., **14**, 794, 1961

3.12   G.K. White, Proc. Roy. Soc. Lond., **A286**, 204, 1965.

3.13   J.E. Rapp and H.D. Merchant, J. Appl. Phys., **44**, 3919, 1973.

3.14   Values quoted without any details by L.G. Van Uitert, H.M. O'Bryan, M.E. Lines, H.J. Guggenheim and G. Zydzik, Mat. Res. Bull., **12**, 261, 1977.

3.15   G.K. White and J.G. Collins, Proc. Roy. Soc. Lond. **A333**, 237, 1973.

3.16   P.P.M. Meincke and G.M. Graham, Can. J. Phys., **43**, 1853, 1965.

3.17   A.C. Bailey and B. Yates, Phil. Mag., **16**, 1241, 1967.

3.18   R.S. Krishnan, *Progress in Crystal Physics*, S. Viswanathan, Madras, India, 1958

3.19   Y.S. Touloukian, R.K. Kirby, R.E. Taylor and T.Y.R. Lee, *Thermophysical Properties of Matter*, **Vol. 13**: *Thermal expansion of Nonmetallic Solids*, Plenum Press, New York, 1977.

3.20   R.S. Krishnan, R. Srinivasan and D. Devanarayanan, *Thermal Expansion of Crystals*, Thomson Press, Faridabad, India, 1979.

3.21   D.B. Sirdeshmukh, J. Chem. Phys. **45**, 2333, 1966.

3.22   A.A. Khan, Acta Cryst. **A30**, 105, 1974.

3.23   R.E. Hanneman and H.C. Gatos, J. Appl. Phys. **36**, 1794, 1965.

3.24   D.B. Sirdeshmukh, J. Appl. Phys. **38**, 4083, 1967.

3.25   Y. Watanabe, A. Sakai and T. Sakurai, J. Phys. Soc. Japan, **66**, 649, 1997.

3.26   O. Urvas, Ann. Acad. Sci. Fenn., **A366**, 7, 1971.

3.27   V. Hovi and J. Pirinen, Ann. Acad. Sci. Fenn., **A390**, 3, 1972.

3.28   P.D. Pathak and N.G. Vasavada, Acta Cryst., **A28**, 30, 1972.

3.29   P.D. Pathak, J.M. Trivedi and N.G. Vasavada, Acta Cryst., **A29**, 477, 1973.

3.30   P.D. Pathak and N.G. Vasavada, Acta Cryst., **A26**, 655, 1970.

3.31   P.D. Pathak and N.M. Pandya, Acta Cryst., **A31**, 155, 1975.

3.32   R. Boehler and G.C. Kennedy, J. Phys. Chem. Solids, **41**, 1019, 1980.

3.33   T. Yagi, J. Phys. Chem. Solids, **39**, 563, 1978.

3.34   M. Kumar, Ind. J. Phys., **69A**, 517, 1995, calculated by using thermodynamic relations

3.35   R. Srinivasan, J. Ind. Inst. Sci., **A37**, 200, 1955.

3.36   *CRC Handbook of Chemistry and Physics*, (a) 60th Ed., (1979–80); (b) 76th Ed., (1995-1996), CRC Press, Boca Raton, Florida, U.S.

3.37   G.A. Slack ,Solid State Phys., **34**, 1, 1979.

3.38   Compilation by Y.S. Touloukian, R.W. Powell, C.Y. Ho and P.G. Klemens, *Thermophysical Properties of Matter*, Vol.**2**, *Thermal conductivity of Nonmetallic Solids*, Plenum Press, New York, 1970, and references therein.

3.39   R.G. Ross, P. Andersson, B. Sundquist and G. Buckstrom, Rept. Progr. Phys., **47**, 1347, 1984 and references therein.

3.40   J.W. Schwartz and C.T. Walker, Phys. Rev., **155**, 959, 1967.

3.41   M.V. Klein in *Physics of Colour Centres*, Ed. W.B. Fowler (Academic Press, New York, 1968) and references therein.

3.42   J.A. Krumhansl, Proc. Int. Conf. Lattice Dynamics, Copenhagen, 1963.

3.43   S.P. Clark Jr., J. Chem. Phys., **31**, 1526, 1959.

3.44   C.W.F.T. Pistorius, J. Chem. Phys., **45**, 3513, 1966.

3.45   C.W.F.T. Pistorius, J. Phys. Chem. Solids, **26**, 1543, 1965.

3.46   C.W.F.T. Pistorius, J. Chem. Phys., **43**, 1557, 1965.

3.47   J. Shanker, W.N. Bhende and M. Kumar, Solid State Commun., **55**, 479, 1985.

3.48   J. Tateno, Solid State Commn., **10**, 61, 1972.

3.49   L. Pietronero, Phys. Rev. **B17**, 3946, 1978.

3.50   A.R. Ubbelohde, *The Molten State of Matter*, John Wiley, New York, 1978.

3.51   J. Shanker and M. Kumar, phys. stat. sol., **(b)158**, 11, 1990.

3.52   L.W. Barr and D.K. Dawson, Proc. Brit. Ceram. Soc., **19**, 152, 1971.

3.53   K. Sangwal, J. Cryst. Growth, **97**, 393, 1989 and references therein.

3.54   K. Furukawa, Discuss. Faraday Soc., **32**, 53, 1941.

3.55   M. Kumar, Ind. J. Pure and Appl. Phys., **31**, 67, 1993.

3.56   S.N. Vaidya and E.S.R. Gopal, J. Phys. Chem. Solids, **28**, 1074, 1967.

3.57   C.J. Martin and D.A. O' Connor, J. Phys. C: Solid State, **10**, 3521, 1977.

3.58   M.A. Viswamitra and A. Jayalakshmi Ramanuja, Acta Cryst., **A28**, S 189, 1972.

3.59   M.S. Kushwaha, Nuovo Cimento, **65B**, 285, 1981.

3.60   Compilation by N.M. Butt, J. Bashir and M. Nasir Khan, Acta Cryst., **A49**, 171, 1993 and references therein.

3.61   O. Inkinen and M. Jarvinen, Phys. Kondens. Materie, **7**, 372, 1968.

3.62   P. Geeta Krishna, K.G. Subhadra and D.B. Sirdeshmukh, Acta Cryst., **A54**, 253, 1998.

3.63   R.K. Gupta, Phys. Rev., **B12**, 4452, 1975 ($\overline{u^2}$ values converted into B by multiplying with $8\pi^2/3$); Lattice dynamical calculations based on a 11–parameter shell model).

3.64   M.V. Linkaoho, Acta Cryst., **A25**, 450, 1969.

3.65   G.J. McIntyre, G. Moss and Z. Barnea , Acta Cryst., **A36**, 482, 1980.

3.66   M. Merisalo, Ann. Acad. Sci. Fenn., **A VI**, 245, 1967.

3.67   M. Merisalo and T. Paakkari, Acta Cryst., **23**, 1107, 1967.

3.68   C.W. Tompson, Private Communication, 1978.

3.69   R.W. James and G.W. Brindley, Proc. Roy. Soc., **A122**, 155, 1928.

3.70   V. Ganesan and K.S. Girirajan, Pramana, **30**, 331, 1988.

3.71   V. Ganesan and K.S. Girirajan, Pramana, **30**, 337, 1988.

3.72   T.J. Bastow, S.L. Mair and S.W. Wilkins, J. Appl. Phys., **48**, 494, 1977.

3.73   C.W. Tompson, private communication to T.J. Bastow.

3.74   L.K. Patomaki and M.V. Linkoaho, Acta Cryst., **A25**, 304, 1969.

3.75   D.I. Vadets and Ya.L. Giller, Ukr. Fiz. Zh., **14**, 348, 1969.

3.76   N.M. Butt and A.K. Cheetham, Acta Cryst., **A29**,727,1973.

3.77   H.X. Gao, L.M. Peng and J.M. Zuo, Acta Cryst., A55, 1014, 1999 for NaCl type crystals and Acta Cryst., (in press), for CsCl type crystals.

3.78   A.J.F. Boyle and G.J. Perlow, Phys. Rev., 151, 211, 1966.

3.79   N.T. Padial, L.M. Brescansin and M.M. Shukla, Acta Phys. Pol., A57, 129, 1980.

3.80   K.G. Subhadra and D.B. Sirdeshmukh, Ind. J. Pure and Appl. Phys., **16**, 693, 1978.

3.81   Calculated from Blackman's formula by S. Narain, phys. stat. sol., **(b)182**, 273, 1994.

3.82   P.D. Pathak and J.M. Trivedi, Acta Cryst., **A30**, 231, 1974.

3.83   K.G. Bansigir, J. Appl. Phys., **39**, 4024, 1968.

3.84   H. Hoinks, H. Nahr and H. Wilsh, Surface Sci., **33**, 516, 1972; **40**, 457, 1973.

3.85   J.C. Ho and D.P. Dandekar, Phys. Rev., **B30**, 2117, 1984 and references therein.

3.86   A. Konti and Y.P. Varshni, Can. J. Phys., **49**, 3115, 1971 and references therein.

3.87   W.T. Berg and J.A. Morrison, Proc. Roy. Soc. Lond., **A242**, 467, 1957.

3.88   T.H.K. Barron, W.T. Berg and J.A. Morrison, Proc. Roy. Soc. Lond., **A242**, 478, 1957.

3.89   K. Clusius, J. Goldman and B. Perlick, Z. Naturforsch., A4, 424, 1949 quoted by C.V. Briscoe and C.F. Squire, Phys. Rev., 100, 1175, 1957 and references therein.

3.90   A. R. Taylor, Jr., T.E. Stelle Gardner and D.F. Smith, U.S. Bureau of Mines Report No. 6157, 1963.

3.91   A. A. A. Z. Ahmad, H.G. Smith, N. Vakabayashi and M.K.Wilkinson, Phys. Rev., B6, 3956, 1972.

3.92   R.J.Rollefson and P.P. Peressini, J. Appl. Phys., **43**, 727, 1972.

3.93   A.D. Redmond and B. Yates, J. Phys. C: Solid State, **5**, 1589, 1972.

3.94   P.D. Pathak and J.M. Trivedi, Acta Cryst., **A29**, 45, 1973.

3.95   P.D. Pathak and N.M. Pandya, Acta Cryst., **A31**, 185, 1975.

3.96   D.B. Sirdeshmukh and K.G. Subhadra, phys. stat. sol., **(b)150**, K11, 1988.

3.97   M. Kumari and N. Dass, phys. stat. sol., **(b)133**, 101, 1986.

3.98   C.M. Kachava and S.C. Saxena, J. Appl. Phys., **39**, 2973, 1968.

3.99   Calculated by B. Kameshwara Rao, Ph.D. Thesis, Kakatiya University, Warangal, India, 1980.

3.100   R.W. Roberts and C.S. Smith, J. Phys. Chem. Solids, **31**, 2397, 1970.

3.101   J.J. Fontanella and D.E. Schuele, J. Phys. Chem. Solids, **31**, 647, 1970.

3.102   H.H. Demarest Jr., J. Phys. Chem. Solids, **35**, 1393, 1973.

3.103   D. Gerlich, J. Phys. Chem. Solids, **53**, 865, 1992.

3.104   R. Nava, J. Phys. Soc. Japan, **59**, 1537, 1998.

3.105   R.P. Lowndes and A. Rastogi, Phys. Rev., **B14**, 3598, 1976.

3.106   S.S. Mitra, C. Postmus and J.R. Ferraro, Phys. Rev. Lett., **18**, 455, 1967.

3.107   J. Fontanella, C. Andeen and D. Schuele, Phys. Rev., **B6**, 582, 1972.

3.108   C. Postmus, J.R. Ferraro and S.S. Mitra, Phys. Rev., **174**, 983, 1968.

3.109   J.R. Ferraro, S.S. Mitra and A. Quattrochi, J. Appl. Phys., **42**, 3677, 1971.

3.110   S.S. Mitra (Proc Int. Colloq. on *Physical Properties of Solids under Pressure*, CNRS, Paris, 1969.

3.111   G.R. Barsch and B.N.N. Achar, phys. stat. sol., **35**, 881, 1969.

3.112   A.V. Singh and J. Shanker, phys. stat. sol., **(b)103**, 733, 1981.

3.113   J. Shanker and A.V. Singh, Pramana, **11**, 411, 1978.

3.114   J. Shanker and K. Singh, phys. stat. sol., **(b)103**, 151, 1981.

3.115   V.K. Mathur and S.P. Singh, J. Phys. Chem. Solids, **29**, 959, 1968.

3.116   M.N. Sharma and R. Jain, J. Phys. Soc. Japan, **35**, 194, 1973.

3.117   M.N. Sharma and S.R. Tripathi, J. Phys. Chem. Solids, **36**, 45, 1975.

3.118   R.K. Pandey, J. Phys. Chem. Solids, **59**, 1157, 1998.

4 Optical Properties

4.1 Refractive Index

4.1.1 Refractive Index at RT

Table 4.1 Values of n (the refractive index measured at the sodium yellow wavelength 0.5893 μm) and n_∞ (the high frequency refractive index); uncertainty in the last digit given in parenthesis, wherever reported

Parameter	n		n_∞		
Ref.	[4.1]	[4.2]	[4.1]	[4.2]	[4.3]
Crystal					
NaCl Structure					
LiF	1.3915	1.3917(4)	1.386	1.388	1.3878
LiCl	1.662	-	1.658	-	1.6463
LiBr	1.784	-	1.778	-	1.7520
LiI	1.955	-	1.949	-	1.9065
NaF	1.3258	1.3250(4)	1.3188	1.319	1.3210
NaCl	1.5443	1.5443(4)	1.5252	1.526	1.5278
NaBr	1.6412	-	1.6111	-	1.6126
NaI	1.7745	-	1.706	-	1.7305
KF	1.352	-	1.360	-	1.3545
KCl	1.4904	1.4907(4)	1.4770	1.474	1.4751
KBr	1.5594	1.5596(4)	1.5332	1.535	1.5367
KI	1.6670	1.6668(4)	1.6238	1.626	1.6275
RbF	1.396	-	1.388	-	1.3873
RbCl	1.4936	-	1.4765	-	1.4723
RbBr	1.5528	-	1.5283	-	1.5232
RbI	1.6474		1.6061		1.6050
CsF	1.478	-	1.469	-	1.4672
CsCl Structure					
CsCl	1.6418	-	1.6209	-	1.6190
CsBr	1.6984	-	1.6676	-	1.6686
CsI	1.7876	-	1.7378	-	1.7428

Notes and Comments

1. The value of n_∞ is determined from data on dispersion of refractive index. Different procedures have been followed by different authors:

i)   Tessman et al [4.1] determined n_∞ from refractive index data at only three wavelengths. The n-λ data were fitted to the Cauchy dispersion equation

$$n = A + (B/\lambda^2) \tag{4.1}$$

and the value for $\lambda = \infty$ was taken as n_∞.

ii) Lowndes and Martin [4.2] measured n at eight wavelengths and obtained n_∞ by fitting the dispersion data to the Sellmeier equation

$$n^2 - 1 = \frac{\varepsilon_0 - n_\infty^{\,2}}{1 - (\omega/\omega_0)^2} + \frac{n_\infty^{\,2} - 1}{1 - (\omega/\omega_e)^2} \tag{4.2}$$

where ε_0 is the static dielectric constant, ω the frequency corresponding to the wavelength of measurement of n; ω_0 and ω_e are absorption frequencies.

iii) Johannsen [4.3] determined n_∞ by fitting the dispersion data to the following expression for the refractive index n (as a function of photon energy E) obtained by considering the band structure:

$$n(E) = 1 + [\Delta\varepsilon_1^{\,TO}(E) + \Delta\varepsilon_1^{\,el}(E)]^{1/2} \tag{4.3}$$

The expressions for the two terms in the bracket involve n_∞ and are given by Johannsen [4.3].

2. n_∞ is an important parameter. $n_\infty^{\,2}$ is equal to ε_∞ the electronic dielectric constant. ε_∞ is a basic input parameter in lattice dynamical calculations.

3. Combined with the Clausius-Mosotti equation, n_∞ gives the electronic polarisabilities:

$$(n_\infty^{\,2} - 1)/(n_\infty^{\,2} + 2) = (4\pi/V_m)\alpha_m \tag{4.4}$$

where V_m is the volume per molecule and α_m is the polarisability per molecule.

4. Combined with the static dielectric constant ε_0 and the transverse optical frequency ω_{TO}, n_∞ gives the Szigeti effective ionic charge e^* :

$$(e^*)^2 = (9\pi\mu v_{TO}^{\,2}/N_V e^2)(\varepsilon_0 - n_\infty^{\,2})/(2 + n_\infty^{\,2})^2 \tag{4.5}$$

where μ is the reduced mass v_{TO} the long wavelength TO mode frquency, N_V the number of ion pairs per unit volume and ε_0 the static dielectric constant.

5. Pantelides [4.4] showed that $n_\infty^{\,2}$ and the interionic distance r follow the relation

$$n_\infty^{\,2} = 1 + A\,r^3 \tag{4.6}$$

where the constant A has the values 0.108, 0.059, 0.038 and 0.034 (in Å^{-3}) for the halides of Li, Na, K and Rb respectively.

6. Sangwal and Kucharezyk [4.5] showed that the following correlations hold between the refractive index n, the density ρ and the mean atomic mass $\overline{M}$:

$$(n^2 - 1)/\rho = k_1 \overline{M}^{-0.4} \qquad \text{and}$$

$$(n^2 - 1)/(n^2 + 2)\rho = k_2 \overline{M}^{-0.4} \tag{4.7}$$

The constants (k_1, k_2) have the values (0.520, 0.320), (1.024, 0.581), (0.991, 0.547) and (1.187, 0.634) respectively for the alkali fluorides, chlorides, bromides and iodides.

7. A linear plot with positive slope was obtained between the coefficient of thermal expansion and the refractive index by Toulokian et al. [4.6].
8. Reddy and Rao [4.7] established the following correlation between the refractive index n and the energy gap (E_G):

$$(n^2 + 2)^2 (E_G - 0.365) = 154 \qquad (4.8)$$

where E_G is in eV.

4.1.2 Dispersion Equations for Refractive Index

Table 4.2 Parameters in the dispersion equation

$$n^2 = A + \frac{B_i \lambda^2}{(\lambda^2 - \lambda_i^2)} \qquad (4.9)$$

Wavelengths λ and λ_i are in μm

Crystal	Wavelength range [μm]	Equation for n^2; Ref. [4.8]
LiF	0.10–11.0	$1 + \dfrac{0.92549\ \lambda^2}{\lambda^2 - (0.07376)^2} + \dfrac{6.96747\ \lambda^2}{\lambda^2 - (32.79)^2}$
LiCl	0.17–16.0	$2.51 + \dfrac{0.24\ \lambda^2}{\lambda^2 - (0.137)^2} + \dfrac{9.11\ \lambda^2}{\lambda^2 - (49.26)^2}$
LiBr	0.21–20.0	$2.88 + \dfrac{0.28\ \lambda^2}{\lambda^2 - (0.164)^2} + \dfrac{10.07\ \lambda^2}{\lambda^2 - (57.80)^2}$
LiI	0.25–25	$3.55 + \dfrac{0.25\ \lambda^2}{\lambda^2 - (0.171)^2} + \dfrac{7.23\ \lambda^2}{\lambda^2 - (70.42)^2}$
NaF	0.15–17	$1.41572 + \dfrac{0.32785\ \lambda^2}{\lambda^2 - (0.117)^2} + \dfrac{3.18248\ \lambda^2}{\lambda^2 - (40.57)^2}$
NaCl	0.20–30	$1.00055 + \dfrac{0.19800\ \lambda^2}{\lambda^2 - (0.050)^2} + \dfrac{0.48398\ \lambda^2}{\lambda^2 - (0.100)^2} + \dfrac{0.38696\ \lambda^2}{\lambda^2 - (0.128)^2}$ $+ \dfrac{0.25998\ \lambda^2}{\lambda^2 - (0.158)^2} + \dfrac{0.08796\ \lambda^2}{\lambda^2 - (40.50)^2} + \dfrac{3.17064\ \lambda^2}{\lambda^2 - (60.98)^2}$ $+ \dfrac{0.30038\ \lambda^2}{\lambda^2 - (120.34)^2}$
NaBr	0.21–34	$1.06728 + \dfrac{1.10463\ \lambda^2}{\lambda^2 - (0.125)^2} + \dfrac{0.18816\ \lambda^2}{\lambda^2 - (0.145)^2} + \dfrac{0.00243\ \lambda^2}{\lambda^2 - (0.176)^2}$ $+ \dfrac{0.24454\ \lambda^2}{\lambda^2 - (0.188)^2} + \dfrac{3.7960\ \lambda^2}{\lambda^2 - (74.63)^2}$

Table 4.2 (Continued)

Crystal	Wavelength range [μm]	Equation for n^2; Ref. [4.8]
NaI	0.25–40	$1.478 + \dfrac{1.532\,\lambda^2}{\lambda^2 - (0.170)^2} + \dfrac{4.27\,\lambda^2}{\lambda^2 - (86.21)^2}$
KF	0.15–22	$1.55083 + \dfrac{0.29162\,\lambda^2}{\lambda^2 - (0.126)^2} + \dfrac{3.60001\,\lambda^2}{\lambda^2 - (51.55)^2}$
KCl	0.18–35	$1.26486 + \dfrac{0.30523\,\lambda^2}{\lambda^2 - (0.100)^2} + \dfrac{0.41620\,\lambda^2}{\lambda^2 - (0.131)^2} + \dfrac{0.18870\,\lambda^2}{\lambda^2 - (0.162)^2}$ $+ \dfrac{2.6200\,\lambda^2}{\lambda^2 - (70.42)^2}$
KBr	0.20–42	$1.39408 + \dfrac{0.79221\,\lambda^2}{\lambda^2 - (0.146)^2} + \dfrac{0.01981\,\lambda^2}{\lambda^2 - (0.173)^2} + \dfrac{0.15587\,\lambda^2}{\lambda^2 - (0.187)^2}$ $+ \dfrac{0.17673\,\lambda^2}{\lambda^2 - (60.61)^2} + \dfrac{2.06217\,\lambda^2}{\lambda^2 - (87.72)^2}$
KI	0.25–50	$1.47285 + \dfrac{0.16512\,\lambda^2}{\lambda^2 - (0.129)^2} + \dfrac{0.41222\,\lambda^2}{\lambda^2 - (0.175)^2} + \dfrac{0.44163\,\lambda^2}{\lambda^2 - (0.187)^2}$ $+ \dfrac{0.16076\,\lambda^2}{\lambda^2 - (0.219)^2} + \dfrac{0.33571\,\lambda^2}{\lambda^2 - (69.44)^2} + \dfrac{1.92474\,\lambda^2}{\lambda^2 - (98.04)^2}$
RbF	0.15–25	$1.395 + \dfrac{0.535\,\lambda^2}{\lambda^2 - (0.124)^2} + \dfrac{4.55\,\lambda^2}{\lambda^2 - (63.29)^2}$
RbCl	0.18–40	$1.47558 + \dfrac{0.56600\,\lambda^2}{\lambda^2 - (0.138)^2} + \dfrac{0.14493\,\lambda^2}{\lambda^2 - (0.166)^2} + \dfrac{2.74000\,\lambda^2}{\lambda^2 - (85.84)^2}$
RbBr	0.21–50	$1.45931 + \dfrac{0.16301\,\lambda^2}{\lambda^2 - (0.123)^2} + \dfrac{0.29841\,\lambda^2}{\lambda^2 - (0.146)^2} + \dfrac{0.17198\,\lambda^2}{\lambda^2 - (0.155)^2}$ $+ \dfrac{0.12186\,\lambda^2}{\lambda^2 - (0.178)^2} + \dfrac{0.13039\,\lambda^2}{\lambda^2 - (0.191)^2} + \dfrac{2.520\,\lambda^2}{\lambda^2 - (114.29)^2}$
RbI	0.24–64	$1.60563 + \dfrac{0.00947\,\lambda^2}{\lambda^2 - (0.120)^2} + \dfrac{0.01073\,\lambda^2}{\lambda^2 - (0.134)^2} + \dfrac{0.00136\,\lambda^2}{\lambda^2 - (0.156)^2}$ $+ \dfrac{0.41864\,\lambda^2}{\lambda^2 - (0.179)^2} + \dfrac{0.41771\,\lambda^2}{\lambda^2 - (0.187)^2} + \dfrac{0.13707\,\lambda^2}{\lambda^2 - (0.223)^2}$ $+ \dfrac{2.36091\,\lambda^2}{\lambda^2 - (132.45)^2}$
CsF	0.15–30	$1.60 + \dfrac{0.56\,\lambda^2}{\lambda^2 - (0.121)^2} + \dfrac{5.92\,\lambda^2}{\lambda^2 - (78.74)^2}$
CsCl	0.18–40	$1.33013 + \dfrac{0.98369\,\lambda^2}{\lambda^2 - (0.119)^2} + \dfrac{0.00009\,\lambda^2}{\lambda^2 - (0.137)^2} + \dfrac{0.00018\,\lambda^2}{\lambda^2 - (0.145)^2}$ $+ \dfrac{0.30914\,\lambda^2}{\lambda^2 - (0.162)^2} + \dfrac{4.320\,\lambda^2}{\lambda^2 - (100.50)^2}$

Table 4.2 (Continued)

Crystal	Wavelength range [μm]	Equation for n^2; Ref. [4.8]
CsBr	0.21–55	$1.14600 + \dfrac{1.26628\,\lambda^2}{\lambda^2-(0.120)^2} + \dfrac{0.01137\,\lambda^2}{\lambda^2-(0.146)^2} + \dfrac{0.00975\,\lambda^2}{\lambda^2-(0.160)^2}$ $+ \dfrac{0.00672\,\lambda^2}{\lambda^2-(0.173)^2} + \dfrac{0.34557\,\lambda^2}{\lambda^2-(0.187)^2} + \dfrac{3.76339\,\lambda^2}{\lambda^2-(136.05)^2}$
CsI	0.25–67	$1.27587 + \dfrac{0.68689\,\lambda^2}{\lambda^2-(0.130)^2} + \dfrac{0.26090\,\lambda^2}{\lambda^2-(0.147)^2} + \dfrac{0.06256\,\lambda^2}{\lambda^2-(0.163)^2}$ $+ \dfrac{0.06527\,\lambda^2}{\lambda^2-(0.177)^2} + \dfrac{0.14991\,\lambda^2}{\lambda^2-(0.185)^2} + \dfrac{0.51818\,\lambda^2}{\lambda^2-(0.206)^2}$ $+ \dfrac{0.01918\,\lambda^2}{\lambda^2-(0.218)^2} + \dfrac{3.38229\,\lambda^2}{\lambda^2-(161.29)^2}$

Notes and Comments

1. Dispersion equations for some alkali halides are also given in [4.9].

4.1.3 Temperature Derivative of Refractive Index at Selected Wavelengths

Table 4.3 Values of the temperature derivative of the refractive index (dn/dT) measured at some selected wavelengths (λ); measurements over a range close to room temperature

λ [μm] Crystal ↓	$dn/dT\,[10^{-6}\,(^{\circ}\mathrm{C})^{-1}]$; Ref. [4.9]				
	0.2537	0.3650	0.4358	0.5461	0.5893
NaCl Structure					
LiF	−11.0	−12.0	−12.5	−12.7	−12.7
NaF	-	-	-	−10.0	-
NaCl	2.99	−21.05	−35.4	-	−37.3
KCl	-	-	−35.6	-	−36.4
KBr	-	-	−37.1	−39.2	−39.7
KI	-	-	-	−50.0	-
CsCl Structure					
CsBr	-	−73.3	−60.0	−63.3	−63.3
CsI	-	-	−98.2	−99.4	−100.0

Notes and Comments

1. Compilations of dn/dT have also been given in [4.10, 4.11].

4.1.4 Temperature Derivative of Refractive index – Dispersion Equations (Empirical)

Table 4.4 Parameters in the dispersion equations for the temperature derivative of refractive index

$$2n\frac{\mathrm{d}\,n}{\mathrm{d}\,T} = -A\,(n^2-1) - B + \frac{C_1\,\lambda^4}{\left(\lambda^2-D_1\right)^2} + \frac{C_2\,\lambda^4}{\left(\lambda^2-D_2\right)^2} \qquad (4.10)$$

λ is in µm, D_1 and D_2 in [(µm)^2]; Ref. [4.8].

Crystal	A	B	C_1	D_1	C_2	D_2
NaCl Structure						
LiF	9.96	8.13	12.09	0.00544	184.86	1075.18
LiCl	13.14	12.85	22.75	0.02045	382.62	2426.55
LiBr	14.94	14.18	28.08	0.02993	503.50	3340.84
LiI	17.82	14.90	36.40	0.04494	318.12	4958.98
NaF	9.51	0.92	3.404	0.01369	83.30	1645.92
NaCl	11.91	0.50	6.118	0.02496	199.36	3718.56
NaBr	12.69	0.12	7.36	0.03534	242.94	5569.64
NaI	13.65	−0.57	9.246	0.05198	247.66	7432.16
KF	10.44	0.08	2.465	0.01588	167.90	2657.40
KCl	11.13	−0.19	3.393	0.02624	142.56	4958.98
KBr	11.61	−0.39	3.944	0.03497	182.88	7694.80
KI	12.24	−0.80	4.785	0.04796	165.92	9611.84
RbF	8.25	0.89	1.581	0.01742	227.50	4005.62
RbCl	10.80	0.84	2.006	0.02756	186.32	7368.51
RbBr	11.25	0.89	2.278	0.03648	191.52	13062.20
RbI	12.45	0.85	2.686	0.04973	169.92	17543.00
CsF	9.60	2.54	1.42	0.01850	296.00	6199.99
CsCl Structure						
CsCl	13.89	4.27	1.989	0.02624	276.48	10100.25
CsBr	14.22	4.75	2.172	0.03497	310.40	18509.60
CsI	14.70	5.53	2.464	0.04752	242.76	26014.46

Notes and Comments

1. These equations have been obtained by empirical extrapolation and interpolation of the trends observed in the experimental data on a few of the alkali halides (LiF, NaF, NaCl, KCl and CsI).

4.1.5 Density Derivative of Refractive Index (Experimental)

Table 4.5 Values of the density derivative ($\rho\, dn/d\rho$) of refractive index where ρ is the density

	($\rho\, dn/d\rho$)				
Ref.	[4.12]	[4.13]	[4.14]	[4.15]	[4.3]
Crystal ↓					
NaCl Structure					
LiF	-	0.126	0.129 ± 0.016		-
NaF	-	-	0.124		0.124
NaCl	0.255	-	0.289 ± 0.014		0.276
NaBr	-	-		0.380	0.360
NaI	-	-		-	0.490
KCl	-	0.294	0.330 ± 0.014		-
KBr	-	0.343	0.369 ± 0.014		-
KI	0.423	-	0.443 ± 0.018		-
RbCl	-	-	0.357 ± 0.014		-
RbBr	-	-	0.404 ± 0.014		-
RbI	-	-		0.441	
CsCl Structure					
CsCl	-	-		-	0.351
CsBr	-	-	0.572 ± 0.016		0.416
CsI	-	-		-	0.474

4.1.6 Pressure Derivative of Refractive Index

Table 4.6 Values of the pressure derivative ($\partial n/\partial P)_T$ of refractive index

	$(\partial n/\partial P)_T\,[10^{-12}\,\mathrm{cm}^2\,\mathrm{dyne}^{-1}]$				
	Expt.	Expt.	Expt.	Expt.	Theor.
Ref.	[4.16]	[4.17]	[4.18]	[4.19]	[4.20]
Crystal ↓					
NaCl Structure					
LiF	0.198	-	0.18	-	0.197
LiCl	-	-	-	-	0.358
LiBr	-	-	-	-	0.347
LiI	-	-	-	-	−0.156
NaF	0.272	-	-	-	0.465
NaCl	1.17	-	-	0.97	1.178
NaBr	1.57	-	-	1.61	1.65
NaI	-	-	-	3.68	2.14
KF	-	-	-	-	0.585
KCl	1.82	-	-	1.96	1.87
KBr	2.44	-	-	1.67	2.46
KI	3.85	-	-	2.20	3.68
RbF	-	-	-	-	1.13

Table 4.6 (Continued)

	$(\partial n/\partial P)_T \, [10^{-12} \, cm^2 \, dyne^{-1}]$				
Ref.	Expt. [4.16]	Expt. [4.17]	Expt. [4.18]	Expt. [4.19]	Theor. [4.20]
Crystal ↓					
NaCl Structure					
RbCl	-	2.24	-	2.62	2.26
RbBr	-	3.0	-	-	2.93
RbI	-	-	-	5.0	4.10
CsCl Structure					
CsCl	-	-	-	1.84	-
CsBr	-	-	-	2.9	-
CsI	-	-	-	2.85	-

Notes and Comments

1. The experimental $(\partial n/\partial P)_T$ values are at $P \to 0$.

4.1.7 Pressure Variation of Refractive Index (Experimental)

Table 4.7 Parameters in the polynomial represented by

$$n = a_0 + a_1 P + a_2 P^2 \tag{4.11}$$

for the pressure variation of refractive index n; pressure P in GPa

Crystal	Pressure range [GPa]	a_0	a_1 [(GPa)$^{-1}$]	a_2 [(GPa)$^{-2}$]	Ref.
NaCl Structure					
LiF	0–8	1.378	0.0018	-	[4.18]
NaF	0–10	1.3168	0.004	−0.0002	[4.19]
NaCl	0–10	1.5297	0.0097	−0.0003	[4.19]
NaBr	0–10	1.6168	0.0161	−0.0007	[4.19]
NaI	0–8	1.7116	0.0368	−0.0020	[4.19]
KCl	0–2	1.4816	0.0196	−0.0035	[4.19]
KBr	0–2	1.5495	0.0167	-	[4.19]
KI	0–2	1.642	0.022	-	[4.19]
RbCl	0–0.7	1.4818	0.0262	-	[4.19]
RbI	0.1–0.4	1.614	0.050	-	[4.19]
CsCl Structure					
CsCl	0–11.2	1.6205	0.0184	−0.0002	[4.19]
CsBr	0–9	1.6669	0.029	−0.0008	[4.19]
CsI	0–9	1.7586	0.0285	-	[4.19]
KCl	3.6–8.7	1.5573	0.011	-	[4.19]
KBr	2.3–10	1.6406	0.0102	0.0005	[4.19]
KI	3–5.5	1.7388	0.0196	-	[4.19]
RbCl	1–9	1.5711	0.0209	−0.0007	[4.19]
RbI	0.9–5.7	1.7821	0.0319	−0.0007	[4.19]

Notes and Comments

1. The pressure versus refractive index plots are either linear or sublinear.
2. If the refractive index is plotted as a function of the density change (instead of the pressure), three types of plots are observed: (a) sublinear in LiF and NaF, (b) linear in NaCl, NaBr and NaI and (c) nonlinear in CsCl, CsBr and CsI. The sublinear behaviour is attributed to the opening of the p-s gap and the nonlinear behaviour to the closing of the p-d gap. The two aspects cancel each other resulting in the linear behaviour in systems like NaCl [4.3].

4.2 Photoelasticity

4.2.1 Strain-Optical Constants

Table 4.8 Values of the strain-optical constants P_{11}, P_{12}, P_{44} at room temperature and $\lambda = 0.5893$ µm

Crystal	$(P_{11}-P_{12})$	P_{11}	P_{12}	P_{44}	Ref.
NaCl Structure					
LiF	−0.11	0.02	0.130	−0.045	[4.9]
NaF	−0.101	0.075	0.176	−0.024	[4.21]
NaCl	−0.0408	0.1372	0.178	−0.0108	[4.9]
	−0.047	0.126	0.174	−0.0107	[4.21]
NaBr	−0.0356	-	-	−0.0036	[4.21]
KF	0.061	-	-	−0.027	[4.21]
KCl	0.056	0.215	0.159	−0.024	[4.9]
	0.064	0.231	0.168	−0.027	[4.21]
KBr	0.047	0.212	0.165	−0.022	[4.9]
	0.049	0.193	0.144	−0.020	[4.21]
KI	0.038	0.203	0.164	-	[4.9]
	0.039	0.209	0.170	−0.0098	[4.21]

Notes and Comments

1. The strain-optical constants P_{ij} are defined as

$$\Delta(1/n^2)_{ij} = P_{ijkl}\, \tau_{kl} \tag{4.11}$$

where $\Delta(1/n^2)_{ij}$ are the changes in the coefficients of the index ellipsoid and τ_{kl} is the strain tensor. For cubic crystals with m3m point group (to which NaCl and CsCl structures belong) there are only three strain optical constants P_{11}, P_{12} and P_{44} in the standard contracted form. These can be converted into the stress-optical constants q_{ij} with the relations

$$q_{11} - q_{12} = (P_{11} - P_{12})(S_{11} - S_{12}) \tag{4.12}$$

$$q_{12} = P_{11}S_{12} + P_{12}(S_{11} + S_{12})$$
(4.13)

$$q_{44} = P_{44}S_{44}$$
(4.14)

where S_{ij} are the elastic compliances.

2. P_{11} and P_{12} values are also listed by Coker [4.15].
3. Theoretically calculated values of P_{ij} are given in [4.20, 4.23].
4. The photoelastic constants have considerable dispersion. The combination $(P_{11}-P_{12})$ changes sign at a certain wavelength. The values of these reversal wavelengths are given in the section on dispersion of photoelastic constants.
5. The strain-optical constants are related to the pressure derivative of the refractive index through the relation

$$n^3\psi(P_{11} + P_{12}) = 6(\mathrm{d}n / \mathrm{d}P)$$
(4.15)

where n is the refractive index, ψ is the compressibility and P the pressure.
6. The strain-optical constants $(P_{11}-P_{12})$ and P_{44} decrease in the sequence F–Cl–Br–I in the sodium and potassium halides.

4.2.2 Stress-Optical Constants

Table 4.9 Values of stress-optical constants q_{11}, q_{12} and q_{44} at room temperature and $\lambda = 0.5893$ μm

Crystal	$(q_{11}-q_{12})$	q_{11}	q_{12}	q_{44}	Ref.
	$[10^{-13}\ \mathrm{cm}^2\ \mathrm{dyne}^{-1}]$				
NaCl Structure					
LiF	−1.52	−0.40	1.12	−0.83	[4.22]
LiCl	−1.52	-	-	−0.71	[4.22]
NaF	−1.39	0.05	1.44	−0.85	[4.22]
NaCl	−1.31	1.27	2.58	−0.84	[4.22]
	−1.17	-	-	−0.92	[4.24]
NaBr	−1.21	-	-	−0.36	[4.22]
NaI	−0.66	-	-	0.67	[4.22]
KF	1.22	-	-	−2.16	[4.22]
KCl	1.88	4.75	2.87	−4.32	[4.22]
	1.47	-	-	−4.94	[4.24]
KBr	1.68	4.62	2.93	−3.94	[4.22]
	1.76	-	-	−4.42	[4.24]
KI	1.69	6.18	4.49	−2.66	[4.22]
RbCl	4.04	7.04	2.99	−7.86	[4.22]
	3.83	-	-	−9.40	[4.25]
RbBr	4.40	7.40	3.00	−8.58	[4.22]
	4.07	-	-	−8.99	[4.25]
	3.84	-	-	−9.03	[4.26]
RbI	4.24	9.23	4.99	−7.92	[4.22]
	4.31	-	-	−8.40	[4.25]
	4.10	-	-	−8.53	[4.26]

Table 4.9 (Continued)

Crystal	$(q_{11}-q_{12})$	q_{11}	q_{12}	q_{44}	Ref.
	$[10^{-13}\ \mathrm{cm}^2\ \mathrm{dyne}^{-1}]$				
CsCl Structure					
CsCl	−2.82	-	-	5.52	[4.26]
CsI	−2.48	-	-	3.80	[4.26]

Notes and Comments

1. The stress-optical constants q_{ij} are defined as

$$\Delta(1/n^2)_{ij} = q_{ijkl}\ \sigma_{kl} \tag{4.16}$$

where σ_{kl} is the stress tensor. For NaCl and CsCl structures, there are only three constants viz. q_{11}, q_{12} and q_{44} in standard contracted form.

2. The q_{ij}'s are related to the P_{ij}'s through the relation

$$P_{ij} = \sum_{1}^{6} q_{ik} C_{kj} \tag{4.17}$$

where the C_{kj} are the elastic constants.

3. No systematic trends are observable in $(q_{11}-q_{12})$, q_{11} and q_{12}. However, q_{44} shows an increase in the sequence Li–Na–K–Rb for a given halide. Systematic dependence of q_{44} on the halogen for a fixed alkali ion is not observed.

4.2.3 Dispersion of Photoelastic Constants

The variation of the photoelastic constants with wavelength in different regions of the spectrum is shown in Figs. 4.1–4.3.

Notes and Comments

1. A classical theory for the dispersion of photoelastic constants of NaCl type crystals has been proposed by Bansigir and Iyengar [4.28] and for CsCl type crystals by Ethiraj et al. [4.29] but the expressions are not in closed form.

2. A theory of the dispersion of photoelastic constants based on the band theory of solids has been proposed by Rahman and Iyengar [4.30]. Here, again, the expressions are not in closed form.

3. Bendow et al. [4.20] have developed the following closed form expression for the dispersion of the strain-optical constants:

$$P_{ij} = -n^{-4}\left[\frac{\alpha_{ij}}{\left(1-\omega^2/\omega_0^2\right)} + \frac{\beta_{ij}}{\left(1-\omega^2/\omega_0^2\right)^2} + \gamma_{ij}\right] \tag{4.18}$$

from the Phillips-Van Vechten theory. ω is the frequency of the radiation corresponding to the measurement of P_{ij} and ω_0 is the TO mode frequency. The values of α_{ij}, β_{ij} and γ_{ij} and also the values of P_{ij} calculated at some wavelengths for several alkali halides are given in [4.20].

4. It is observed that the sign of $(P_{11}-P_{12})$ reverses at a certain wavelength λ_{rev} given in Table XI.

Table XI Values of λ_{rev} for some alkali halides:

Crystal	λ_{rev} [µm]	Ref.	Crystal	λ_{rev} [µm]	Ref.
KCl	0.2480	[4.27]	RbBr	0.2020	[4.26]
KBr	0.2800	[4.27]	RbI	0.2420	[4.26]
KI	0.3380	[4.27]	CsCl	0.2240	[4.26]
			CsI	0.2590	[4.26]

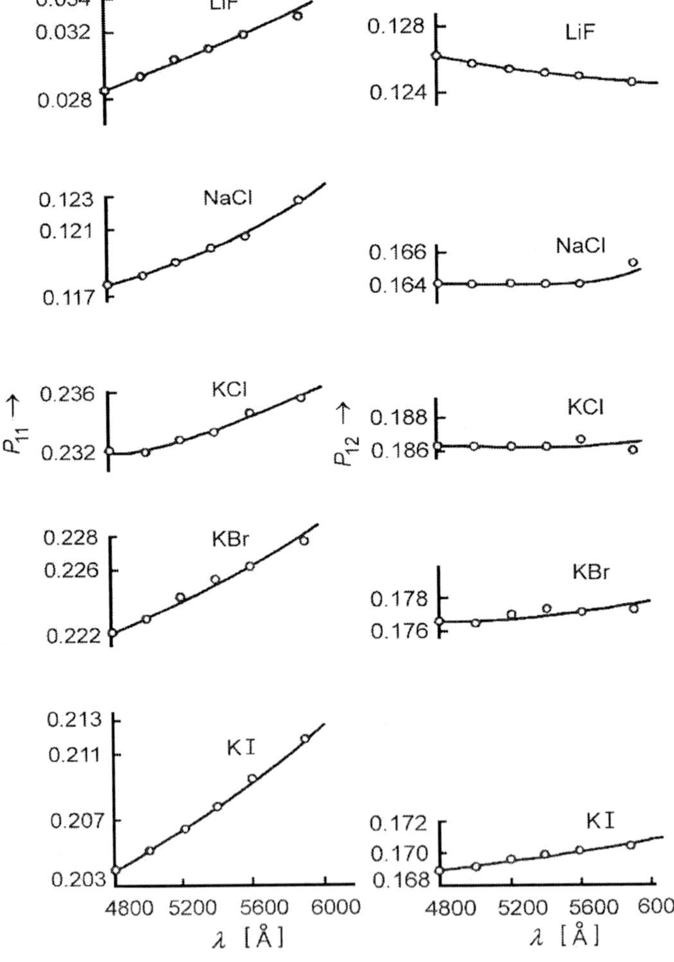

Fig. 4.1 Dispersion of P_{ij} in the visible region (after [4.23])

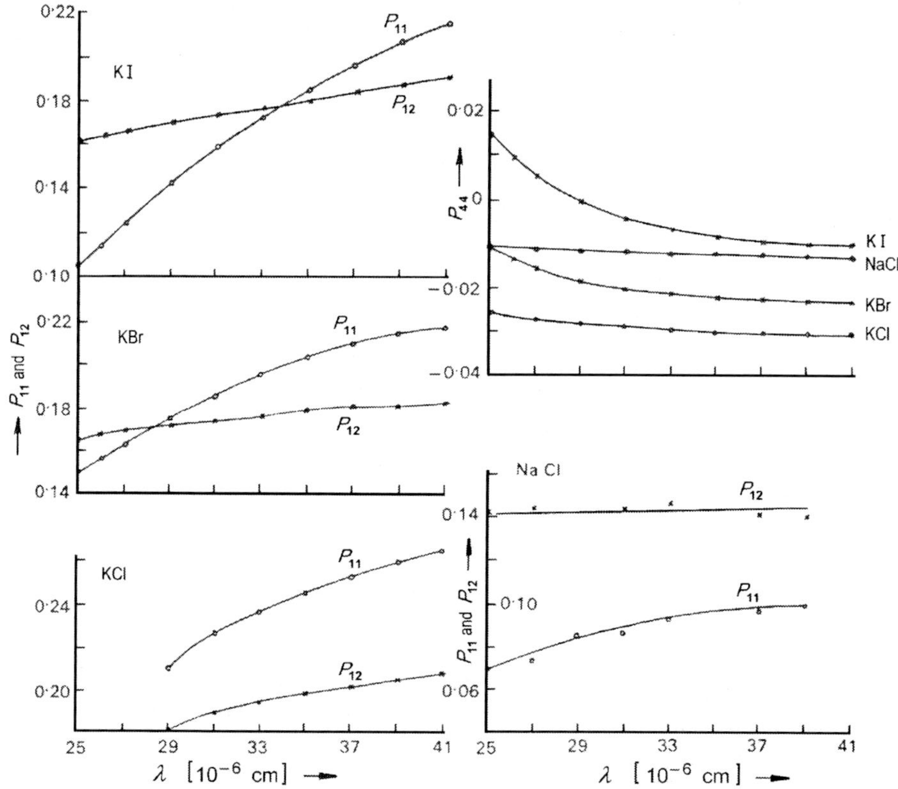

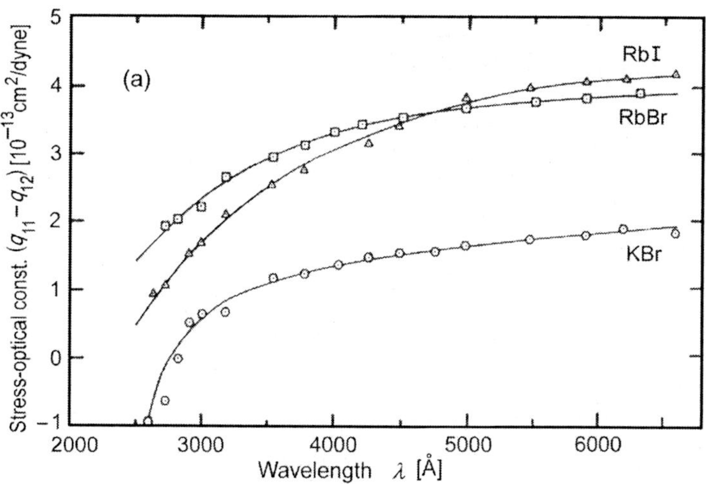

Fig. 4.2 Dispersion of P_{ij} in the UV region (after [4.27])

Fig. 4.3a, b and c Dispersion of q_{ij} in the visible-UV region (after [4.26])

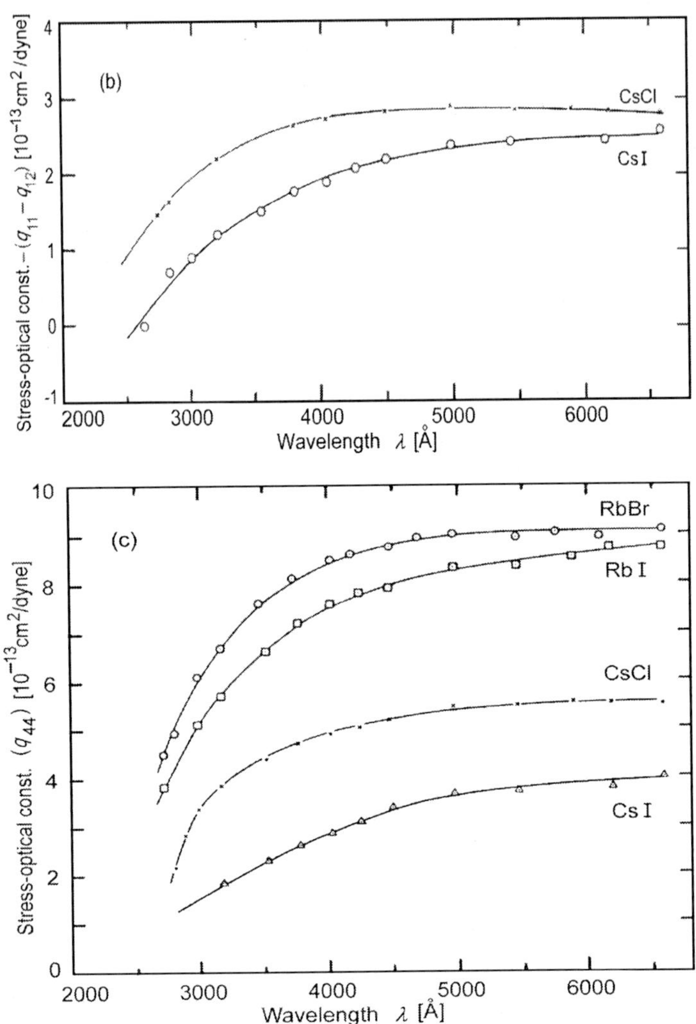

4.2.4 Polycrystalline Photoelastic Constants

Table 4.10 Values of the polycrystalline photoelastic constants $P_{ij}, \overline{q}'_{ij}$ for strain continuity (Voigt) and $\overline{q}_{ij}, \overline{P}'_{ij}$ for stress continuity (Reuss); [4.31]

Crystal	$\overline{P}_{11}$ $\overline{P}'_{11}$	$\overline{P}_{12}$ $\overline{P}'_{12}$	$\overline{P}_{44}$ $\overline{P}'_{44}$	$\overline{q}'_{11}$ $\overline{q}_{11}$	$\overline{q}'_{12}$ $\overline{q}_{12}$	$\overline{q}'_{44}$ $\overline{q}_{44}$
	$[10^{-2}]$			$[10^{-13}\ \text{cm}^2\ \text{dyne}^{-1}]$		
LiF	5.202	16.664	−5.731	−0.077	0.958	−1.035
	5.073	16.729	−5.828	−0.124	0.982	−1.106
NaF	9.926	17.642	−3.858	0.231	1.349	−1.118
	10.339	17.435	−3.548	0.266	1.332	−1.066

Table 4.10 (Continued)

Crystal	$\overline{P}_{11}$ $\overline{P}'_{11}$	$\overline{P}_{12}$ $\overline{P}'_{12}$	$\overline{P}_{44}$ $\overline{P}'_{44}$	$\overline{q}'_{11}$ $\overline{q}_{11}$	$\overline{q}'_{12}$ $\overline{q}_{12}$	$\overline{q}'_{44}$ $\overline{q}_{44}$
	$[10^{-2}]$			$[10^{-13}\,\mathrm{cm}^2\,\mathrm{dyne}^{-1}]$		
NaCl	14.597	18.354	−1.879	1.415	2.507	−1.092
	14.906	18.119	−1.647	1.458	2.486	−1.028
KCl	17.777	18.490	−0.357	3.270	3.610	−0.341
	16.205	19.276	−1.536	2.270	4.110	−1.840
KBr	18.487	18.502	−0.007	3.489	3.496	−0.007
	16.861	19.316	−1.228	2.368	4.056	−1.688
KI	19.945	18.983	0.481	5.429	4.865	0.564
	18.652	19.629	−0.488	4.440	5.360	−0.920
RbCl	25.122	23.862	0.630	4.750	4.135	0.614
	21.408	25.719	−2.155	2.276	5.372	−3.069
RbBr	22.658	20.905	0.877	5.100	4.150	0.950
	18.837	22.815	−1.989	2.208	5.596	−3.388
RbI	26.537	24.479	1.029	7.324	5.943	1.381
	23.420	26.037	−1.309	4.366	7.422	−3.056

Notes and Comments

1. The photoelastic constants P_{ij}, $\overline{q}'_{ij}$ for strain continuity (Voigt) and $\overline{q}_{ij}$, $\overline{P}'_{ij}$ for stress continuity (Reuss) are derived from the single crystal photoelastic constants P_{ij}, q_{ij} using the following working equations [4.31]:

$$\overline{P}_{11} = (1/5)(3R + 2S + 4T) \tag{4.19}$$

$$\overline{P}_{12} = (1/5)(R + 4S - 2T) \tag{4.20}$$

$$\overline{P}_{44} = (1/2)(\overline{P}_{11} - \overline{P}_{12}) \tag{4.21}$$

where

$$R = (1/3)(P_{11} + P_{22} + P_{33}) \tag{4.22}$$

$$S = (1/6)(P_{12} + P_{21} + P_{23} + P_{32} + P_{31} + P_{13}) \tag{4.23}$$

$$T = (1/3)(P_{44} + P_{55} + P_{66}) \tag{4.24}$$

and

$$\overline{q}'_{ij} = \overline{P}_{im}\,\overline{S}'_{mj} \tag{4.25}$$

where S'_{mj} is the polycrystal elastic constant for strain continuity. Similarly,

$$\overline{q}_{11} = (1/5)(3R' + 2S' + 2T')$$
(4.26)

$$\overline{q}_{12} = (1/5)(R' + 4S' - T')$$
(4.27)

$$\overline{q}_{44} = (\overline{q}_{11} - \overline{q}_{12})$$
(4.28)

where

$$R' = (1/3)(q_{11} + q_{22} + q_{33})$$
(4.29)

$$S' = (1/6)(q_{12} + q_{21} + q_{23} + q_{32} + q_{31} + q_{13})$$
(4.30)

$$T' = (1/3)(q_{44} + q_{55} + q_{66})$$
(4.31)

and

$$\overline{P}'_{ij} = \overline{q}_{im}\overline{C}'_{mj}$$
(4.32)

where C'_{mj} is the polycrystal elastic constant for stress continuity.

2. Note that there are large differences in the Voigt and Reuss values of P_{44} and $\overline{q}_{44}$. In fact, the values have different signs in KI and the rubidium halides.
3. The polycrystalline values of the photoelastic constants of alkali halides have not been determined experimentally.

4.3 Faraday Effect

4.3.1 Verdet Constant and its Temperature Coefficient

Table 4.11 Values of Verdet constant (V) measured at room temperature and $\lambda = 0.633$ μm and its temperature coefficient (d log V/dT); uncertainty in last digit is indicated in parenthesis. Temperature derivative measured over 220–310 K; [4.32]

Crystal	V [10^{-4} degree amp^{-1}]	d log V/dT [10^{-6} K^{-1}]
NaCl Structure		
LiF	1.68(2)	275
LiCl	6.7(1)	95
LiBr	10.2(2)	162
NaF	2.00(4)	231
NaCl	6.1(1)	80
NaBr	9.5(2)	141
NaI	16.2(4)	147
KF	1.99(4)	167
KCl	4.81(9)	174
KBr	7.3(1)	62
KI	12.6(2)	177

Table 4.11 (Continued)

Crystal	V $[10^{-4}$ degree amp$^{-1}]$	$d \log V/dT$ $[10^{-6}$ K$^{-1}]$
NaCl Structure		
RbF	2.36(4)	136
RbCl	4.77(9)	171
RbBr	6.7(1)	138
RbI	11.7(3)	227
CsF	3.39(7)	~ 0
CsCl Structure		
CsCl	6.0(1)	24
CsBr	7.8(1)	33
CsI	12.5(3)	200

Notes and comments

1. The Verdet constant is defined as $V = \varphi/LH$ where φ is the rotation of the plane of polarisation caused when the wave travels a distance L in the medium in magnetic field H. V is given here in units of degree amp^{-1}. In older literature, V is given in units of min cm^{-1} oersted^{-1}. Multiply the value in degree amp^{-1} with 47.73 to get the value in min cm^{-1} oersted^{-1}.
2. Compilations of Verdet constants for a few of the alkali halides can be found in [4.9, 4.33]
3. The plots of V versus r are smooth curves, r being the interatomic distance. However, the halides of each alkali ion lie on a different curve.
4. The V values increase in the sequence F–Cl–Br–I. On the other hand, the V values decrease in the sequence Li–Na–K–Rb except for the fluorides for which there is not much variation from alkali to alkali.
5. Haussuhl and Effgen [4.32] calculated the molar Faraday rotation $\Omega_m = V(V_m)\rho^{-1}$ where V_m is the molar volume and ρ is the density. Assigning an ionic molar Faraday rotation Ω' and assuming additivity i.e., $\Omega_m = \Sigma \Omega'_k$ Haussuhl and Effgen [4.32] obtained the values of Ω'_k (given in Table XII) for the ions.

Table XII Values of Ω'_k

Ion	Ω'_k	Ion	Ω'_k
Li$^+$	0.00015	F$^-$	0.0015
Na$^+$	0.0013	Cl$^-$	0.0152
K$^+$	0.0030	Br$^-$	0.029
Rb$^+$	0.0050	I$^-$	0.065
Cs$^+$	0.0095		

Further, Haussuhl and Effgen [4.32] found the following relation between Ω'_k and the corresponding molar electronic polarisability α_k:

$$\Omega'_k \approx 0.00145\,\alpha_k^2 \qquad\qquad (\alpha_k^2 \text{ in Å}^6) \qquad\qquad (4.33)$$

4.3.2 Dispersion of Verdet Constant

Table 4.12 Values of the Verdet constant V at some selected wavelengths

	V [min cm^{-1} oersted^{-1}]; Ref. [4.34]			
λ [µm]	LiF	KCl	KBr	KI
0.5780	0.00935	0.02938	0.0442	0.0753
0.5461	0.01055	0.03330	0.0503	0.0864
0.4916	0.01313	0.04239	0.0648	0.1224
0.4358	0.01702	0.05638	0.0869	0.1558
0.4047	0.01992	0.0670	0.1055	0.1925
0.3650	0.02485	0.0872	0.1390	0.2668
0.3132	0.03531	0.1338	0.2232	-
0.3022	0.03839	0.1490	0.2520	-

Notes and Comments

1. To express V in degree amp^{-1}, the value in min cm^{-1} oersted^{-1} may be multiplied with 2.095×10^{-2}
2. Sivaramakrishnan [4.34] fitted the available data on dispersion of refractive index n and Verdet constant V to Eq. (4.34)

$$n^2 - 1 = A + \sum_i \frac{A_i \lambda^2}{\lambda^2 - \lambda_i^2} \quad \text{and} \quad V = \frac{1}{n}\left(\frac{e}{2mc^2}\right)\sum_i \frac{\bar{\gamma}\, A_i \lambda^2 \lambda_i^2}{(\lambda^2 - \lambda_i^2)^2} \tag{4.34}$$

where A_i's are constants, λ_i's are absorption wavelengths and $\bar{\gamma}$ is called the magnetic anomaly factor. The values of the parameters are given in Table XIII.

Table XIII Values of parameters in Eq. (4.34); λ and λ_i in µm

Crystal	A	A_1	A_2	A_3	λ_1	λ_2	λ_3	$\bar{\gamma}$
LiF	0.2650	0.6596	-	-	0.0865	-	-	0.836
NaCl	0.1560	0.8554	0.3177	-	0.1107	0.1563	-	0.93
KCl	0.2434	0.3573	0.3761	0.1981	0.1000	0.1310	0.1620	0.854
KBr	0.2424	0.4116	0.4265	0.2808	0.1100	0.1460	0.1800	0.850
KI	0.4532	0.2150	0.8027	0.1780	0.1290	0.1805	0.2190	0.871

3. The magnetic anomaly factor $\bar{\gamma}$ would be unity if the ions were in the inert gas configuration. Hence the deviation of $\bar{\gamma}$ (from unity) could be taken as a measure of the deviation from the inert gas structure; the greater the deviation, the lower is the value of $\bar{\gamma}$ [4.35].

4.4 Quadratic Electro-optic Effect

Table 4.13 Values of the quadratic electro-optic coefficients g_{ijkl}

Crystal	$g_{ijkl}\ [10^{-18}\ cm^2\ V^{-2}]$				
	Expt.; [4.36]		Theor.; [4.37]		
	$(g_{1111}{-}g_{1122})$	g_{1212}	g_{1111}	g_{1122}	g_{1212}
	uncertainty: 15 %				
NaCl Structure					
LiF	−3.53	−0.70	−33.4	−29.9	2.34
LiCl	7.51	−2.31	-	-	-
LiBr	9.80	−3.20	-	-	-
LiI	-	-	-	-	-
NaF	−1.15	−0.63	− 9.7	−8.6	1.06
NaCl	1.30	−1.67	−41.4	−42.7	5.56
NaBr	2.51	−2.93	−64.9	−67.4	7.88
NaI	4.81	−5.00	-	-	-
KF	−1.12	1.50	-	-	-
KCl	1.03	1.46	−31.3	−32.3	2.87
KBr	1.03	1.00	−43.5	−44.5	1.84
KI	1.26	−0.82	−69.8	−71.0	0.96
RbF	−0.50	1.61	-	-	-
RbCl	−0.11	3.90	−36.6	−36.5	5.26
RbBr	−0.20	3.52	−49.1	−48.9	3.97
RbI	−0.74	2.15	−65.9	−65.2	0.26

Notes and Comments

1. The quadratic electro-optic effect is defined by the relation

$$\Delta a_{ij} = g_{ijkl}\ E_k E_l \tag{4.35}$$

   where Δa_{ij} are the polarisation constants, E_k and E_l are the electric field components and g_{ijkl} are the components of the quadratic electro-optic tensor. For the NaCl symmetry, there are only three constants g_{1111}, g_{1122} and g_{1212}.
2. Individual experimental values $g_{1111}= -29.0$ and $g_{1122}= -33.9$ (in $10^{-18}\ cm^2\ V^{-2}$) are available only for LiF [4.38].
3. The theoretical values given by Kucharczyk [4.37] are based on the Phillips Van Vechten theory. Theory predicts that all g_{1111} and g_{1122} coefficients have negative sign.

4.5 Laser-Related Properties

4.5.1 Optical Transmittance

Plots of percent transmittance vs wavelength are shown in Figs. (4.4 a-i):

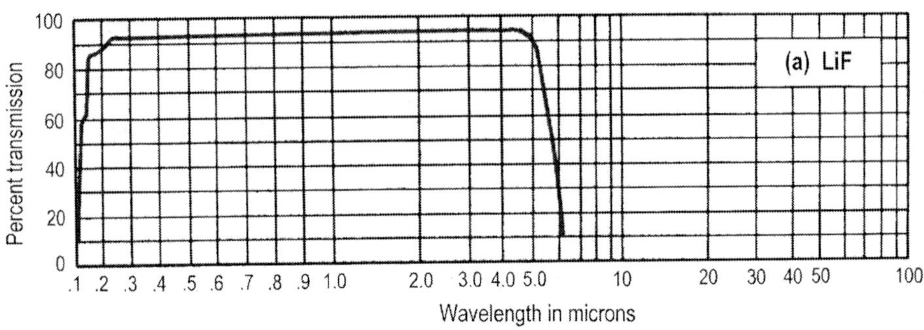

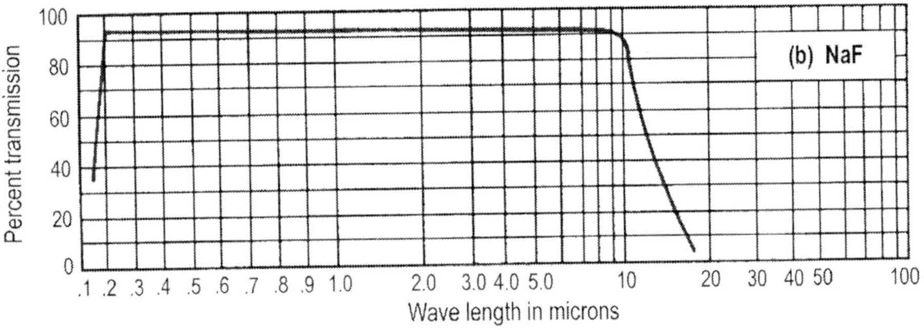

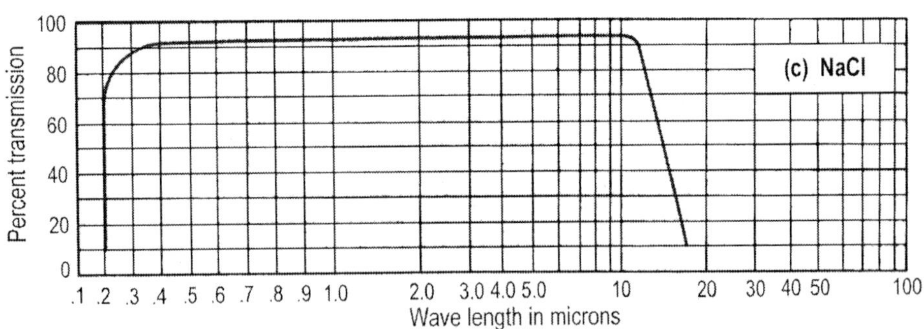

Fig. 4.4 Plots of percent transmittance vs wavelength (in μm) for **(a)** LiF, **(b)** NaF, **(c)** NaCl, **(d)** KCl, **(e)** KBr, **(f)** RbCl, **(g)** RbBr, **(h)** RbI, **(i)** CsBr and **(j)** CsI (after [4.39])

Fig. 4.4 (Continued)

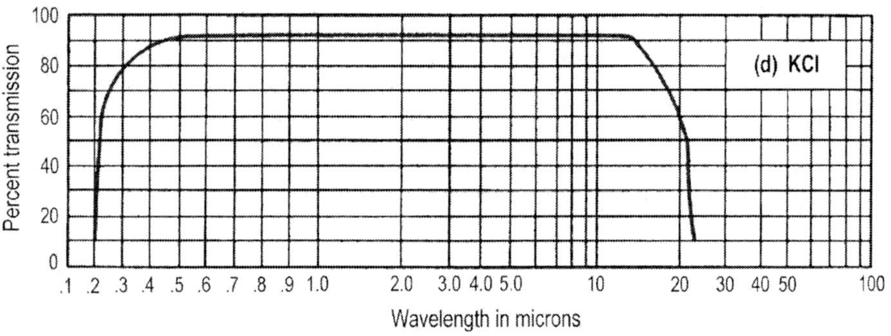

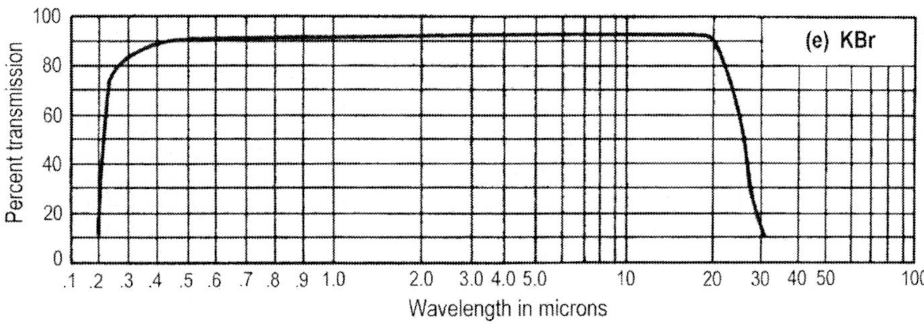

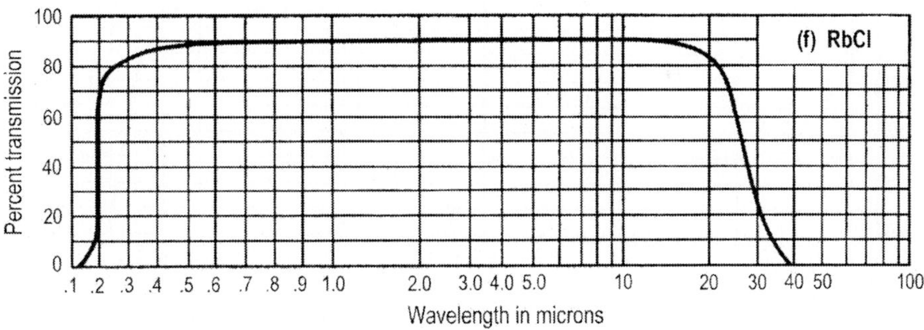

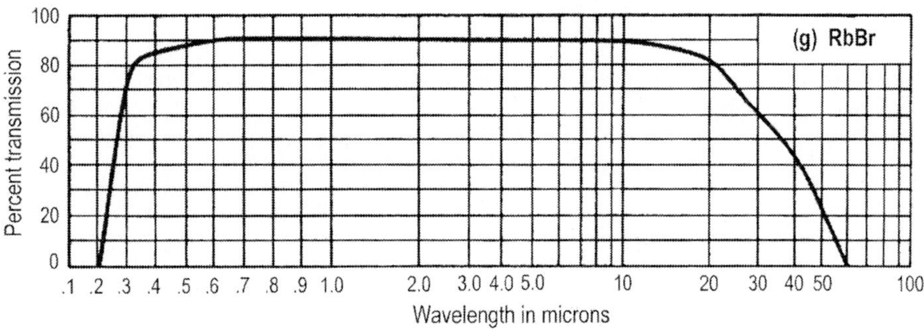

Fig. 4.4 (Continued)

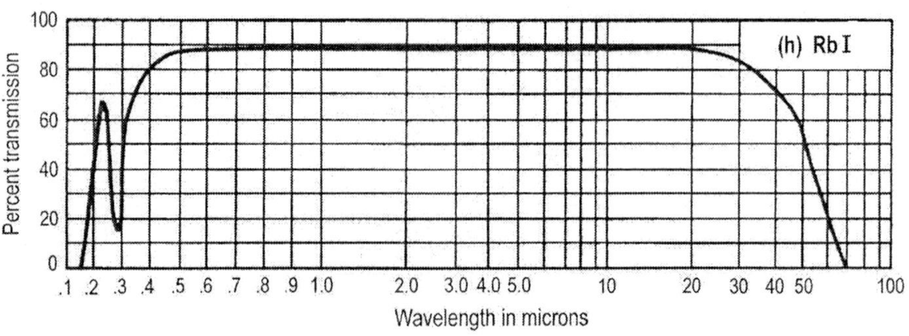

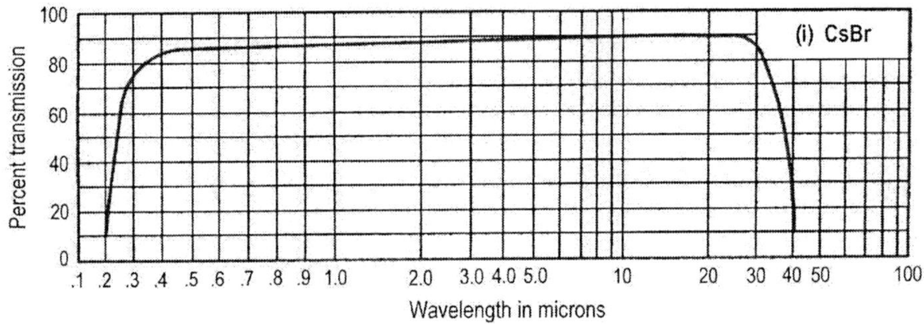

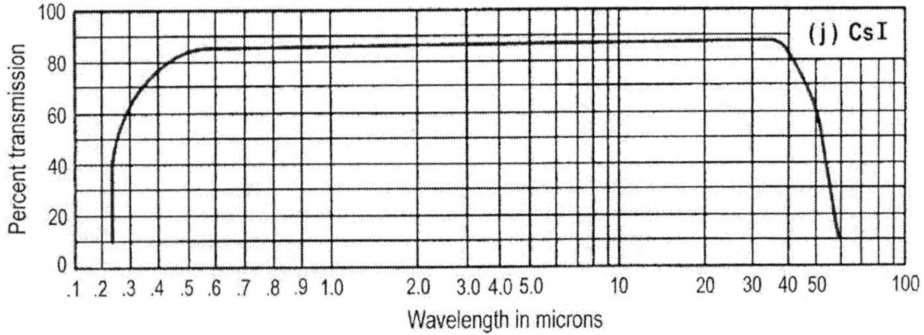

4.5.2 Linear Absorption Coefficient

Table 4.14 Values of the linear absorption coefficient α [cm^{-1}] at selected wavelengths (λ in µm); Ref. [4.40]

LiF		NaF		KCl	
λ	α	λ	α	λ	α
0.1068	1.0	0.161	1.0	0.177	1.0
0.3638	0.009	0.3638	0.004	0.351	0.0002
0.4880	0.003	0.4579	0.002	0.4880	< 0.00002

Table 4.14 (Continued)

LiF		NaF		KCl	
λ	α	λ	α	λ	α
1.3188	0.0005	0.5145	0.001	1.06	0.000007
4.35	0.006	2.7	0.0004	2.7	0.001
5.0	0.045	3.8	0.0002	3.8	0.0003
5.5	0.155	8.0	0.17	5.5	0.00001
6.0	0.44	9.0	0.32	9.8	0.0001
7.0	2.0	10.0	0.54	12.5	0.0015
8.0	7.3	11.0	1.1	15	0.05

Table 4.14 (Continued)

KBr		CsI	
λ	α	λ	α
1.06	0.000003	5.5	0.0001
2.7	0.00012	10.6	0.000009
3.8	0.0002	40	0.01
5.25	0.0002	45	0.3
10.6	0.0001	50	0.5
18	0.02	55	0.9
20	0.09	60	2.0
24	0.24		
28	0.76		
32	2.04		

4.5.3 Two-Photon Absorption Coefficient

Table 4.15 Values of the two-photon absorption coefficient β at select wavelengths

Wavelength λ [μm]	$\beta\,[10^{-11}\ \mathrm{m\ W^{-1}}];\ [4.40]$				
	LiF	KCl	KBr	KI	CsI
0.2484	< 0.0013	-	-	-	-
0.2661	< 0.02	1.7	2.0	3.8	-
0.3471	-	-	5.1	8	-
0.3547	-	-	-	6	6
0.36	-	-	4	9.1	-

4.5.4 Three-Photon Absorption Coefficient

Table 4.16 Values of the three-photon absorption coefficient (γ') at selected wavelengths

Wavelength λ [µm]	γ' [10^{-26} m^3 W^{-2}] [4.40]	
	LiF	KI
0.2484	0.0016	-
0.6943	-	0.18

4.5.5 Nonlinear Refractive Index

Table 4.17 Values of the nonlinear refractive index (γ'') at selected wavelengths (λ)

λ [µm]	γ'' [10^{-20} m^2 W^{-1}]; Ref. [4.40]				
	LiF	NaF	KCl	KBr	KI
0.2661	1.3	-	-	-	-
0.3547	0.6	-	-	-	-
0.5321	0.6	3.0	-	12.7	-
1.0642	0.8	1.1	5.7	7.9	29

4.5.6 Bulk Damage Threshold Parameters

Table 4.18 Values of the pulse width τ_p [ns] and bulk damage threshold I_{thr} [10^{12} W m^{-2}] at selected wavelengths (λ); Ref. [4.40]

λ [µm]	LiF		NaF		KCl		KBr		KI		CsI	
	τ_p	I_{thr}	τ_p	I_{thr}	τ_p	I_{thr}	τ_p	I_{thr}	τ_p	I_{thr}	τ_p	I_{thr}
0.1933	30	1.3	30	0.3-1.0	-	-	-	-	-	-	-	-
0.2484	20	5-20	20	7.5-12.5	-	-	-	-	-	-	-	-
0.5321	0.02	26000	0.02	13300	8	250	8	600	-	-	8	96
0.6943	10	3600	14	1300	14	110	14	130	14	130	10	130
1.06	10	3600	10	1400	31	380-1240	10	500	10	200	10	150
2.76	85	1200	85	500	85	200	85	80	85	20	85	40
3.8-3.9	75	27	75	> 40	80	200-1030	80	8-90	-	-	-	-
10.6	-	-	-	-	-	0.0003	140	40	140	20	60	5

4.6 Harmonic Generation

4.6.1 Second Harmonic Generation Parameters at Crystal-Glass Boundary

Table 4.19 Values of the parameters: $P_\perp$, $P_{||}$, P_γ, α, β, γ; P's in 10^{17}; α, β, γ in 10^{16} esu; [4.41]

Crystal	$P_\perp$	P_γ	$P_{||}/P_\perp$	α/β	β	γ
LiF	8	~ 1 or less	0.8	3.6	0.5	< 0.1
NaF	8.5	~ 1 or less	0.96	3.4	0.5	< 0.1
NaCl	9.5	~ 1 or less	0.4	3.9	0.55	< 0.1
KCl	12	~ 1 or less	0.61	4.0	0.6	< 0.1
KBr	6.5	~ 1 or less	0.73	4.6	0.45	< 0.1
KI	25	1.4	0.31	4.5	1.0	< 0.2

Notes and Comments

1. α, β, γ are parameters entering into the equation for the nonlinear polarisation (denoted by $P_i(2\omega)$) at the second harmonic frequency 2ω; $P_i(2\omega)$ is given by:

$$P_i(2\omega) = (\alpha - \beta)E_j(\omega)\nabla_j E_i(\omega) + \beta E_i(\omega)\nabla_j E_j(\omega) + \gamma \nabla_i[E_j(\omega)E_j(\omega)]$$

(4.36)

The components of the second harmonic power polarised parallel and perpendicular to the plane of reflection corresponding to the fundamental beam polarised at 45° to the plane of incidence are $P_{||}$ and $P_\perp$. The second harmonic power generated with fundamental beam polarised $\perp$ to the plane of incidence and always polarised in the plane of reflection is denoted by P_γ.

2. For a centrosymmetric crystal in bulk, for a transverse electromagnetic wave, $P_i(2\omega)$ becomes zero. However, it becomes nonzero in an experiment involving a discontinuity, such as the crystal-glass boundary.

4.6.2 Third Harmonic Generation Parameters

Table 4.20 Values of the third harmonic generation parameters C_{11} and C_{18}

Crystal	Nonlinear optical process	C_{11} [10^{-20} m^2 V^{-2}]	C_{18}/C_{11}	λ [μm]
		Ref. [4.42]		
NaF		0.0035	-	1.06
NaCl	$(-3\omega, \omega, \omega, -\omega)$	0.0168	0.4133	1.06
KCl		0.0168	0.28	1.06
KBr		−0.0392	0.3667	1.06

Notes and Comments

1. In third harmonic generation (THG), the expression for polarisation is

$$P_j\left(\omega_1,\omega_2,\omega_3\right)= g\,\psi_{jklm}\,E_k^{\omega_1}\,E_l^{\omega_2}\,E_m^{\omega_3} \tag{4.37}$$

The new frequency produced is $\omega_4 = \omega_1 + \omega_2 + \omega_3$ When $\omega_1 = \omega_2 = \omega_3 = \omega$, new frequency is $\omega_4 = 3\omega$. In this case, $g = 1/4$. Further, the THG coefficient C_{jklm} is

$$C_{jklm} = \psi_{jklm}/4 \tag{4.38}$$

Because of symmetry, the coefficients are denoted with two indices only i.e. C_{jn}. For the alkali halides only two coefficients are measured viz. C_{11} and C_{18}. Note that the C's in this section are not to be confused with the C's used to represent elastic constants (Sec. 4.2).

4.7 Polarisability

4.7.1 Electronic Polarisabilities

Table 4.21 Experimental room temperature values of the electronic polarisabilities (α^+ and α^-)

Ref. Ion ↓	α^+, α^- [Å^3]	
	[4.1]	[4.43]
Li$^+$	0.029	0.029
Na$^+$	0.255	0.285
K$^+$	1.201	1.149
Rb$^+$	1.797	1.707
Cs$^+$	3.137	2.789
F$^-$	0.759	0.876
Cl$^-$	2.974	3.005
Br$^-$	4.130	4.168
I$^-$	6.199	6.294

Notes and Comments

1. The experimental values of the molecular polarisabilities $\alpha\ (=\alpha^- +\alpha^+)$ are obtained from refractive index (n) data at $\lambda = \infty$ and molecular volumes (V_m) using the Clausius-Mosotti equation:

$$\frac{4\pi\alpha}{3} = V_m\,\frac{(n_\infty^2 - 1)}{(n_\infty^2 + 2)} \tag{4.39}$$

Tessman et al. [4.1] have used the values of refractive index from a compilation of refractive index by Winchell [4.46]. Jaswal and Sharma [4.43] have used

More recent refractive index data from Lowndes and Martin [4.2]. The polarisabilities for the ions (α^-, α^+) are obtained from the molecular polarisability α using the additivity rule: $\alpha = \alpha^- + \alpha^+$.

2. Tessman et al. [4.1] have given values of polarisability corresponding to the sodium D wavelength also.
3. Jaswal and Sharma [4.43] have given values of polarisability at 4 K also.
4. Polarisabilities calculated by theoretical methods have been given in [4.22, 4.44, 4.45].
5. Polarisability values are extremely useful in the analysis of various physical parameters. Thus, Hanlon and Lawson [4.47] showed that the values of Szigeti charge e^* decrease as the values of the polarisability difference ($\alpha^- - \alpha^+$) increase and the plot between the two is a smooth curve.
6. The polarisabilities are used as input parameters in lattice dynamical calculations.
7. Anderson [4.48] showed that a linear correlation with positive slope exists between ψ^{-1} $(n^2 - 1)^{-1}$ and the polarisability sum ($\alpha^- + \alpha^+$) where ψ is the compressibility.

4.7.2 Strain Derivative of Polarisability

Table 4.22 Values of the strain derivative of polarisability (Λ)

| | Λ at 0.5893 µm; [4.15] | | |
| | Experimental | | Theoretical |
Method* Crystal ↓	a	b	c
NaCl Structure			
LiF	0.706	0.719	0.745
LiCl	-	-	0.648
LiBr	-	-	0.600
LiI	-	-	0.616
NaF	0.69	0.599	0.581
NaCl	0.559	0.610	0.582
NaBr	-	0.529	0.556
NaI	-	-	0.593
KF	-	-	0.326
KCl	0.428	0.490	0.432
KBr	0.456	0.495	0.433
KI	0.479	0.502	0.491
RbF	-	-	0.244
RbCl	0.384	0.396	0.371
RbBr	0.396	0.381	0.374
RbI	-	0.457	0.442
CsF	-	-	0.178

Table 4.22 (Continued)

	Λ at 0.5893 µm; [4.15]		
	Experimental		Theoretical
Method*	a	b	c
Crystal $\downarrow$			
CsCl Structure			
CsCl	-	-	0.349
CsBr	0.365	-	0.380
CsI	-	-	0.461

*See Notes and Comments

Notes and Comments

1. Λ is defined as $- (\partial \log \alpha / \partial \log \rho)_T$ where α is the molecular polarisability and ρ the density. Λ is sometimes referred to as the strain polarisability constant.
2. The values (a) in the table are obtained from the experimental values of $\rho (\partial n / \partial \rho)_T$ using the relation:

$$\Lambda = 1 - \frac{6\,n\,\rho}{(n^2 + 2)\,(n^2 - 1)} \left(\frac{\partial n}{\partial \rho} \right)_T \tag{4.40}$$

The values (b) are obtained from the experimental values of P_{11} and P_{12} using the relation

$$\Lambda = 1 - [n^4 (P_{11} + 2 P_{12}) / (n^2 - 1) (n^2 + 2)] \tag{4.41}$$

where P_{11} and P_{12} are the strain-optical constants. The theoretical values (c) are obtained from the Wilson-Curtis model .
3. Burstein and Smith [4.49] suggested that the strain polarisability constant is a measure of the ionicity of the bond (larger ionicity, smaller Λ). Bansigir [4.50] showed that there is a linear relation between Λ and $[1 - (e^*/e)]$, e^* being the effective ionic charge.

4.7.3 Wavelength and Temperature Variation of the Strain Polarisability Constant

The variation of the strain polarisability constant (Λ) with wavelength and temperature is shown in Fig. 4.5.

Notes and Comments

1. The strain polarisability constants at different wavelengths are calculated using the dispersion data on n and the strain-optical constants P_{ij}'s.
2. The strain polarisability constants at different temperatures are calculated using data on dn/dT and the P_{ij}'s at different temperatures.

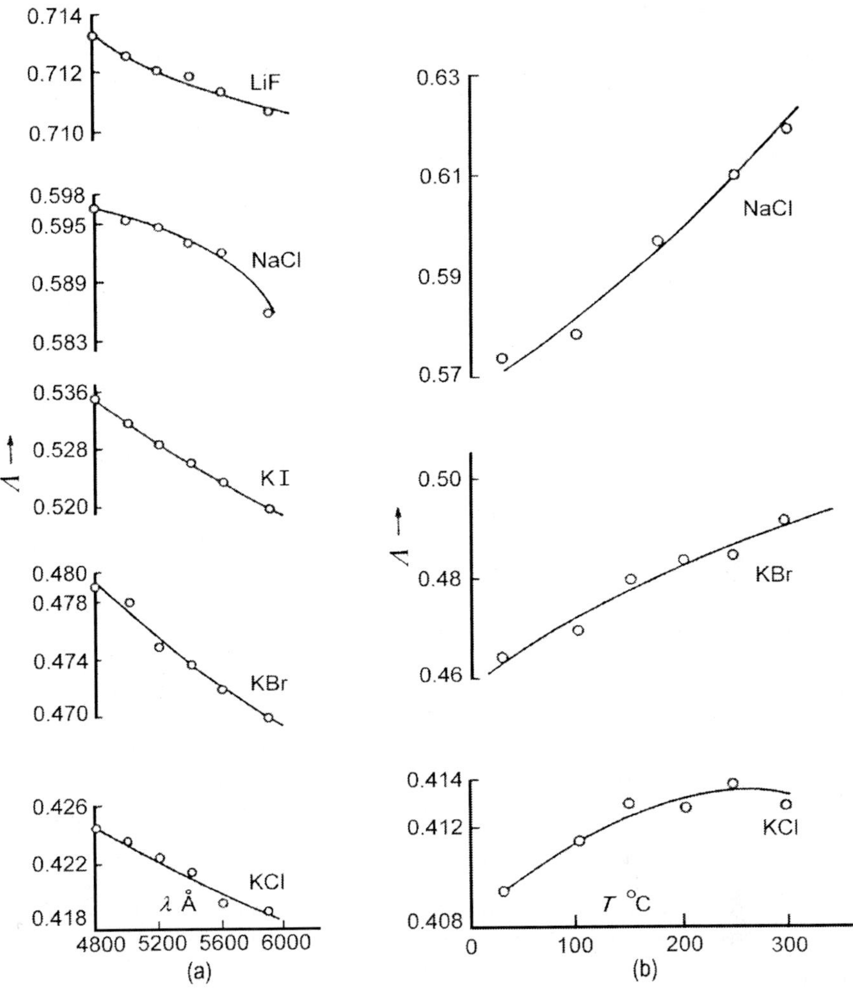

Fig. 4.5 Strain polarisability constant Λ vs **(a)** wavelength λ and **(b)** temperature (after [4.23])

4.7.4 Pressure Variation of Polarisability

Table 4.23 Values of the pressure derivatives of the molecular polarisability $(d\alpha/dP)$ and of the anion polarisability $(d\alpha^-/dP)$

Crystal	$-(d\alpha/dP)\ [\text{Å}^3\,(\text{atm})^{-1}]$	$-(d\alpha^-/dP)\ [\text{Å}^3\,(\text{atm})^{-1}]$
	Ref. [4.51]	
NaCl Structure		
LiF	0.9	1.2
LiCl	7.5	14.7
LiBr	14.2	27.3

Table 4.23 (Continued)

Crystal	$-(\mathrm{d}\alpha/\mathrm{d}P)\,[\text{Å}^3\,(\text{atm})^{-1}]$	$-(\mathrm{d}\alpha_-/\mathrm{d}P)\,[\text{Å}^3\,(\text{atm})^{-1}]$
	Ref. [4.51]	
NaCl Struceture		
LiI	29.7	55.4
NaF	1.5	2.0
NaCl	7.3	15.9
NaBr	12.5	28.6
NaI	26.2	57.5
KF	3.3	2.5
KCl	9.7	18.0
KBr	14.9	31.5
KI	25.4	62.6
RbF	5.1	2.8
RbCl	11.7	19.2
RbBr	17.5	33.5
RbI	31.8	65.9

Notes and Comments

1. The pressure derivative of the molecular polarisability $\mathrm{d}\alpha/\mathrm{d}P$ given above is estimated from the experimental values of the pressure derivatives of the various quantities appearing in the Clausius-Mosotti equation.
2. Assuming the pressure derivative of the cation polarisability to be negligible, the pressure derivative of anion polarisability is expressed as

$$\mathrm{d}\alpha^-/\mathrm{d}P = \left(\partial\alpha^-/\partial r_-\right)\left(\mathrm{d}r_-/\mathrm{d}P\right) + \left(\partial\alpha^-/\partial r_+\right)\left(\mathrm{d}r_+/\mathrm{d}P\right) \tag{4.42}$$

where r_- and r_+ are the anion and cation radii. The values of $(\partial\alpha^-/\partial r_-)$ and $(\partial\alpha^-/\partial r_+)$ are obtained from the plots of α^- and the interionic distance r for fixed cation and fixed anion. The ionic compressibilities $(\mathrm{d}r_-/\mathrm{d}P)$ and $(\mathrm{d}r_+/\mathrm{d}P)$ are estimated from the experimental compressibility by a minimisation procedure.
3. The pressure derivatives of the anion polarisability are negative, consistent with the idea that increasing pressure restricts the ability of the electron cloud to polarise.
4. The calculated pressure derivatives $(\mathrm{d}\alpha^-/\mathrm{d}P)$ are systematically larger than the experimental derivatives $(\mathrm{d}\alpha/\mathrm{d}P)$ typically by a factor of 2.

4.7.5 Quadrupole and Octupole Polarisabilities

Table 4.24 Values of the quadrupole and octupole polarisabilities for the alkali and halogen ions; [4.52]

Halogen ion Alkali ion $\downarrow$	F^-	Cl^-	Br^-
Quadrupole Polarisability [$\mathring{A}^5$]			
Li^+	-	0.89	4.83	7.84
Na^+	0.057	1.34	6.08	9.48
K^+	0.50	1.62	6.70	10.2
Rb^+	1.02	1.84	7.26	11.0
Octupole Polarisability [$\mathring{A}^7$]			
Li^+	-	1.84	16.8	32.4
Na^+	0.042	3.25	23.2	42.28
K^+	0.73	4.35	27.0	47.1
Rb^+	1.90	5.22	30.3	52.5

Notes and Comments

1. The quadrupole and octupole polarisabilities of the alkali ions are independent of the companion halogen ion.
2. On the other hand, the quadrupole and octupole polarisabilities of the anions vary with the companion alkali ion. The anion quadrupole and octupole polarisabilities increase in the sequence Li–Na–K–Rb and also in the sequence F–Cl–Br–I.
3. Mahan [4.52] has excluded cesium halides and alkali iodides as the assumptions in the calculation were not valid for the Cs^+ and I^- ions.

References

4.1   J.R. Tessman, A.H. Kahn and W. Shockley, Phys. Rev., **92**, 890, 1953.

4.2   R.P. Lowndes and D.H. Martin, Proc. Roy. Soc. Lond. **A308**, 473, 1969.

4.3   P.G. Johannsen, Phys. Rev. **B55**, 6856, 1997; P.G. Johannsen, G. Reiß, U. Bohle, J. Magiera, R. Muller, H. Spiekermann and W.B. Holzapfel, Phys. Rev., **B55**, 6865, 1997; P.G. Johannsen, G. Reiß, U. Bohle and W.B. Holzapfel, phys. stat. sol., **(b) 198**, 93, 1996.

4.4   S.T. Pantelides, Phys. Rev. Lett., **35**, 250, 1975.

4.5   K. Sangwal and W. Kucharezyk, J. Phys. D: Appl. Phys.,**20**, 522, 1987.

4.6   Y.S. Toulokian, R.K. Kirby, R.E. Taylor and T.V.R. Lee, *Thermophysical Properties of Matter*, **Vol.13**, *Thermal expansion – Non-metallic Solids*, Plenum Press, New York, 1977.

4.7   R.R. Reddy and T.V.R. Rao, Infrared Physics and Tech. **36**, 825, 1995.

4.8   H.H. Li, J. Phys. Chem. Ref. Data, **5**, 329, 1976 and references therein (equations fitted to available data).

4.9   R.S. Krishnan, *Progress in Crystal Physics*, S. Viswanathan, Madras, India, 1958 and references therein.

4.10   *Handbook of American Institute of Physics*, 2nd Ed., McGraw Hill, New York, 1963.

4.11   Y. Tsay, B. Bendow and S.S. Mitra, Phys. Rev. **B8**, 2688, 1973 and references therein.

4.12   E. Burstein and P.L. Smith, Phys. Rev., **74**, 229, 1948.

4.13   K.S. Iyengar, Nature, **176**, 1119, 1955.

4.14   E.D. Schmidt, Ph.D. Thesis, Pennsylvania State University, USA, 1972, quoted in [4.15].

4.15   H. Coker, J. Phys. Chem. Solids, **40**, 1079, 1988.

4.16   Experimental values by K.Vedam and coworkers quoted by J. Fontanella, C. Andeen and D. Schuele, Phys. Rev., **B6**, 582, 1972.

4.17   K. Vedam and E.D.D. Schmidt, J. Mater. Sci., 1, 310 ,1966.

4.18   N.M. Balzaretti and J.A.H.Da. Jornada, J. Phys. Chem. Solids, **57**, 179, 1996.

4.19   Calculated by D.B. Sirdeshmukh from raw n-P data at 0.65 eV communicated by P.G. Johannsen (1997); also see [4.3].

4.20   B. Bendow, P.D. Gianino, Y. Tsay and S.S. Mitra, Appl. Optics, **13**, 2382, 1974.

4.21   H.E. Pettersen, J. Opt. Soc. Am., **63**, 1243, 1973, q_{ij} values in [4.22] converted into P_{ij}.

4.22   H. Leibssle, Zeit. fur Krist., **114**, 457, 1960.

4.23   K.G. Bansigir and K.S. Iyengar, Acta Cryst., **14**, 727, 1961.

4.24   K.G. Bansigir and K.S. Iyengar, Proc. Phys. Soc., **71**, 225, 1958.

4.25   T.S. Narasimhamurthy, J. Opt. Soc. Am., **51**, 914, 1961.

4.26   R. Laiho and A. Korpela, Ann. Acad. Sci. Fenn., **A6**, 272, 1968.

4.27   A. Rahman and K.S. Iyengar, Acta Cryst., **A26**, 128, 1970.

4.28   K.G. Bansigir and K.S. Iyengar, Acta Cryst. **14**, 670, 1961.

4.29   R. Ethiraj, V.G. Krishnamurthy and K.G. Bansigir, Acta Cryst., **A29**, 636, 359, 1973.

4.30   A. Rahman and K.S. Iyengar, Acta Cryst., **A26**, 359, 1970.

4.31   G.S. Ranganath and S. Ramaseshan, Pramana, **1**, 78, 1973.

4.32   S. Haussuhl and W. Effgen, Zeit. fur Krist., **183**, 153, 1988.

4.33   *Landolt-Bornstein*, 6th Ed. **Vol. I**, Part 1, Springer-Verlag, Berlin, 1950.

4.34   V. Sivaramakrishnan, J. Ind. Inst. Sci. **A39**, 1, 1957 quoted in [4.9].

4.35   S. Ramaseshan, Proc. Ind. Acad. Sci., **A24**, 104, 1946.

4.36   S. Haussuhl and H. Hesse, phys. stat. sol., **30**, 209, 1968.

4.37   W. Kucharczyk, Zeit. fur Krist., **191**, 277, 1990.

4.38   S. Haussuhl and G. Walda, phys. stat. sol., **(a) 5**, K163, 1971.

4.39   OPTOVAC Optical Crystal Handbook, OPTOVAC, Brookfield, MA USA, 1993.

4.40   D.N. Nikogosyan, *Properties of Optical and Laser-related Materials–A Handbook*, John Wiley, Chichester, U.K., 1997, and references therein.

4.41   C.C. Wang and A.N. Duminski, Phys. Rev. Lett., **20**, 668, 1968.

4.42   CRC Handbook of Chemistry and Physics, 76th Ed., CRC Press, Florida, 1995-1996.

4.43   S.S. Jaswal and T.P. Sharma, J. Phys. Chem. Solids, **34**, 509, 1973.

4.44   A.R. Ruffa, Phys. Rev., **130**, 1412, 1963.

4.45   J. Shanker, A.K.G. Lashkari and V.P. Gupta, phys. stat. sol., **(b) 91**, 263, 1979.

4.46   A. Winchell, *Microscopic Characters of Inorganic Substances or Artificial Minerals*, John Wiley and Sons, New York, 1931.

4.47   J.E. Hanlon and A.W. Lawson, Phys. Rev., **113**, 472 1959.

4.48   O.L. Anderson, Am. Min., **51**, 1001, 1966.

4.49   E. Burstein and P.L. Smith, Proc. Ind. Acad. Sci., **A28**, 377, 1948.

4.50   K.G. Bansigir, Nature, **216**, 256, 1967.

4.51   A. Batana, J. Bruno and R.W. Munn, Molecular Physics, **92**, 1029, 1997.

4.52   ab initio calculations by G.D. Mahan, Phys. Rev., **B34**, 4235, 1986.

5 Dielectric and Electrical Properties

5.1 Static Dielectric Constant

5.1.1 Static Dielectric Constant at Room Temperature

Table 5.1 Values of the static dielectric constant (ε_0) at room temperature.

Ref.	[5.1]	[5.2]	[5.3]	[5.4]	
Frequency	1.6 kHz	1 kHz	500 Hz–20 MHz	1 MHz	
Uncertainty	1 %	0.01 %	0.3 %	0.6 %	
Temp.	25 °C	27 °C	27 °C		
Crystal $\downarrow$	ε_0	ε_0	ε_0	t °C	ε_0
NaCl Structure					
LiF	8.42	9.0354	9.03	19.7	8.94
LiCl	-	-	11.89	-	-
LiBr	-	-	13.28	-	-
LiI*					
NaF	4.88	5.0719	5.10	-	-
NaCl	5.72	5.8940	5.92	16	5.93
NaBr	6.44	6.3957	6.29	10	6.34
NaI	-	-	7.31	15	7.28
KF	-	-	5.51	-	-
KCl	4.69	4.8112	4.86	21	4.80
KBr	4.73	4.8735	4.92	10	4.87
KI	4.89	-	5.11	22	5.12
RbF	-	-	6.50	-	-
RbCl	4.81	-	4.90	29	4.94
RbBr	4.64	-	4.87	36	4.85
RbI	4.69	-	4.93	16	4.83
CsF	-	-	8.08	-	-
CsCl Structure					
CsCl	-	-	6.98	-	-
CsBr	6.38	-	6.68	-	-
CsI	6.31	-	6.56	-	-

*see Notes and Comments

Notes and Comments

1. Data on static dielectric constants of alkali halides can also be found in Refs. [5.5–5.7].

2. Though the static dielectric constant is defined as the dielectric constant at zero frequency (D.C. field), values of dielectric constant measured with A.C. fields up to 1 MHz are taken as static values.
3. There is very little work on LiI, perhaps due to its highly hygroscopic nature. There is a single value of 11.03 at 2 MHz by Hojendahl [5.5].
4. Andeen et al. [5.2] in their accurate measurement on well-characterised crystals observed variations in ε_0 determined from different samples which were in excess of the claimed error of 0.01%. They attributed these variations to the presence of trace-level impurities.
5. The static dielectric constant ε_0 is related to the high frequency dielectric constant ε_∞ and the transverse and longitudinal optic mode frequencies v_{TO} and v_{LO} through the Lyddane-Sachs-Teller relation [5.8]

$$(\varepsilon_0 / \varepsilon_\infty) = (v_{LO} / v_{TO})^2 \tag{5.1}$$

6. The validity of this relation has been verified in most of the alkali halides.
7. ε_0 is one of the basic input parameters for lattice dynamical calculations.

5.1.2 Static Dielectric Constant at Low Temperatures

Table 5.2 Values of static dielectric constant (ε_0) at low temperatures (1.5–350 K)

	ε_0; uncertainty 1 % for RbF and CsF; 0.3 % for others; Ref. [5.3]								
Temp. [K] Crystal ↓	1.5	25	50	75	100	150	200	250	300	350
NaCl Structure										
LiF	8.50	8.52	8.55	8.57	8.60	8.70	8.81	8.92	9.03	9.14
LiCl	10.92	10.98	11.05	11.12	11.21	11.37	11.56	11.71	11.89	12.08
LiBr	11.95	11.99	12.12	12.22	12.32	12.56	12.80	13.03	13.28	13.54
NaF	4.73	4.74	4.76	4.78	4.80	4.86	4.92	5.00	5.10	5.22
NaCl	5.45	5.47	5.50	5.54	5.58	5.67	5.75	5.84	5.92	6.02
NaBr	5.78	5.80	5.85	5.89	5.93	6.02	6.11	6.20	6.29	6.39
NaI	6.62	6.66	6.72	6.77	6.82	6.93	7.05	7.17	7.31	7.46
KF	5.11	5.13	5.17	5.20	5.24	5.31	5.39	5.45	5.51	5.58
KCl	4.49	4.51	4.53	4.55	4.58	4.65	4.72	4.79	4.86	4.94
KBr	4.52	4.54	4.58	4.61	4.64	4.71	4.78	4.85	4.92	5.00
KI	4.68	4.71	4.76	4.79	4.83	4.90	4.97	5.04	5.11	5.18
RbF	5.99	6.01	6.04	6.09	6.14	6.24	6.32	6.41	6.50	6.59
RbCl	4.53	4.56	4.59	4.61	4.64	4.71	4.77	4.84	4.90	4.96
RbBr	4.51	4.53	4.56	4.58	4.61	4.67	4.73	4.80	4.87	4.95
RbI	4.55	4.58	4.61	4.63	4.66	4.72	4.79	4.86	4.93	5.00
CsF	7.27	7.30	7.35	7.41	7.48	7.62	7.77	7.92	8.08	8.22
CsCl Structure										
CsCl	6.75	6.76	6.78	6.79	6.81	6.85	6.89	6.92	6.98	7.03
CsBr	6.39	6.41	6.43	6.45	6.47	6.52	6.57	6.62	6.68	6.74
CsI	6.29	6.30	6.32	6.34	6.36	6.41	6.46	6.50	6.56	6.62

5.1.3 Temperature Coefficient of the Static Dielectric Constant at Low Temperatures

Table 5.3 Values of the temperature coefficient $[(1/\varepsilon_0)(\partial\varepsilon_0/\partial T)_P]$ of the static dielectric constant at low temperatures (25–300 K)

	$(1/\varepsilon_0)(\partial\varepsilon_0/\partial T)_P$ in $[10^{-5}\,\mathrm{K}^{-1}]$; uncertainty ± 0.7 [5.3]							
Temp. [K] Crystal ↓	25	50	75	100	150	200	250	300
NaCl Structure								
LiF	2.3	3.8	7.9	12.5	23.1	26	27.5	28.2
LiCl	20.8	25.0	30.0	31.0	32.1	34.0	36.7	37.7
LiBr	32.6	39.2	40.0	36.5	37.9	38.5	39.7	40.9
NaF	3.0	10.5	16.5	20.9	23.7	26.8	28.8	30.0
NaCl	17.5	23.7	25.5	27.5	29.0	30.2	31.0	31.7
NaBr	23.0	26.6	27.6	28.7	31.0	32.0	32.9	33.4
NaI	25.0	30.2	30.6	31.2	32.0	32.7	33.4	34.2
KF	21.6	28.3	30.6	27.0	27.7	28.1	28.5	28.9
KCl	8.5	17.9	22.5	25.8	26.5	27.2	27.9	28.8
KBr	16.9	22.7	24.5	26.1	27.0	27.7	29.9	30.7
KI	18.0	23.4	26.3	27.9	29.0	30.4	31.7	33.0
RbF	19.5	27.2	27.8	29.1	30.2	32.4	33.7	34.9
RbCl	13.0	20.0	22.0	24.0	25.5	26.9	27.8	28.7
RbBr	15.5	22.5	23.7	24.6	25.9	27.0	28.0	29.0
RbI	17.5	23.3	24.5	25.8	26.9	28.1	28.8	29.9
CsF	20.0	28.8	33.3	37.0	39.6	40.2	41.0	42.3
CsCl Structure								
CsCl	6.4	8.6	9.3	11.0	12.3	13.0	17.0	19.9
CsBr	7.9	12.1	12.6	13.5	14.2	14.9	15.6	16.2
CsI	7.4	11.6	12.6	13.4	14.1	14.9	15.8	16.6

Notes and Comments

1. Values of the temperature coefficient of static dielectric constant of some of the alkali halides are reported in Refs. [5.9–5.11].
2. The temperature variation of ε_0 and $(1/\varepsilon_0)(\partial\varepsilon_0/\partial T)_P$ is nonlinear. The nonlinear temperature variation of ε_0 can be expressed in terms of the first and second derivatives of ε_0 (see Sec. 5.1.4).

5.1.4 First and Second Temperature Derivatives of the Static Dielectric Constant

Table 5.4 Values of the first and second temperature derivatives [$(1/\varepsilon_0)(\partial\varepsilon_0/\partial T)_P$ and $(1/\varepsilon_0)(\partial^2\varepsilon_0/\partial T^2)_P$] of the static dielectric constant over the temperature range 200–308 K

Parameter	$(1/\varepsilon_0)(\partial\varepsilon_0/\partial T)_P$ [10^{-5} K^{-1}]	$(1/\varepsilon_0)(\partial^2\varepsilon_0/\partial T^2)_P$ [10^{-7} K^{-2}]
Uncertainty	0.2 %	20 %
Ref.	[5.12]	[5.12]
Crystal $\downarrow$		
NaCl Structure		
LiF	26.96	0.90
NaF	28.25	2.00
NaCl	31.59	2.28
NaBr	32.78	2.38
KCl	29.48	2.77
KBr	29.82	2.79

5.1.5 High Temperature Data on Static Dielectric Constant

Data on temperature variation of the static dielectric constant (ε_0) and its temperature derivative [$(1/\varepsilon_0)(\partial\varepsilon_0/\partial T)_P$] at high temperatures up to 700°C are shown in Figs. 5.1 and 5.2.

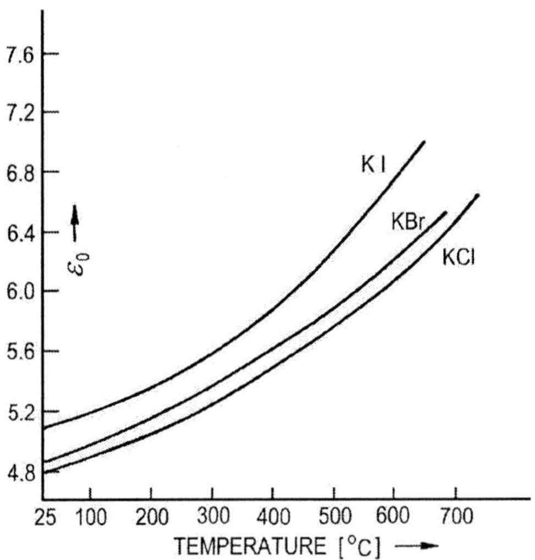

Fig.5.1 Static dielectric constants (ε_0) of KCl, KBr and KI as a function of temperature: data on KCl, KBr from [5.13], data on KI from [5.14]. Here, ε_0 is the real part of the dielectric constant measured at 23.5 GHz

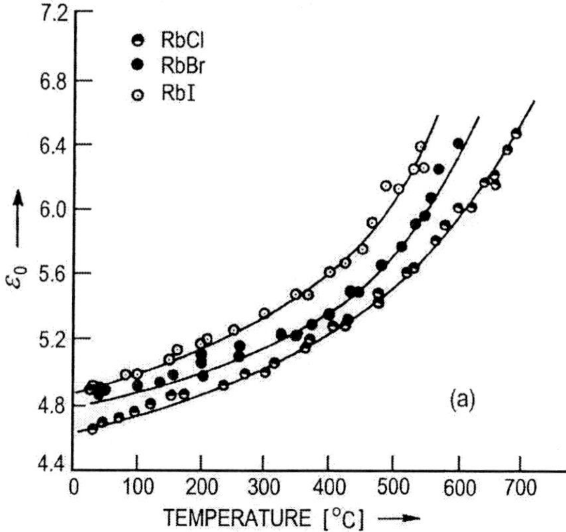

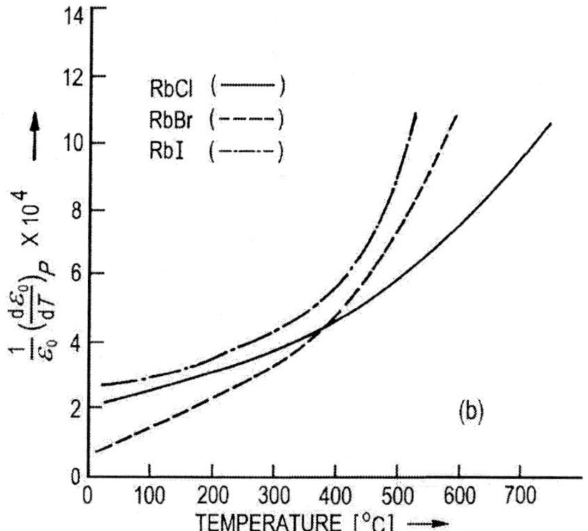

Fig. 5.2 **(a)** Static dielectric constant (ε_0) and **(b)** temperature derivative $[(1/\varepsilon_0)(\partial\varepsilon_0/\partial T)_P]$ for rubidium halides as a function of temperature (after [5.15]). ε_0 is the real part of the dielectric constant measured at 24.6 GHz

5.1.6 Pressure Coefficient of Static Dielectric Constant

Table 5.5 Pressure coefficient of static dielectric constant $[(1/\varepsilon_0)\,(\partial\varepsilon_0/\partial P)_T]$ at 1 bar and 3.5 kbar at room temperature

	$-(1/\varepsilon_0)(\partial\varepsilon_0/\partial P)_T$ in 10^{-6} bar^{-1}				
Pressure	1 bar	1 bar	1 bar	1 bar	3.5 kbar
Ref.	[5.3]	[5.11]	[5.1]	[5.12]	[5.3]
Uncertainty	5 %		1 %	0.2 %	5 %
Crystal ↓					
NaCl Structure					
LiF	3.8	4.23	4.87	5.085	3.8
LiCl	8.6	-	-	-	8.6
LiBr	12.8	-	-	-	12.1
NaF	5.3	-	5.06	5.398	5.2
NaCl	9.2	9.24	10.00	10.388	9.5
NaBr	11.9	-	12.39	12.704	11.3
NaI	15.1	-	-	-	14.6
KF	7.6	-	-	-	7.2
KCl	10.0	9.92	10.57	11.006	10.2
KBr	11.5	9.91	11.84	12.488	11.9
KI	13.7	-	12.60	-	12.8
RbF	8.5	-	-	-	8.3
RbCl	10.2	10.86	11.30	-	9.4
RbBr	11.8	12.90	12.60	-	10.0
RbI	13.4	-	13.20	-	13.3
CsCl Structure					
CsCl	12.5	12.90	-	-	12.0
CsBr	13.9	13.70	13.90	-	12.6
CsI	13.6	14.70	15.60	-	12.7

Notes and Comments

1. Note −ve sign of the pressure coefficients.
2. The pressure coefficient of ε_0 is useful in the estimation of the volume derivatives of several other parameters like polarisability, transverse optical mode frequency and effective ionic charge etc. [5.1].

5.1.7 Higher-Order Pressure Derivatives of Static Dielectric Constant

Table 5.6 Values of the pressure derivatives $\varepsilon' = (1/\varepsilon_0)(\partial \varepsilon_0/\partial P)_T$ in 10^{-6} bar^{-1}; $\varepsilon'' = (1/\varepsilon_0)(\partial^2 \varepsilon_0/\partial P^2)_T$ in 10^{-12} bar^{-2} and $\varepsilon''' = (1/\varepsilon_0)(\partial^3 \varepsilon_0/\partial P^3)_T$ in 10^{-18} bar^{-3}

Ref.	[5.1]		[5.12]		
Pr. range	1–7 kbar		1–2.5 kbar		
Parameter	ε'	ε''	ε'	ε''	ε'''
Uncertainty	1 %	10 %	0.2 %	2 %	20 %
Crystal ↓					
NaCl Structure					
LiF	−4.87	40	−5.085	86.6	−1610
NaF	−5.06	70	−5.398	120.8	−2630
NaCl	−10.00	290	−10.388	491.3	−32200
NaBr	−12.39	560	−12.704	812.6	−85600
KCl	−10.57	470	−11.006	691.2	−47400
KBr	−11.84	600	−12.488	978.1	−110000
KI	−12.60	880	-	-	-
RbCl	−11.30	600	-	-	-
RbBr	−12.60	540	-	-	-
RbI	−13.20	980	-	-	-
CsCl Structure					
CsBr	−13.90	870	-	-	-
CsI	−15.60	920	-	-	-

Notes and Comments

1. The signs of the derivatives: ε' and ε''' are −ve for all crystals while ε'' is +ve.
2. The values of ε' reported by Fontanella et al. [5.12] are systematically higher than those by Jones [5.1] by 2–5%. The ε'' values from the two sources differ by a factor of about 1.5. These differences are due to i) difference in accuracy in the two measurements, ii) difference in the degree of nonlinearity observed in the two sets and iii) difference in the compression data used for converting capacitance changes into dielectric constant changes. These aspects are discussed by Fontanella [5.16].
3. Dutt et al. [5.17] theoretically calculated the pressure derivatives ε', ε'' and ε''' for the alkali halides and found good agreement with the experimental results of Fontanella et al. [5.12].
4. The pressure variation of static dielectric constant of RbCl and RbBr has been studied through the Fm3m→Pm3m transition pressure. Jones [5.1] failed to observe any discontinuity in the dielectric constant at the transition pressure. Havinga and Bosman [5.11] observed a sharp change in the dielectric constant of RbBr from 4.6 to 6.5 at a pressure of 4.9 kbar. Samara [5.18] observed a sudden increase in dielectric constant of RbCl from 4.6 to 6.1 at a pressure of 7.8 kbar.

5.2 Electronic Dielectric Constant

Table 5.7. Values of the electronic dielectric constant (ε_∞) and its temperature and pressure derivatives [$(1/\varepsilon_\infty)\,(\partial\varepsilon_\infty/\partial T)_P$ and $(1/\varepsilon_\infty)\,(\partial\varepsilon_\infty/\partial P)_T$].

Parameter	ε_∞		$(1/\varepsilon_\infty)(\partial\varepsilon_\infty/\partial T)_P$ $[10^{-6}\,\mathrm{K}^{-1}]$	$(1/\varepsilon_\infty)(\partial\varepsilon_\infty/\partial P)_T$ $[10^{-6}\,\mathrm{bar}^{-1}]$	
Ref.	[5.19]		[5.20]	[5.21]	[5.22]
Temp. Crystal ↓	2 K	290 K			
NaCl Structure					
LiF	1.93	1.93	−18.30	0.259	-
LiCl	2.79	2.75	-	-	-
LiBr	3.22	3.16	-	-	-
LiI	3.89	3.80	-	-	-
NaF	1.75	1.74	−15.14	0.606	-
NaCl	2.35	2.33	−48.83	1.269	-
NaBr	2.64	2.60	-	1.997	-
NaI	3.08	3.01	-	4.254	-
KF	1.86	1.85	-	-	-
KCl	2.20	2.17	−49.35	2.657	-
KBr	2.39	2.36	−51.67	2.174	-
KI	2.68	2.65	−61.44	3.109	-
RbF	1.94	1.93	-	-	-
RbCl	2.20	2.18	-	3.559	-
RbBr	2.36	2.34	-	-	3.939
RbI	2.61	2.58	-	6.23	-
CsF	2.17	2.16	-	-	-
CsCl Structure					
CsCl	2.67	2.63	-	2.273	-
CsBr	2.83	2.78	−75.87	3.476	-
CsI	3.09	3.02	−114.75	3.041	-

Notes and Comments

1. The electronic dielectric constant ε_∞ is equal to the square of the refractive index. For accurate values of ε_∞, the refractive index n_∞ obtained from dispersion equations (Table 4.1.1) is used.
2. ε_∞ is one of the input parameters in most lattice dynamical calculations.
3. The pressure coefficient of ε_∞ is useful in estimating mode Gruneisen parameters and the pressure dependence of effective ionic charge.

5.3 Dielectric Polarisability

Table 5.8. Values of the dielectric polarisability of ions [α_D (i)]

Ion	α_D (i) [Å^3]	
	Ref. [5.23] *	Ref. [5.24]
Li$^+$	0.93	1.20
Na$^+$	1.57	1.80
K$^+$	3.29	3.83
Rb$^+$	4.56	5.29
Cs$^+$	5.69	7.43
F$^-$	2.05	1.62
Cl$^-$	5.00	-
Br$^-$	6.45	-
I$^-$	8.71	-

* The original values have been divided by 4π for uniformity with units in [5.24]

Notes and Comments

1. The dielectric polarisability α_D per molecule is calculated from the static dielectric constant ε_0 from the relation

$$\alpha_D = \left(\frac{3}{4\pi}\right) V_m \left(\frac{\varepsilon_0 - 1}{\varepsilon_0 + 2}\right) \tag{5.2}$$

where V_m is the molecular volume and α_D the sum of the dielectric polarisabilities of all the ions $\alpha_D(i)$:

$$\alpha_D = \sum \alpha_D(i) \tag{5.3}$$

2. The dielectric polarisability has contributions from the optical polarisability ($\alpha_{opt.}$) and the infrared polarisability α_{IR}. Thus $\alpha_D = \alpha_{opt.} + \alpha_{IR}$.
3. The dielectric polarisabilities of the alkali ions (as a group) and the halogen ions (as a group) when plotted against the cube of the ionic radii result in straight line plots [5.23, 5.24].

5.4 Effective Ionic Charge

5.4.1 Effective Ionic Charge (Szigeti Charge)

Table 5.9 Values of the effective ionic charge e^* expressed as a fraction of the electron charge e

	e^*/e				
Ref.	[5.25]	[5.19]		[5.16]	
Temp. Crystal ↓	RT	2 K	290 K	0 K	RT
NaCl Structure					
LiF	0.87 (0.83)	0.80	0.81	0.790	0.807
LiCl	-	0.77	0.79	-	-
LiBr	-	0.73	0.73	-	-
NaF	0.93	0.82	0.83	0.815	0.827
NaCl	0.74	0.76	0.77	0.753	0.765
NaBr	0.69	0.73	0.74	0.731	0.753
NaI	0.71	0.71	0.73	-	-
KF	-	0.88	0.91	-	-
KCl	0.80	0.79	0.81	0.773	0.810
KBr	0.76	0.75	0.78	0.745	0.776
KI	0.69	0.72	0.74	-	-
RbF	-	0.92	0.95	-	-
RbCl	0.84	0.81	0.83	-	-
RbBr	0.82	0.78	0.80	-	-
RbI	0.89 (0.79)	0.74	0.77	-	-
CsF	-	0.93	0.95	-	-
CsCl Structure					
CsCl	0.84	0.85	0.85	-	-
CsBr	0.79	0.82	0.82	-	-
CsI	-	0.77	0.78	-	-

Notes and Comments

1. The effective ionic charge is defined as:

$$(e^*/e)^2 = (9\pi\mu v_{\mathrm{TO}}^2 / N_V e^2)(\varepsilon_0 - \varepsilon_\infty)/(\varepsilon_\infty + 2)^2 \tag{5.4}$$

    where μ is the reduced mass, v_{TO} the long wavelength TO mode frequency, N_V the number of ion pairs per unit volume and ε_0 and ε_∞ the static and high frequency dielectric constants ($\varepsilon_\infty = n^2$, n being the refractive index).
2. An error in the values of (e^*/e) for LiF and RbI was noticed and the corrected values (in parenthesis) were given by Hardy [5.26].
3. In the literature, the quantity (e^*/e) is sometimes represented by 's' or q^*.

4. Values of (e^*/e) are also given in [5.27–5.29].
5. Values of e^*/e from different sources for a given crystal show variations of 2–5%. These differences are mostly due to differences in the values of ε_0 employed. The most accurate values are those by Fontanella [5.16].
6. Values of e^*/e for the alkali halides range from 0.69 (for NaBr) to 0.95 (for CsF).
7. The deviation of e^*/e from its formal value of unity indicates the extent of covalent bonding [5.30]. A material with a higher Szigeti charge is considered to be more ionic and vice versa [5.31].
8. Lowndes and Martin [5.19] found that the values of e^*/e when plotted against the radius ratio r_+/r_- lie on smooth curves with a separate curve for each alkali ion (or each halogen ion).
9. Niwas et al. [5.32] obtained linear plots for e^*/e and $r^{0.5}$ with a different slope for each common halogen ion. Here, r is the interionic distance.
10. Koh [5.31] found a linear plot for e^*/e and r_-/r with a different slope and intercept for each common alkali ion.
11. Lawaetz [5.33] showed that (e^*/e) equals $[C/(\omega_P)]$ where C is the Phillips electronegativity difference and ω_P the plasma frequency.
12. Hanlon and Lawson [5.34] observed that e^*/e decreases with increasing values of polarisability difference $(\alpha^- - \alpha^+)$ and that the plot between the two is a smooth curve.
13. Hardy [5.26] derived a linear relation between $[1 - (e^*/e)]$ and the parameter

$$[\lambda(r_-^2 - r_+^2)\{(1/\rho)-(2/r)\}\exp(-r/\rho)]$$

where $\lambda\exp(-r/\rho)$ is the short range repulsion interaction term. The plot between the two parameters is linear but does not pass through the origin as required by theory.
14. Sirdeshmukh [5.35] showed that the product αe^{*2} is a constant for the alkali halides where α is the linear coefficient of thermal expansion.
15. Mitra [5.36] obtained a linear correlation between (e^*/e) and the mode Gruneisen parameter γ_{TO}.
16. Bansigir [5.37] obtained a linear plot between $(1-e^*/e)$ and the strain polarisability constant Λ.
17. Sirdeshmukh et al. [5.38] showed that the product $(e^*/e)^2 (H/C_{44})$ is a constant where H is the microhardness and C_{44} the shear elastic constant.

5.4.2 Temperature and Volume Derivatives of Effective Ionic Charge

Table 5.10 Values of the temperature and volume derivatives $(1/s)(\partial s / \partial T)_P$ and $(\partial \log s / \partial \log V)_T$, where $s = (e^*/e)$

Parameter	$(1/s)(\partial s / \partial T)_P$ $[10^{-5}\,\mathrm{K}^{-1}]$	$(\partial \log s / \partial \log V)_T$				
		Experimental			Model calculations	
Ref. Crystal ↓	[5.39]	[5.1]	[5.40]	[5.16]	[5.41a]	[5.41b]
NaCl Structure						
LiF	4.29	−0.7	-	0.57 ± 0.4	0.53	0.25
LiCl	8.79	-	-	-	-	-
LiBr	4.69	-	-	-	-	-
NaF	4.18	−1.6	-	0.14 ± 0.4	0.49	0.29
NaCl	4.51	−0.8	0.61	0.74 ± 0.4	0.91	0.32
NaBr	4.69	−0.4	0.45	1.68 ± 0.2	1.06	0.41
NaI	9.51	-	-	-	-	-
KF	11.4	-	-	-	-	-
KCl	8.57	−0.9	-	-0.06 ± 0.4	0.73	0.32
KBr	13.4	−0.2	0.35	0.02 ± 0.4	0.92	0.43
KI	9.38	−0.9	-	-	1.15	0.46
RbF	10.9	-	-	-	-	-
RbCl	8.37	−0.04	0.24	-	0.71	0.46
RbBr	8.68	−0.05	0.16	-	0.85	0.50
RbI	13.5	0.2	0.43	-	1.09	0.48
CsF	7.31	-	-	-	-	-
CsCl Structure						
CsCl	4.04	-	-	-	0.80	0.45
CsBr	4.18	−0.4	0.19	-	1.00	0.50
CsI	4.45	−0.4	0.21	-	1.42	0.54

Notes and Comments

1. As mentioned, the coefficient $(1/s)(\partial s / \partial T)_P$ has been computed for the temperature interval 2–290 K; the value is +ve for all the crystals. Kim et al. [5.42] made calculations at different temperatures up to 320 K for 4 crystals (LiF, NaF, KCl, KBr) using temperature derivatives of ε_0, ε_∞ and v_{TO}. They found a complex temperature variation, the coefficient $(1/s)(\partial s / \partial T)_P$ taking +ve and −ve values in different temperature regions.

2. For the strain derivative of s, the experimental values are obtained using pressure derivatives of ε_0, ε_∞ and v_{TO}. It can be seen that there are large differences in the values of the strain derivatives of s for a given salt given by different sources and, even, difference in sign. This is so because the expression for $(\partial \log s / \partial \log V)$ is a difference between two large quantities and hence inaccuracies in experimental data have a large effect on the resulting value of

$(\partial \log s / \partial \log V)$. However, the model calculations yield a consistent +ve sign i.e. s increases as V increases.

3. For the same cation, the value of $(\partial \log s / \partial \log V)$ tends to increase with increasing anion size.

5.5 Electrostriction

Table 5.11 Values of the quadratic electrostrictive coefficients d_{ijkl} at room temperature

Crystal ↓	d_{ijkl} [10^{-21} m^2 V^{-2}]; uncertainty given in parenthesis; Ref.[5.43]		
	d_{1111}	d_{1122}	d_{1212}
NaCl Structure	**Experimental Values**		
LiF	5.2(3)	−1.7(3)	−0.6(2)
NaF	2.4(1)	−0.51(9)	−0.57(6)
NaCl	3.5(2)	−0.7(3)	0.0(3)
KCl	4.0(3)	−1.0(3)	0.3(2)
KBr	4.2(3)	−0.9(3)	0.4(2)
KI	5.3(3)	−1.2(2)	1.0(2)
RbCl	4.0(3)	−0.9(2)	0.5(2)
RbBr	5.6(4)	−1.3(2)	0.8(2)
RbI	6.2(4)	−1.5(3)	1.0(3)
	Estimated Values		
LiCl	15.4	−3.4	6.2
LiBr	30.1	−6.5	14.2
NaBr	6.3	−1.5	1.3
NaI	11.3	−2.5	4.0
KF	3.3	−0.8	−0.35
RbF	4.4	−1.1	0.26

Notes and Comments

1. The quadratic electrostrictive coefficients d_{ijkl} are defined by the relation

$$\varepsilon_{ij} = d_{ijkl}\, E_k E_l \tag{5.5}$$

where ε_{ij} are the components of the strain tensor and E_k, E_l the electric field vector components. For the m3m symmetry, only d_{1111}, d_{1122} and d_{1212} exist.

2. Values of d_{ijkl} for some alkali halides have been reported in Refs. [5.44 –5.47], but the values given in [5.43] are the most accurate. A detailed comparison of the values is given in [5.43].

3. The coefficient d_{1111} is positive, d_{1122} is negative while d_{1212} has positive values for some alkali halides and negative values for the others.

4. The quadratic electrostrictive coefficients are related to the pressure derivative of the static dielectric constant ε_0 as follows:

$$2(d_{1111} + 2d_{1122}) = -(\partial \varepsilon_0 / \partial P) \tag{5.6}$$

The experimental values of d_{ijkl} in [5.43] and the literature values of $(\partial \varepsilon_0/\partial P)$ satisfy Eq. (5.6).

5. Kucharezyk [5.48] proposed a linear relation of the form:

$$d_{ijkl} = a_{ijkl} + b_{ijkl}\,\Omega \tag{5.7}$$

where a_{ijkl} and b_{ijkl} are constants and the quantity Ω is given by

$$\Omega = [9\varepsilon_\infty^{\,2}(\varepsilon_0 - \varepsilon_\infty)^2 a^4 / 8s^2(\varepsilon_\infty + 2)^2] \tag{5.8}$$

where ε_0 and ε_∞ are static and high frequency dielectric constants, s the effective ionic charge and a the lattice constant respectively. Schreuer and Haussühl [5.43] fitted their experimental data to these equations and with the values of a_{ijkl} and b_{ijkl}, thus obtained, estimated the values of d_{ijkl} for the hygroscopic alkali halides for which there is no experimental data; these 'estimated values' are given in Table 5.11.

5.6 Electric Breakdown

Table 5.12 Values of the electric breakdown voltage F at room temperature; electric field applied in <100> direction

Crystal	$F\,[10^5$ volts cm$^{-1}]$ [5.49, 5.50]
LiF	31
NaF	24
NaCl	15
NaBr	8.1
KF	19
KCl	10
KBr	7.0
KI	5.7
RbCl	8.3
RbBr	6.3
RbI	4.9

Notes and Comments

1. Although uncertainties in values are not clearly mentioned, Von Hippel [5.51] pointed out the possibility of large error in the LiF value.
2. The breakdown voltage smoothly decreases in the sequence F–Cl–Br–I and Li–Na–K–Rb.
3. The electric breakdown voltage has some preferential crystallographic directions. Thus with the electric field applied in the <100> direction, breakdown

takes place at the lowest breakdown voltage in the <110> direction and at a higher voltage in the <111> and / or the <100> direction (Von Hippel [5.51]).

4. Callen [5.52] derived the following equation for the breakdown field at 0 K:

$$F = 134 \times 10^6 (h\nu_{TO})_{eV} \, [(\varepsilon_0 - \varepsilon_\infty)/(\varepsilon_0 \varepsilon_\infty^{\,3})^{1/2}](m_e^*/m_e) \ \text{volts cm}^{-1} \qquad (5.9)$$

where ν_{TO} is the transverse optical IR frequency, ε_0 and ε_∞ are the static and electronic dielectric constants and (m_e^*/m_e) the ratio of the effective mass to free electron mass. With $(m_e^*/m_e) = 1$, the theoretical values agree with the experimental values.

5. The breakdown voltage value increases with temperature up to a certain temperature T_C and then decreases with increasing temperature (KBr: [5.53]; NaCl: [5.54]).

6. The low temperature results (F increasing with temperature) were accounted for by the following theoretical expression (Frohlich and Simpson [5.55]).

$$F = 1.64 \times 10^5 (\rho/M)^{1/3} (\varepsilon_0 - \varepsilon_\infty)(\lambda_0/\lambda^{3/2})[1 + \{\exp(h\nu/kT) - 1\}^{-1}]^{1/2} \qquad (5.10)$$
$$\text{volts cm}^{-1}$$

where ρ is the density, M the molecular weight, ε_0 and ε_∞ the static and electronic dielectric constant, λ the IR absorption wavelength and λ_0 the wavelength of the first ultraviolet absorption band (λ in Å) and ν the characteristic frequency.

References

5.1   B.W. Jones, Phil. Mag., **16**, 1085, 1967.

5.2   C. Andeen, J. Fontanella and D. Schuele, Phys. Rev., **B2**, 5068, 1970.

5.3   R.P. Lowndes and D.H. Martin, Proc. Roy. Soc. Lond., **A316**, 351, 1970.

5.4   K. Kamiyoshi and Y. Nigara, phys. stat. sol., **(a) 3**, 735, 1970.

5.5   K. Hojendahl, Kgl. Danske Videnskabe. Selskab. Mat. Fys. Medd., **16**, 1, 1938.

5.6   S. Haussuhl, Z. Naturforsch., **12A**, 445, 1957.

5.7   K.F. Young and H.P.R. Fredrikse, J. Physical and Chemical Ref. Data, **2**, 313, 1973.

5.8   R.H. Lyddane, R.G. Sachs and E. Teller, Phys. Rev., **59**, 673, 1941.

5.9   S. Mayburg, Phys. Rev., **79**, 375, 1950.

5.10   A.J. Bosman and E.E. Havinga, Phys. Rev., **129**, 1593, 1963.

5.11   E.E. Havinga and A.J. Bosman, Phys. Rev.,**140**, A292, 1965.

5.12   J. Fontanella, C. Andeen and D. Schuele, Phys. Rev., **B6**, 582, 1972.

5.13   G.C. Smith, Rept. No. 51, Mat.Sci. Centre, Cornell Univ., Ithaca (US), 1962.

5.14   S. Chandra, Canad. J. Phys., **47**, 969, 1969.

5.15   S. Chandra and J. Prakash, Canad. J. Phys., **50**, 1053, 1972.

5.16   J. Fontanella, AEC Tech. Rept., No. 69, Case Western Reserve University, Cleveland, 1971.

5.17   N. Dutt, G.G. Agrawal and J. Shanker, phys. stat. sol., **(b)132**, 99, 1985.

5.18   G.A. Samara, Phys. Rev.,**165**, 959, 1967.

5.19   R.P. Lowndes and D.H. Martin, Proc. Roy. Soc. Lond., **A308**, 473, 1969.

5.20   R.S. Krishnan, *Progress in Crystal Physics*, S. Viswanathan, Madras, India, 1958.

5.21   Raw data from P.G. Johannsen communicated to D.B. Sirdeshmukh, 1998.

5.22   G.R. Barsch and B.N.N. Achar, phys. stat. sol., **35**, 881, 1969.

5.23   S. Roberts, Phys. Rev., **81**, 865, 1951.

5.24   R.D. Shannon, J. Appl. Phys., **73**, 348, 1993.

5.25   B. Szigeti, Trans. Faraday Soc., **45**, 155, 1949.

5.26   J.R. Hardy, Phil. Mag., **6**, 27, 1961.

5.27   G.O. Jones, D.H. Martin, P.A. Mawer and C.H. Perry, Proc. Roy. Soc. Lond., A261, 10, 1961.

5.28   D.H. Martin, Adv. Phys., **14**, 39, 1965.

5.29   B. Szigeti, Proc. Roy. Soc. Lond. **A204**, 51, 1950.

5.30   S.S. Mitra and K.V. Namjoshi, J. Chem. Phys., **55**, 1817, 1971.

5.31   A.K. Koh, J. Phys. Chem. Solids, **50**, 39, 1989.

5.32   R. Niwas, S.C. Goyal and J. Shanker, J. Phys. Chem. Solids, **38**, 219, 1977.

5.33   P. Lawaetz, Phys. Rev. Lett., **12**, 697, 1971.

5.34   J.K. Hanlon and A.W. Lawson, Phys. Rev., **113**, 472, 1959.

5.35   D.B. Sirdeshmukh, J. Chem. Phys. **45**, 2366, 1966.

5.36   S.S. Mitra, Proc. Int. Colloq. Phys. Prop. of Solids under Pressure, CNRS, Paris, 1969.

5.37   K.G. Bansigir, Nature, **216**, 256, 1967.

5.38   D.B. Sirdeshmukh, K.G. Subhadra, K. Kishan Rao and T. Thirmal Rao, Cryst. Res. Tech., **30**, 861,1995.

5.39   J. Shanker and R. Sunderaj, phys. stat. sol., **(b)101**, 303, 1980 calculated from the 2 K and 290 K values of e* given by Lowndes and Martin [5.19].

5.40   A. Batana and J. Faour, J. Phys. Chem. Solids, **45**, 571, 1984.

5.41   V.T. Gupta, H.P. Sharma and J. Shanker, Solid State Comm., **24**, 739, 1977 calculated from (a) Deformation Dipole model, (b) Phillips-Lawaetz model.

5.42   C.K. Kim, A. Feldman, D. Horowitz and R.M. Waxler, Solid State Comm., **25**, 397, 1978.

5.43   J. Schreuer and S. Haussühl, J. Phys. D: Appl. Phys.,**32**, 1263, 1999.

5.44   J.S. Zheludev and and A.A. Fotchenkov , Sov. Phys. Crystallogr., **3**, 312, 1958.

5.45   H. Burkard, W. Kanzig and M. Rossinelli, Helv. Phys. Acta, **49**, 13,1976.

5.46   L. Bohaty and S. Haussühl, Acta Cryst., **A33**, 114, 1977.

5.47   P. Preu and S. Haussühl, Solid State Commun., **45**, 619, 1983.

5.48   W. Kucharczyk, Z. Kristallogr., **176**, 319, 1986.

5.49   A. Von Hippel, Zeit. fur Physik, **75**, 145, 1932 (quoted in [5.52]).

5.50   A. Von Hippel, Zeit. fur Physik, **88**, 358, 1934 (quoted in [5.52]).

5.51   A. Von Hippel , J. Appl. Phys., **8**, 815, 1937.

5.52   H.B. Callen, Phys. Rev., **76**, 1394, 1949.

5.53   A. Von Hippel and R.J. Maurer, Phys. Rev., **59**, 820, 1941: KBr.

5.54   A.Von Hippel and G.M. Lee, Phys. Rev., **59**, 824, 1941: NaCl.

5.55   H. Frohlich and J.H. Simpson, Adv. Electronics, **2**, 185, 1950.

6 Phonon Spectra

6.1 IR Spectra

6.1.1 Transverse Optical and Longitudinal Optical Frequencies (RT)

Table 6.1 $k \sim 0$ transverse optical (ν_{TO}) and longitudinal optical (ν_{LO}) frequencies [cm^{-1}] and damping constant (γ) at room temperature

Crystal	Method A*; Ref. [6.1] (uncertainty 0.5 %)		Method B*			γ* [6.5]
	ν_{TO}	ν_{LO}	ν_{TO}	ν_{LO}	Ref.	
NaCl Structure						
LiF	305	659	320	660	[6.2]	0.053
LiCl	203	422	-	-	-	-
LiBr	173	354	-	-	-	-
LiI	142	-	-	-	-	-
NaF	246.5	421	247	419	[6.3]	0.065
NaCl	164	261	161	260	[6.2]	0.055
NaBr	134	200	-	-	-	0.067
NaI	116	180	-	-	-	0.180
KF	194	335	-	-	-	-
KCl	142	212	143	203	[6.4]	0.064
KBr	114	164	113	158	[6.4]	0.034
KI	102	141	103.5	135	[6.4]	0.048
RbF	158	290	-	-	-	-
RbCl	116.5	174	-	-	-	0.076
RbBr	87.5	126	-	-	-	0.056
RbI	75.5	104	75	99	[6.4]	0.058
CsF	127	246	-	-	-	-
CsCl Structure						
CsCl	99.5	162	-	-	-	0.096
CsBr	73.5	114	77	112	[6.2]	0.042
CsI	62	91	-	-	-	0.055

* For methods A and B and definition of γ see Notes and Comments

Notes and Comments

1. **Method A:** ν_{TO} from normal incidence thin film transmission IR spectra and ν_{LO} from Lyddane-Sachs-Teller relation: $\nu_{LO} = (\varepsilon_0 / \varepsilon_\infty)^{1/2} \, \nu_{TO}$ (where ε_0 and ε_∞ are the static and high frequency dielectric constants respectively).

Method B: ν_{TO} and ν_{LO} from Kramers-Kronig (KK) analysis or damped oscillator dispersion analysis of IR reflectivity curves.

Definition of γ: The dimensionless damping constant γ appears in the equation for the dielectric function $\varepsilon(\omega)$:

$$\varepsilon(\omega) = \varepsilon_\infty + \frac{\varepsilon_0 - \varepsilon_\infty}{1 - (\omega/\omega_0)^2 - i(\omega/\omega_0)\gamma} \tag{6.1}$$

where $\omega_0 = 2\pi\nu_{\text{TO}}$. Details of Kramers-Kronig analysis, damped oscillator analysis and determination of damping constant from reflectivity are given by Martin [6.5] and Mitra [6.6].

2. Data on IR transmission spectra for some of the alkali halides are reported in [6.7–6.12].
3. For a diatomic lattice, for every wave vector, there are six modes of vibration. These are an LO and two TO modes and an LA and two TA modes. The two TO and the two TA modes are degenerate, thus leaving four branches. At $k = 0$, the two acoustic modes (TA and LA)→0. Further, in first order, the LO mode is inactive. Thus in the IR transmission spectrum, only one line is observed corresponding to $k \to 0$, TO. However, the LO mode contributes to the reflectivity and is hence derived from KK analysis. Further, the LO mode shows up in oblique incidence IR transmission curves of thin films (see [6.2, 6.13]); ν_{TO} and ν_{LO} values can also be determined from hyper-Raman scattering spectra and neutron inelastic scattering (see relevant sections).
4. Plendl [6.14] has shown that ν_{TO} decreases with increasing values of interionic distance r and the ν_{TO} vs r plot is a smooth curve.
5. Mitra and Marshall [6.15] showed that for the alkali halides with NaCl structure there is a linear relationship between ν_{TO} and the parameter $(r/\mu\psi)$ where μ is the reduced mass and ψ the compressibility. A similar linear plot for alkali halides with CsCl structure including the high pressure phases of KCl and KBr was obtained by Postmus et al. [6.16].
6. Szigeti [6.17] derived the following formula connecting the transverse optic mode frequency ν_{TO} and the compressibility ψ:

$$\nu_{\text{TO}}^2 = \left(\frac{\varepsilon_\infty + 2}{\varepsilon_0 + 2}\right)\left(\frac{6r}{\mu\psi}\right) \tag{6.2}$$

Hass [6.8] found that the ν_{TO} values calculated for KCl and NaCl from this formula agree fairly with experimental values. Mitra and Marshall [6.15] calculated ψ for several alkali halides from the formula using experimental values of ν_{TO} and obtained fair agreement with experimental values of ψ.

7. Szigeti [6.17] also derived the following equation for the effective ionic charge (e^*) in terms of ν_{TO}:

$$(e^*/e)^2 = (9\pi\mu\nu_{\text{TO}}^2 / N_\text{v}\, e^2)[(\varepsilon_0 - \varepsilon_\infty)/(\varepsilon_\infty + 2)^2] \tag{6.3}$$

where μ is the reduced mass and N_V the number of ion pairs per unit volume.

8. For ionic crystals with NaCl structure Brout [6.18] derived the relation:

$$\sum_{i=1}^{6} 4\pi^2 v_i^2(k) = (18r / \mu \, \psi) \tag{6.4}$$

This is known as the Brout sum rule. Mitra and Marshall [6.15] calculated ψ for the alkali halides at $k = 0$ using experimental values for v_{TO} and v_{LO} but found that the agreement with experimental values was not very satisfactory.

6.1.2 Temperature Variation of TO Frequencies (Low Temp.)

Table 6.2 Values of $k \sim 0$ transverse optical frequencies v_{TO} at selected low temperatures

Temp. [K] Crystal ↓	v_{TO} [cm^{-1}]; uncertainty ~ 0.5%; Ref. [6.12]			
	2	90	200	290
NaCl Structure				
LiF	318.0	316.0	310.0	305.0
LiCl	221.0	217.5	210.0	203.0
LiBr	187.0	183.5	179.0	173.0
LiI	151.5	150.0	146.0	142.0
NaF	262.0	260.0	254.0	246.5
NaCl	178.0	177.0	172.0	164.0
NaBr	146.0	143.5	140.0	134.0
NaI	123.0	121.5	118.5	115.2
KF	201.5	200.5	197.5	194.0
KCl	151.0	150.0	146.0	142.0
KBr	123.0	122.0	118.0	114.0
KI	109.5	108.0	105.5	102.0
RbF	163.0	162.0	160.0	158.0
RbCl	126.0	125.0	121.0	116.5
RbBr	94.5	93.3	90.5	87.5
RbI	81.6	80.4	77.0	75.5
CsF	134.0	132.0	130.0	127.0
CsCl Structure				
CsCl	106.5	105.0	102.0	99.5
CsBr	78.5	78.0	76.5	73.5
CsI	65.8	65.5	64.0	62.1

Notes and Comments

1. Data on v_{TO} for some of the alkali halides at low temperatures have been reported in [6.1, 6.8, 6.9].
2. Lowndes and Rastogi [6.12] determined v_{TO} from normal incidence transmission spectra of thin films.
3. The temperature variation of v_{TO} in the low temperature region is nonlinear.

6.1.3 Temperature Variation of TO and LO Frequencies (High Temp.)

The temperature variation of ν_{TO} and ν_{LO} at high temperatures is shown in Fig. 6.1.

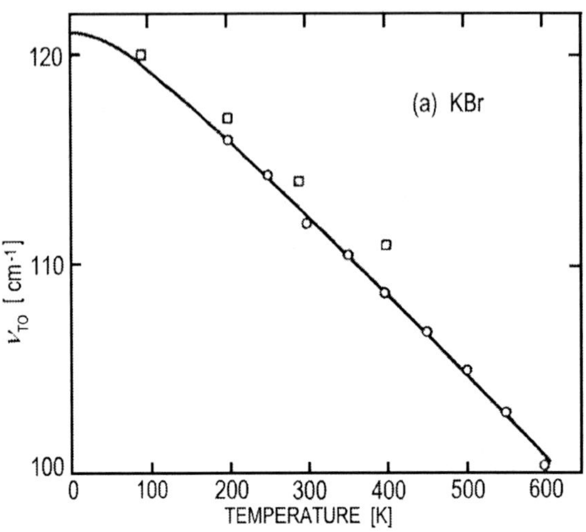

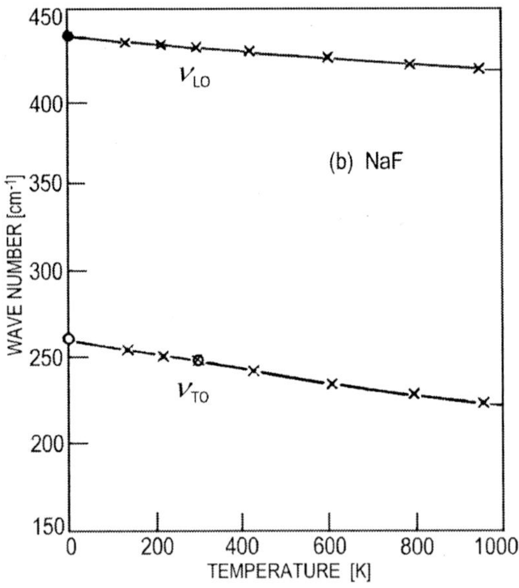

Fig 6.1 Temperature variation of **(a)** ν_{TO} of KBr and **(b)** of ν_{TO} and ν_{LO} of NaF (after [6.19]; data from [6.9, 6.20] are also included)

6.1.4 Temperature Variation of Damping Constant

Table 6.3 Values of the damping constant γ of LiF at selected temperatures [6.21]

Temp. [K]	ν_{TO} [cm^{-1}]	(γ/ν_{TO})
7.5	320	0.0100
85	315	0.0225
295	306	0.0600
420	301	0.1000
605	293	0.1700
840	282	0.2750
1060	271	0.3850

Notes and Comments

1. The parameter (γ/ν_{TO}) used by Jasperse et al. [6.21] is the same as the parameter γ used by Martin [6.5].
2. (γ/ν_{TO}) increases only slightly at lower temperatures but increases at a faster rate at temperatures above 300 K.
3. Maradudin and Wallis [6.22] theoretically proposed an equation for the temperature variation of the parameter (γ/ν) which, for $\nu = \nu_{TO}$, reduces to

$$\left[\frac{\gamma}{\nu}\right]_{\nu = \nu_{TO}} = \frac{\text{constant}}{\nu_{TO}^4}\left[\left(\exp\frac{h\nu_{TO}}{k_B T} - 1\right)^{-1} + \frac{1}{2}\right] \tag{6.5}$$

Comparison of the LiF data with this equation shows very good agreement at high temperatures but shows deviations at low temperatures.

6.1.5 Pressure Derivative of $k \sim 0$ ν_{TO} Frequency

Table 6.4 Values of the pressure derivative $(d\nu_{TO}/dP)$; uncertainty in last digit given in parenthesis [6.12]

Crystal	$(d\nu_{TO}/dP)$ [cm^{-1} (kbar)$^{-1}$]
NaCl Structure	
LiF	1.12(6)
NaF	1.06(5)
NaCl	1.54(7)
NaBr	1.49(9)
KCl	1.80(10)
KBr	1.59(7)
KI	1.88(1)
RbCl	1.63(4)
RbBr	1.63(7)

Table 6.4 (Continued)

Crystal	$\left(d\,v_{\mathrm{TO}}\,/\,dP\right)$ $[\mathrm{cm}^{-1}\,(\mathrm{kbar})^{-1}]$
NaCl Structure	
RbI	1.48(8)
CsCl Structure	
CsCl	1.70(4)
CsBr	1.26(1)
CsI	1.20(3)

Notes and Comments

1. Lowndes and Rastogi [6.12] made measurements up to 7 kbar. Over this range they observed a linear pressure variation; see Sec. 6.1.6 for data at higher pressures.
2. Lowndes and Rastogi [6.12] calculated the mode Gruneisen parameters from the pressure derivative of v_{TO}.

6.1.6 Pressure Variation of k ~ 0 v_{TO}, v_{LO} Frequencies at High Pressures

The pressure variation of v_{TO} and v_{LO} mode frequencies is shown in Figs. 6.2, 6.3.

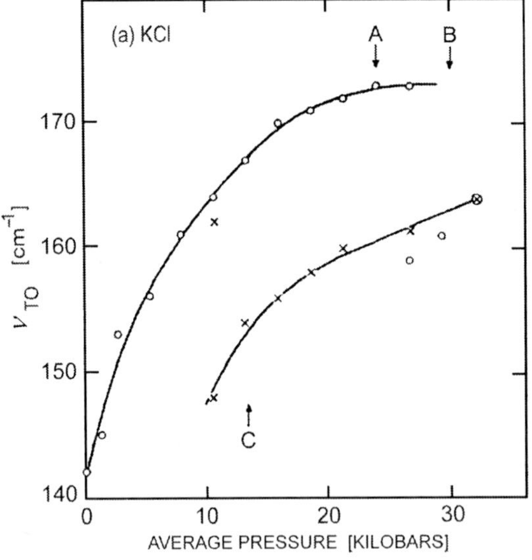

Fig. 6.2 $k \sim 0$ v_{TO} modes of **(a)** KCl and **(b)** KBr (see next page) as a function of pressure upto 35 kbar. o, increasing pressure values; ×, decreasing pressure values; A, first appearance of high pressure phase; B, complete conversion to high pressure phase; C, first appearance of low pressure phase upon release of pressure [after 6.16]

Fig. 6.2 (Continued)

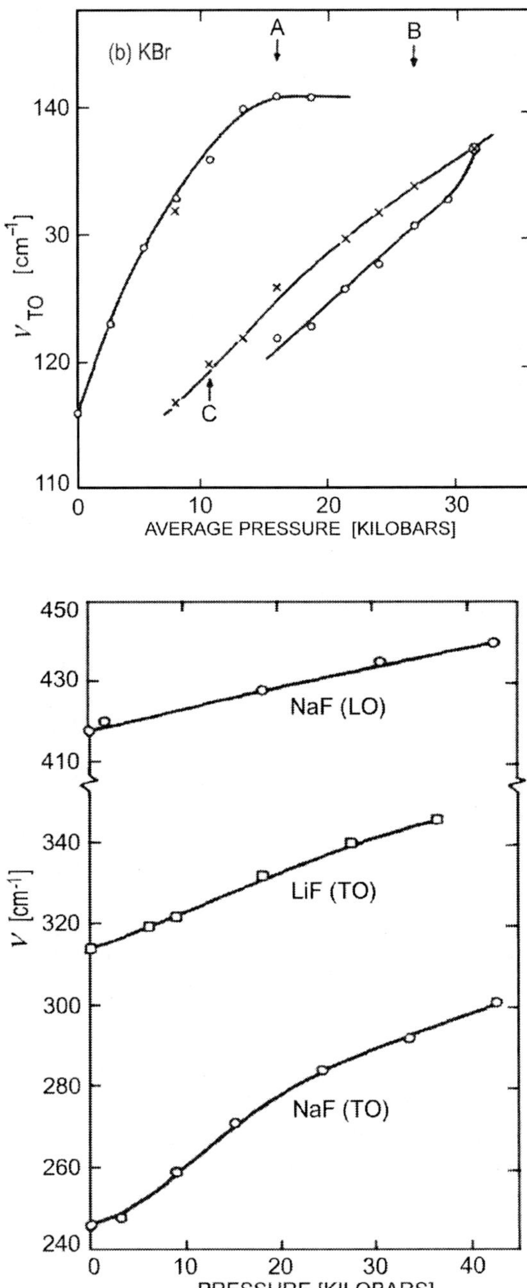

Fig. 6.3 $k \sim 0$ ν_{TO} and ν_{LO} modes of NaF and ν_{TO} mode of LiF as a function of pressure up to 40 kbar (after [6.2])

Notes and Comments

1. At the Fm3m→Pm3m transition pressures of 20 and 13.5 kbars for KCl and KBr respectively, the ν_{TO} phonon frequencies for the Fm3m and Pm3m phases are 173 and 160 cm^{-1} for KCl and 141 and 126 cm^{-1} for KBr. The ratio of the frequencies for the two phases is 1.08 and 1.12 which is close to 1.15, the square root of the ratio of the co-ordination numbers.
2. The pressure variation of ν_{TO} of KCl, KBr and NaF is nonlinear whereas it is linear for ν_{TO} of LiF and for ν_{LO} of NaF.

6.2 Raman Spectra

6.2.1 Second-Order Raman Spectra

Table 6.5 Raman frequency shifts ($\Delta \nu$)

Crystal	Raman frequency shifts $\Delta \nu$ [cm^{-1}]	Ref.
NaCl Structure		
LiCl	49, 86, 116, 128, 147, 159, 168, 207, 227, 274, 292, 298, 307, 337, 357, 375, 405, 432, 444, 453, 472, 498, 522, 540, 558, 618	[6.23]
NaCl	85, 135, 140, 162, 184, 199, 202, 220, 235, 258, 270, 276, 286, 300, 314, 320, 326, 343, 350	[6.24]
NaBr	31, 64, 116, 152, 181, 254	[6.25]
NaI	19, 42, 58, 88, 103, 120, 132, 200, (200 – 250), (310–370)	[6.23]
KCl	122, 212, 237, 257, 331, 349	[6.25]
KBr	46, 85, 126, 146, 170, 186, 216, 228, 232, 242	[6.24]
KI	63, 89, 101, 102, 175, 210, 252	[6.24]
RbI	83.5, 125, 136.5, 154	[6.26]
CsCl Structure		
CsBr	79, 107, 135, 155, 190	[6.24]
CsI	19, 22, 44, 61, 91, 94, 106, 110, 124, 137, 155, 181	[6.23]

Notes and Comments

1. With the exception of RbI, the spectral data given in Table 6.5 have been obtained by the photographic-densitometric method.
2. A good discussion of Raman spectra of solids is given by Menzies [6.27].

6.2.2 Second-Order Laser Raman Spectra

Table 6.6 Second-order laser Raman spectra ; frequency shifts Δv [cm^{-1}] given along with the assignments in terms of the principal phonon modes (see Notes and Comments); [6.26]

NaCl		KBr		KI	
Δv	Assignment	Δv	Assignment	Δv	Assignment
55	LA $-$ TA (X)	46.5	TO $-$ LA (X)	62	2 TA (X)
60.5	$W_3^0 - W_3^A$	61	LO $-$ LA (X)	70	TO $-$ TA (X)
87	TO $-$ TA (X)	76	TO $-$ TA (X)	77	$W_3^0 - W_3^A$
104	LO $-$ TO (Γ)	86	2 TA (X)	90	$W_1 + W_3^A$
174	2 TA (X)	116	LA $+$ TA (X)	102	2 LA (X)
233	LA $+$ TA (X)	138	2 TA (L)	129	2 LA (L)
248	2 TA {XY}	145	2 LA (X)	136	$W_1 + W_2'$
251	$W_1 + W_3$	150	$W_1 + W_3^A$	53	TO $+$ LA (X)
258	TO $+$ TA (X)	165	LA $+$ TA (L)	166	$W_1 + W_3^0$
273	$W_2' + W_3^A$	193	$W_3^0 + W_3^A$	186	2 TO (L)
280	2 TO (L)	200	$W_2' + W_1$	205	2 TO (X)
286	2 LA (X)	208	LO $+$ LA (X)	216	2 LO (X)
316	TO $+$ LA (X)	221	2 W_2'	218	TO$-$LO (L)
333	$W_2' + W_3^0$	235	2TO (X)	229–232	2 W_3^0
350	$2W_3^0$	251	LO $+$ TO (X)	250	2 LO (L)
360	2 LO (X)	259	2 W_3^0	270–300	2 LO (Γ)
530	2 LO (Γ)	316–336	2 LO (Γ)		

Notes and Comments

1. The principal phonon frequencies corresponding to the points in the reduced Brillouin zone are given in Table XIV

Table XIV Principal phonon frequencies [6.26]

Crystal	v [cm^{-1}]							
	Γ		L		X		W	
NaCl	LO	264	LO	226	LO	182.5	W_3^0	174
	TO	162	TO	140	TO	174	W_2'	158.5
			LA	173	LA	142.5	W_1	138
			TA	118	TA	87.5	W_3^A	115
KBr	LO	163	LO	144	LO	133.5	W_3^0	130.5
	TO	113	TO	96.5	TO	118	W_2'	111

Table XIV (Continued)

Crystal	$\nu\,[\mathrm{cm}^{-1}]$							
	Γ		L		X		W	
KBr			LA	92	LA	73	W_1	89
			TA	70	TA	42	$W_3{}^A$	61
KI	LO	139	LO	125	LO	108	$W_3{}^0$	116
	TO	101	TO	93	TO	102.5	$W_2{}'$	86
			LA	64.5	LA	51	W_1	51
			TA	51.5	TA	31	$W_3{}^A$	39

6.2.3 Second-Order Raman Spectra of Alkali Fluorides (Expt. and Theor.)

Experimentally observed and theoretically calculated second-order Raman spectra
of alkali fluorides are given in Fig. 6.4.

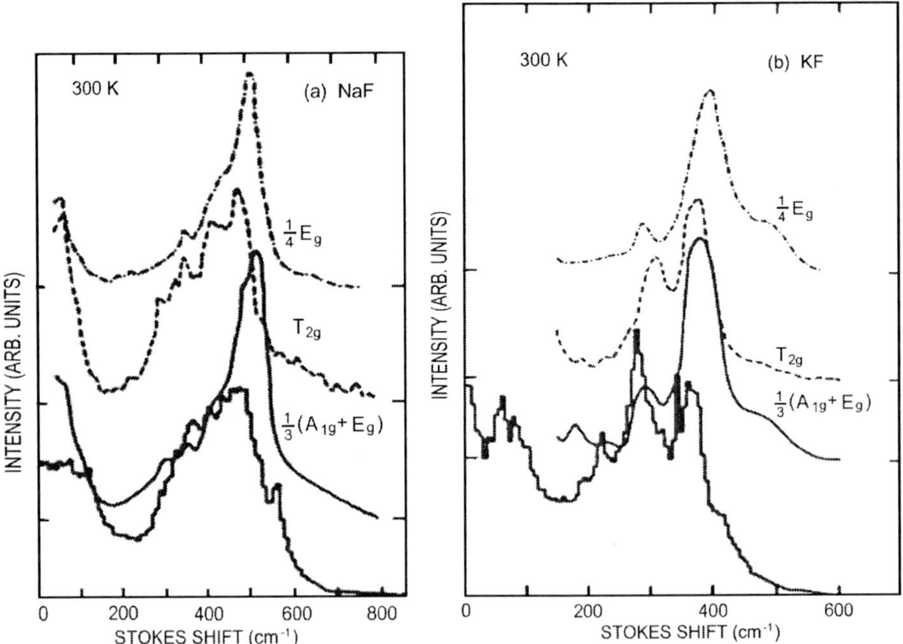

Fig. 6.4 Experimental and theoretical Raman spectra of **(a)** NaF, **(b)** KF, **(c)** RbF and **(d)**
CsF (after [6.28]). In these spectra, the experimental spectra (continuous and dotted lines)
are compared with the theoretical spectrum (histogram) calculated from the theory of Born
and Bradburn [6.29]

Fig. 6.4 (Continued)

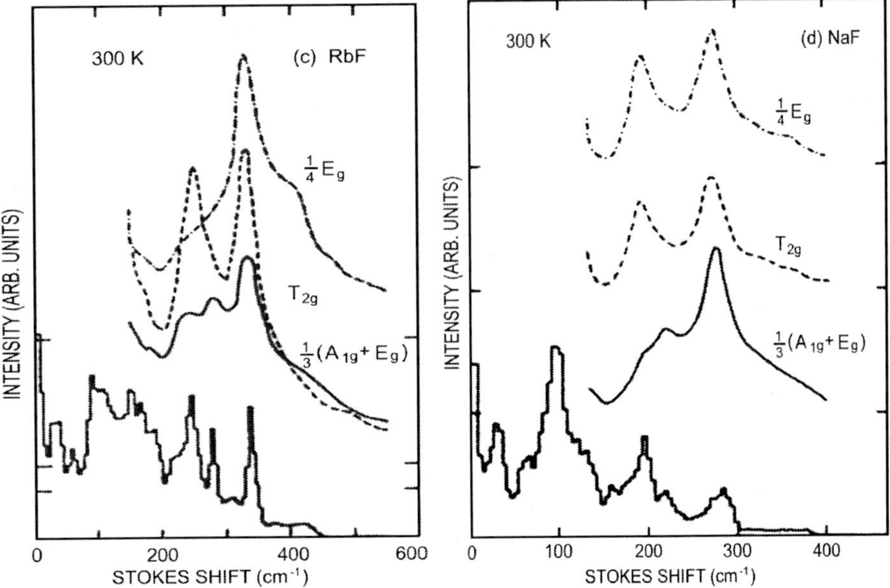

Notes and Comments

1. Cunningham et al [6.28] state that in their work they "calculate the measured spectra rather than reduce the spectra to spectral components".
2. For additional information, see [6.30, 6.31].

6.2.4 Hyper-Raman Spectra (HRS) and Electric Field Induced First Order Raman Spectra (EFIRS)

Table 6.7 Values of the frequencies (ν_{TO}, ν_{LO}) and damping constant (γ) corresponding to the hyper-Raman spectra (HRS) and of frequencies (ν_{TO}, ν_{LO}) and the scattering cross-section corresponding to the electric field induced first-order Raman spectra (EFIRS)

Crystal	HRS [6.32]			EFIRS [6.33]			
	ν_{TO} [cm^{-1}]	γ [cm^{-1}]	ν_{LO} [cm^{-1}]	ν_{TO} [cm^{-1}]	ν_{LO} [cm^{-1}]	EFIRS scattering cross-section [10^{-37} cm^2 sr^{-1}]	
						TO	LO
NaCl Structure							
NaCl	164.7	5.0	270	-	-	-	-
KCl	142.5	5.1	212	151	212	32	15
KBr	114.8	3.8	166	119	-	46	-
KI	100.8	4.5	138	-	-	-	-
RbI	75.9	2.7	106	-	-	-	-

Notes and Comments

1. The values of ν_{TO} and ν_{LO} observed in HRS and EFIRS are close to those obtained from IR spectra. However, the ν_{LO} phonons in HRS are broad and have a structure.
2. For constant laser intensity, the intensity of HR TO phonons increases in the sequence chloride-bromide-iodide i.e. HR efficiency increases with decreasing energy gap.

6.3 Neutron Inelastic Scattering

6.3.1 Phonon Dispersion Relations from Neutron Inelastic Scattering

The phonon dispersion relations obtained from neutron inelastic scattering are given in Fig. 6.5

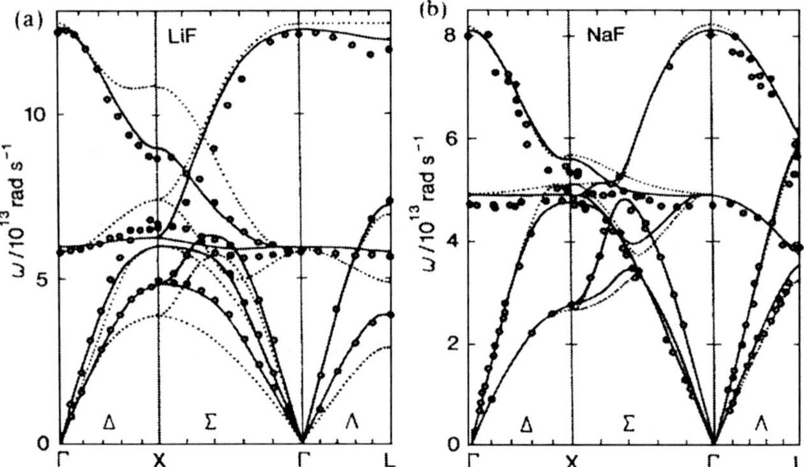

Fig. 6.5 Phonon dispersion relations; experimental data obtained from neutron inelastic scattering denoted by open and full circles or triangles; frequencies ω in 10^{13} rad s^{-1} (THz in the case of CsCl). Continuous curves are from model calculations (not discussed here). Γ point is the centre of the zone, X and L are points at $[(2\pi/a)\,(1,0,0)]$ and $[(2\pi/a)\,(\frac{1}{2},\frac{1}{2},\frac{1}{2})]$ in the Brillouin zone; Δ, Σ and Λ represent points along <100>, <110> and <111> directions; (a–m, after [6.34]; n–q, after [6.35])

Fig. 6.5 (Continued)

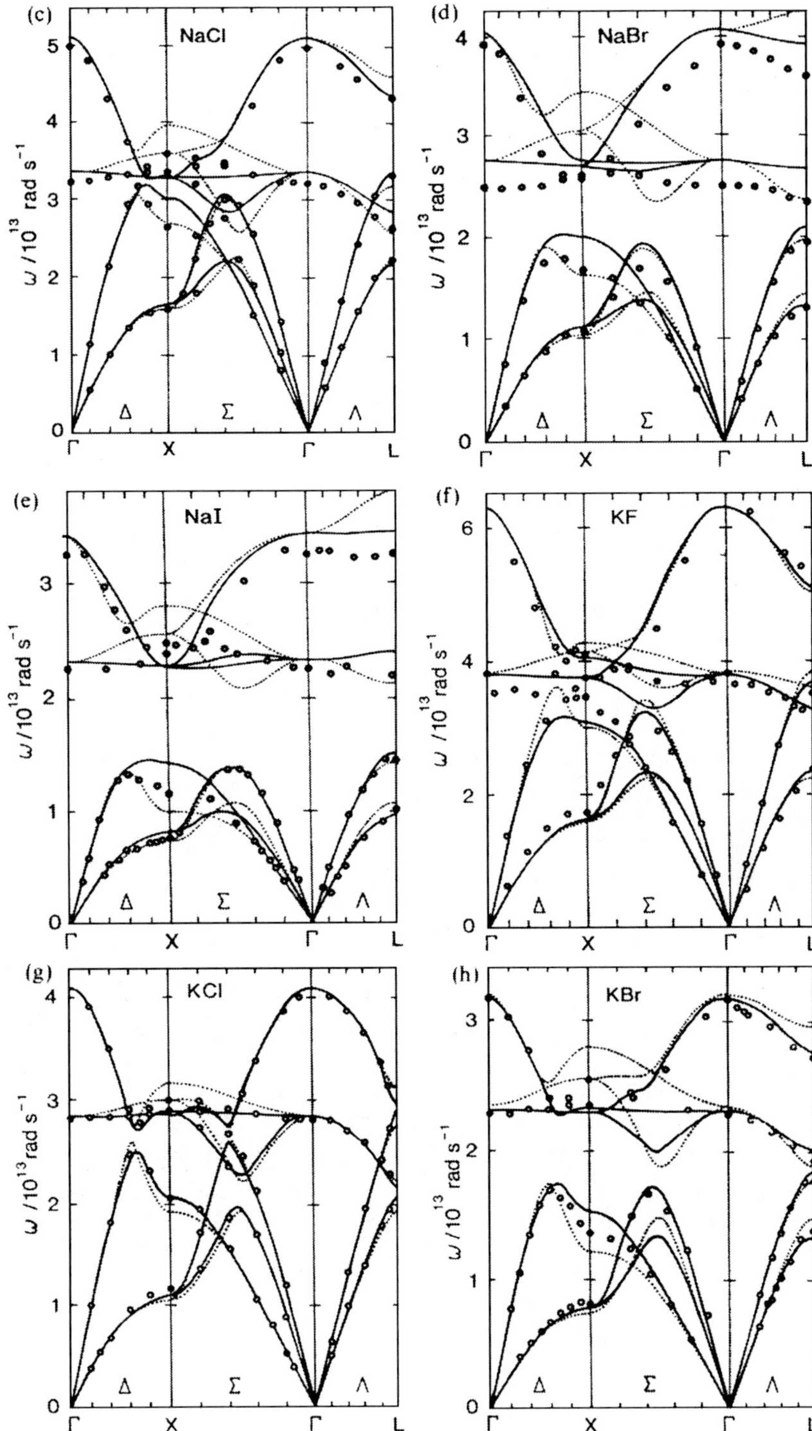

Fig. 6.5 (Continued)

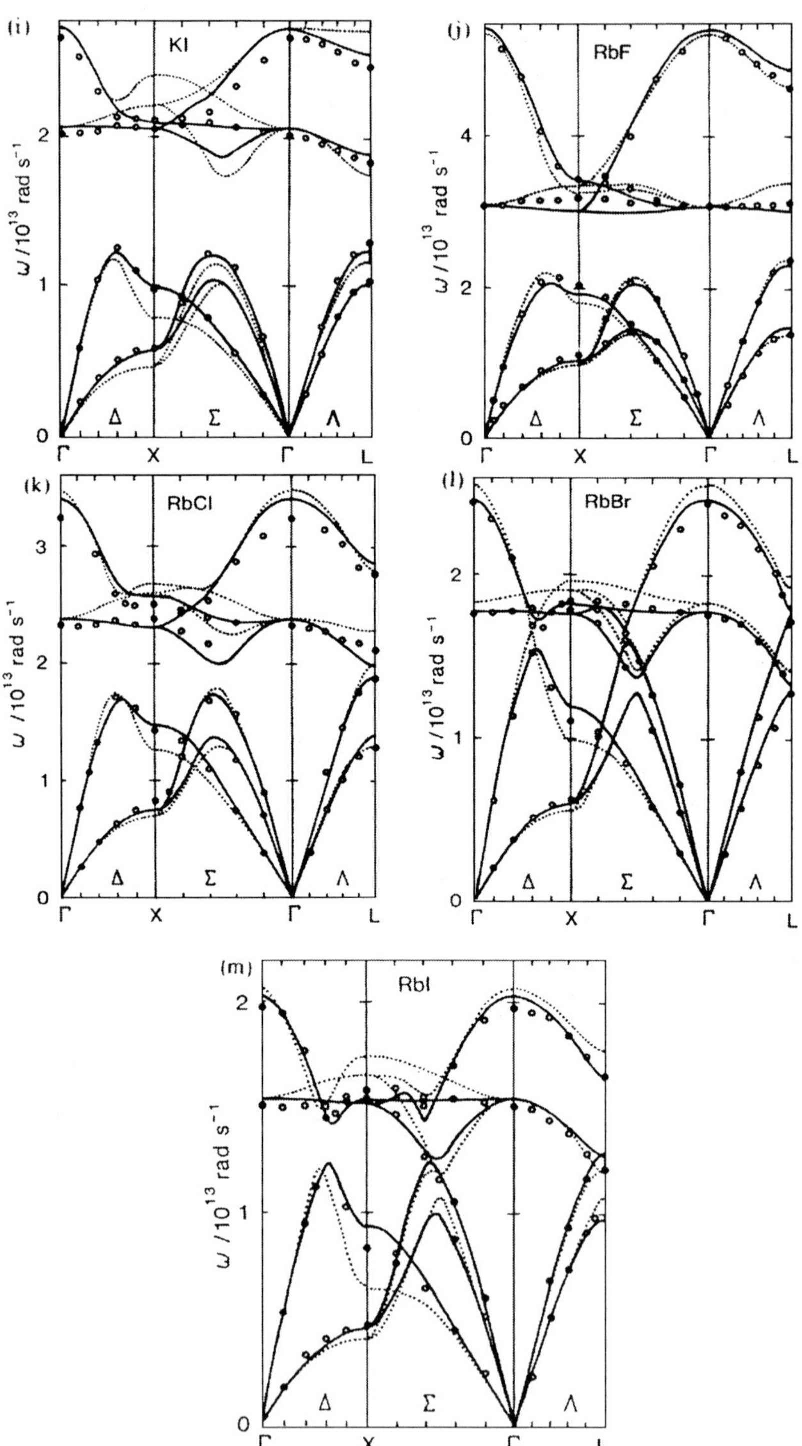

Fig. 6.5 (Continued)

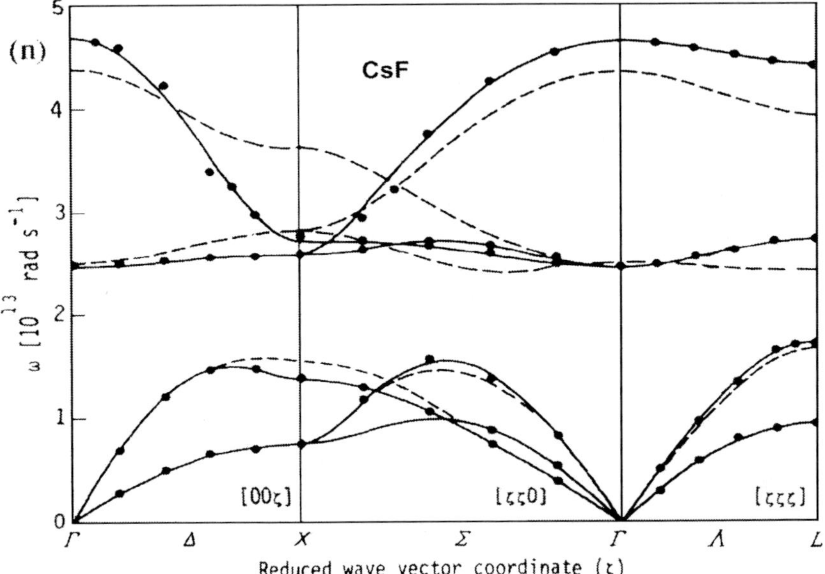

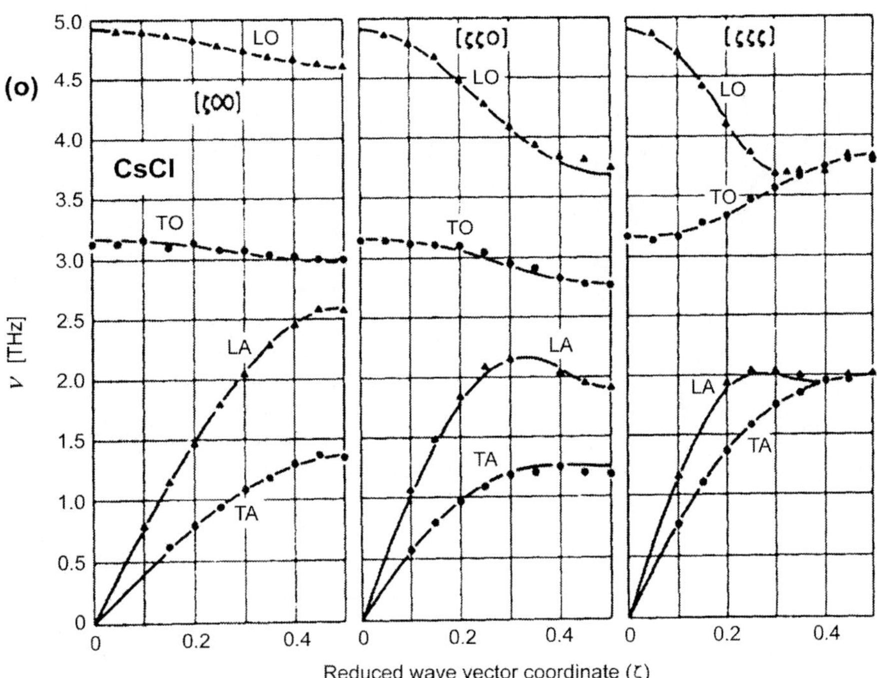

Fig. 6.5 (Continued)

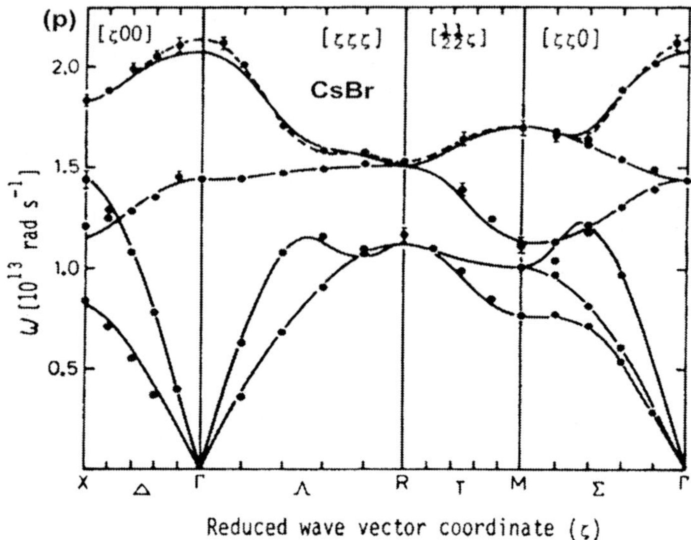

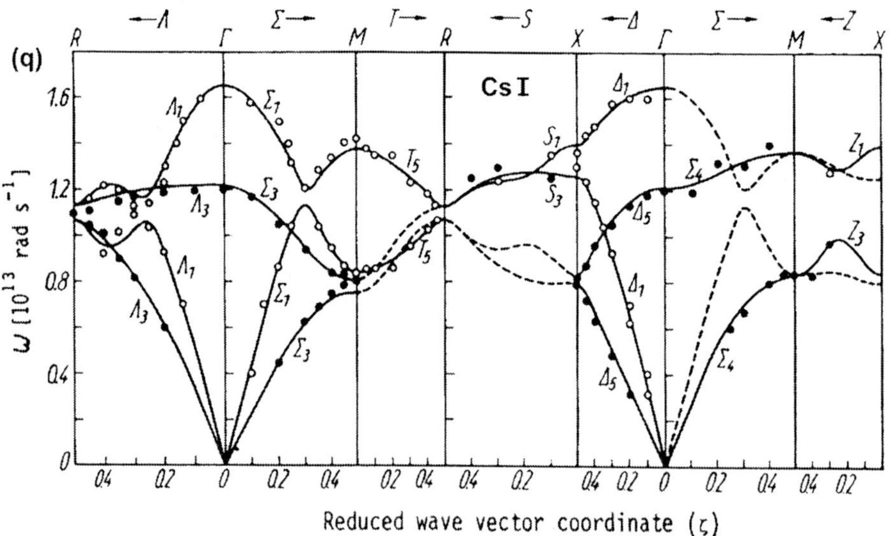

6.3.2 Phonon Frequencies at Zone Centre and Zone Boundary

Table 6.8 Phonon frequencies at zone center ($q = 0$) and zone boundary ($q = 2\pi/a$) measured along <001> direction determined from neutron inelastic scattering ; [6.35] for Cs halides and [6.34] for other crystals

Crystal	Temp. [K]	Frequencies in cm^{-1}; uncertainty $1 - 3\,\%$					
		Wave vector $q = 0$		Wave vector $q = (2\pi/a)\,(0,0,1)$			
		ν_{TO}	ν_{LO}	ν_{TO}	ν_{LO}	ν_{TA}	ν_{LA}
NaCl Structure							
LiF	300	305	657	345	457	257	350
NaF	295	251	422	-	284	146	265
NaCl	80	173	264	179	192	87	142
	300	-	-	173	186	85	143
NaBr	295	135	207	139**	141**	56.7**	97**
NaI	100	120	170	127	130	41.4	62.4
KF	300	202	316*	212*	218	91	185
KCl	80	149	218*	154	159	62	109
	300	141	-	147	151	58	109
KBr	90	120	167	124	134	41.9	71.7
KI	95	107	142	109	113	31.4	52.0
RbF	80	163	286	169	182	57.8	107
	300	156	-	-	180	57.3	105
RbCl	80	124	173	127	134	43.5	75.9
	300	118	176	121	128	40.9	76.4
RbBr	80	94	130	96	99	33.2	59.4
	300	89	-	-	95	32.1	59.9
RbI	80	80	105	82	84	24.6	44.0
	300	75	100	78	80	23.7	44.1
CsF	80	103*	249*	138	148	40.5	73.7
	300	-	-	133	141	-	-
CsCl Structure*							
CsCl	78	106	164*	127	127	67.3	67.3
	298	100	-	124	124	65.7	65.7
CsBr	80	76	114*	64	97	44.8	76.4
	300	74	-	58.4	93	41.6	–
CsI	RT	63.7	87	43.5	72.2	41.9	70.7

* Extrapolated

** These values are at $q = (2\pi/a)\,(0,0,0.8)$

*** In these crystals, instead of zone boundary values, values at $q = (2\pi/a)\,(0,0,0.5)$ are given.

References

6.1 ε_0 and ε_∞ (needed for calculation of ν_{LO} and ν_{TO} from R.P. Lowndes and D.H. Martin, Proc. Roy. Soc. Lond., **A308**, 473, 1969.

6.2   E. Burstein in *Lattice Dynamics*, Edited by R.F. Wallis, Pergamon Press, New York, 1965.

6.3   S.S. Mitra, C. Postmus and J.R. Ferraro, Phys. Rev. Lett., **18**, BV502, 1967.

6.4   J.H. Fertel and C.H. Perry, Phys. Rev., **184**, 874, 1969.

6.5   D.H. Martin, Adv. Phys., **14**, 39, 1965.

6.6   S.S. Mitra, *Optical Properties of Solids*, Plenum Press, New York, 1969.

6.7   R.B. Barnes, Z. Phys., **75**, 723, 1932.

6.8   M. Hass, Phys. Rev., **119**, 633, 1960.

6.9   G.O. Jones, D.H. Martin, P.A. Mawer and C.H. Perry, Proc. Roy. Soc. Lond., **A261**, 10, 1961

6.10   M. Hass, J. Phys. Chem. Solids, **24**, 1159, 1963.

6.11   C. M. Randall, R.M. Fuller and D.J. Montgomery, Solid State Commun., **2**, 273, 1964 .

6.12   R.P. Lowndes and A. Rastogi, Phys. Rev., **B14**, 3598, 1976.

6.13   D.W. Berreman, Phys. Rev.,**130**, 2193, 1963.

6.14   J.N. Plendl, Applied Optics, **9**, 2768, 1970.

6.15   S.S. Mitra and R. Marshall, J. Chem. Phys., **41**, 3158, 1964.

6.16   C. Postmus, J.R. Ferraro and S.S. Mitra, Phys. Rev., **174**, 983, 1968.

6.17   B. Szigeti, Trans. Faraday Soc., 45, 155, 1949; Proc. Roy. Soc. Lond., **A204**, 51, 1950.

6.18   R. Brout, Phys. Rev., **113**, 43, 1959.

6.19   I. F. Chang and S.S. Mitra, Phys. Rev., **B5**, 4094, 1972.

6.20   J. E. Mooij, Phys. Lett., **24A**, 249, 1967.

6.21   J.R. Jasperse, A. Kahan, J.N. Plendl and S.S. Mitra, Phys. Rev., **146**, 526, 1966.

6.22   A.A. Maradudin and R.F. Wallis, Phys. Rev., **125**, 4, 1962.

6.23   R.S. Krishnan in *Lattice Dynamics*, Edited by R.F. Wallis, Pergamon Press, New York, 1965.

6.24   R.S. Krishnan and P.S. Narayanan, Proc. Ind. Acad. Sci., **39**, 85, 1957.

6.25   A.C. Menzies and J. Skinner, J. Phys. Radium, **9**, 93, 1948.

6.26   M. Krauzman, *Light Scattering Spectra in Solids*, Ed. G.B. Wright, Springer-Verlag, New York, 1969.

6.27   A.C. Menzies, Reports on Progress in Physics, **16**, 83, 1953.

6.28   S.L. Cunningham, T.P. Sharma, S.S. Jaswal, M. Hass and J.R. Hardy, Phys. Rev., **B10**, 3500, 1974.

6.29   M. Born and M. Bradburn, Proc. Roy. Soc. Lond., **A188**, 161, 1947.

6.30   M. Krauzman, Solid State Comm., **12**, 157, 1973.

6.31   A. Pasternak, E. Cohen and G. Gilat, Phys. Rev., **B9**, 4584, 1974.

6.32   H. Vogt and H. Presting, Phys. Rev., **B31**, 6731, 1985.

6.33   M. Kranz and F. Luty, Phys. Rev., **B31**, 2599, 1985.

6.34   K. Motida, J. Phys. Soc. Japan, **55**, 1636, 1986 and references therein.

6.35   H. Bilz and W. Kress, in *Phonon Dispersion Relations in Insulators*, Springer-Verlag, Berlin,1979 and references therein.

7 Chemical-Bond-Related Parameters

7.1 Lattice Energy

7.1.1 Interatomic Potentials and Expressions for Theoretical Lattice Energy

The main potentials in use are [7.1, 7.2]:

$$\phi = -\frac{\alpha_r\,e^2}{r} + B\,\exp\left(-\frac{r}{\rho}\right) \tag{7.1}$$

$$\phi = -\frac{\alpha_r\,e^2}{r} + \frac{A}{r^n} \tag{7.2}$$

$$\phi = -\frac{\alpha_r\,e^2}{r} + B\,\exp\left(-\frac{r}{\rho}\right) - \frac{C}{r^6} - \frac{D}{r^8} \tag{7.3}$$

Use of either Hildebrand equation of state or Mie-Gruneisen equation of state leads to the following expressions for lattice energy U_{calc} :

Static Crystal

$$U_{\mathrm{calc}} = N\left[-\frac{\alpha_r\,e^2}{r}\left(1-\frac{1}{r/\rho}\right)\right] \tag{7.4}$$

Hildebrand

$$U_{\mathrm{calc}} = N\left[-\frac{\alpha_r\,e^2}{r}\left(1-\frac{1}{r/\rho}\right) - \frac{3V_{\mathrm{m}}T\beta}{\psi\,r/\rho} + \phi_0\right] \tag{7.5}$$

$$U_{\mathrm{calc}} = N\left[-\frac{\alpha_r\,e^2}{r}\left(1-\frac{1}{r/\rho}\right) - \frac{3V_{\mathrm{m}}T\beta}{\psi\,r/\rho} - \frac{C}{r^6}\left(1-\frac{6}{r/\rho}\right) - \frac{D}{r^8}\left(1-\frac{8}{r/\rho}\right) + \phi_0\right] \tag{7.6}$$

$$U_{\mathrm{calc}} = N\left[-\frac{\alpha_r\,e^2}{r}\left(1-\frac{1}{n}\right) - \frac{3V_{\mathrm{m}}T\beta}{\psi\,r/\rho} - \frac{C}{r^6}\left(1-\frac{6}{n}\right) - \frac{D}{r^8}\left(1-\frac{8}{n}\right) + \phi_0\right] \tag{7.7}$$

Mie-Gruneisen

$$U_{\mathrm{calc}} = N\left[-\frac{\alpha_r\,e^2}{r}\left(1-\frac{1}{r/\rho}\right) - \frac{3V_{\mathrm{m}}\beta\,W_{\mathrm{vib}}}{\psi\,r/\rho\,C_V} - \frac{C}{r^6}\left(1-\frac{6}{r/\rho}\right) - \frac{D}{r^8}\left(1-\frac{8}{r/\rho}\right) + \phi_0\right] \tag{7.8}$$

Huggins-Mayer Form

$$U_{\text{calc}} = N\left[-\frac{\alpha_r\, e^2}{r} - \frac{C}{r^6} - \frac{D}{r^8} + M'\,\beta_{+\,-}\,b\,\exp\left(\frac{r_+ + r_- - r}{\rho}\right)\right.$$

$$\left.+\frac{1}{2}M''\,b\left\{\beta_{+\,+}\,\exp\left(\frac{2r_+}{\rho}\right) + \beta_{-\,-}\,\exp\left(\frac{2r_-}{\rho}\right)\right\}\exp\left(\frac{-a'\,r}{\rho}\right) + \phi_0\right] \tag{7.9}$$

Here,

α_r = Madelung constant

e = Electronic charge

r = Interionic distance

C = Dipole-dipole interaction constant

D = Dipole-quadrupole interaction constant

$\beta_{+\,-}$, $\beta_{+\,+}$ and $\beta_{-\,-}$ = Pauling's overlap constants

b = Repulsion constant

r_+, r_- = Basic radii

ρ = Repulsion parameter; hardness parameter

a' = Ratio of next-nearest to nearest neighbour distance

M', M'' = Coordination numbers of unlike and like ions

ϕ_0 = Zero point energy = $(9/4)\,k_B\,\theta_D$

θ_D = Debye temperature

k_B = Boltzmann constant

V_m = Molar volume

T = Temperature

β = Volume coefficient of thermal expansion

ψ = Compressibility

n = Born index in repulsion term

C_V = Specific heat

W_{vib} = Vibrational energy

N = Avogadro number

7.1.2 Structural Parameters in Lattice Energy Formulae for Alkali Halides

Table 7.1 Values of the structural parameters for the alkali halides (RX) occurring in Eq. (7.9); Ref. [7.2]

Crystals Parameter ↓	LiX	NaX, KX, RbX, CsF	CsCl, CsBr, CsI
α_r	1.7476	1.7476	1.7626
$\beta_{+\,-}$	1.375	1.00	1.00
$\beta_{+\,+}$	2.00	1.25	1.25
$\beta_{-\,-}$	0.75	0.75	0.75
M'	6	6	8
M''	8	8	6
a'	1.414	1.414	1.155

7.1.3 Van der Waal Constants

Table 7.2 Values of the Van der Waal constants C [10^{60} erg cm^6] and D [10^{76} erg cm^8]

Crystal	C	D	C	D
	[7.3]		[7.4]	
NaCl Structure				
LiF	18	11	81	243
LiCl	113	104	265	736
LiBr	183	190	359	1002
LiI	363	470	504	1389
NaF	46	31	176	343
NaCl	180	180	455	996
NaBr	271	300	580	1339
NaI	482	630	763	1817
KF	167	150	514	1084
KCl	452	560	1060	2619
KBr	605	800	1274	3325
KI	924	1420	1566	4235
RbF	278	290	772	1690
RbCl	691	960	1485	3850
RbBr	898	1340	1757	4810
RbI	1330	2240	2119	6024
CsF	495	600	1189	2647
CsCl Structure				
CsCl	1530	2600	3294	9034
CsBr	2070	3600	3807	11020
CsI	2970	5800	4496	13588

Notes and Comments

1. In most of the work on lattice energies of alkali halides (including Tosi [7.1]) the Mayer values for C and D have been used. Ladd [7.4] recalculated the C and D values and revised the lattice energy values using his values of C and D. Though there is a large difference between the two sets of Van der Waal constants given in Table 7.2, the lattice energies calculated with the two sets agree within 1 % [7.1, 7.4].
2. Murti and Selvarajan [7.5] obtained the dipole-dipole constant C by a novel procedure of fitting expressions to lattice dynamical formulae for elastic constants but obtained a range of values for each crystal.

7.1.4 Born Repulsion Parameters

Table 7.3 Values of the Born repulsion parameters; for each salt, the first four columns report the values of the parameters of the single exponential form for the Born repulsive energy: the upper value is the pre-exponential parameter B (in 10^{-8} erg molecule^{-1}) and the lower number is the hardness parameter ρ (in 10^{-8} cm). The fifth column reports the Born repulsive energy of Eq. (7.2) (in 10^{-12} erg molecule^{-1}) and the parameter n. Equation numbers as in Sec. 7.1.1

Crystal	Born repulsion parameters [7.1]				
	Eq. (7.4)	Eq. (7.5)	Eq (7.6)	Eq. (7.8)	Eq.(7.7)
NaCl Structure					
LiF	0.296	0.351	0.391	0.341	3.40
	0.291	0.280	0.280	0.285	6.20
LiCl	0.490	0.836	0.843	0.814	2.38
	0.330	0.303	0.310	0.310	7.30
LiBr	0.591	1.23	1.14	1.11	2.13
	0.340	0.305	0.316	0.316	7.71
LiI	0.599	0.537	0.583	0.490	2.23
	0.366	0.366	0.375	0.384	7.00
NaF	0.641	0.371	0.413	0.335	2.88
	0.290	0.312	0.313	0.322	6.41
NaCl	1.05	2.39	1.99	2.02	1.88
	0.321	0.288	0.301	0.300	8.38
NaBr	1.33	2.07	1.72	1.69	1.82
	0.328	0.308	0.322	0.323	8.27
NaI	1.58	1.56	1.31	1.26	1.76
	0.345	0.341	0.358	0.360	8.03
KF	1.31	0.918	0.903	0.811	2.33
	0.298	0.310	0.319	0.323	7.39
KCl	2.05	3.04	2.21	2.19	1.76
	0.326	0.309	0.330	0.330	8.55
KBr	2.30	5.40	3.34	3.36	1.60
	0.336	0.303	0.328	0.328	9.05
KI	2.85	6.44	3.61	3.60	1.47
	0.348	0.317	0.346	0.346	9.21
RbF	1.78	2.20	1.76	1.71	2.12
	0.301	0.291	0.308	0.309	8.14
RbCl	3.19	10.7	5.31	5.40	1.59
	0.323	0.283	0.313	0.312	9.52
RbBr	3.03	9.71	4.78	4.81	1.53
	0.338	0.298	0.330	0.329	9.45
RbI	3.99	21.7	7.91	7.96	1.36
	0.348	0.293	0.332	0.331	10.07
CsF	4.89	32.7	11.3	12.2	1.67
	0.285	0.236	0.268	0.266	10.22

Table 7.3 (Continued)

Crystal	Born repulsion parameters [7.1]				
	Eq. (7.4)	Eq. (7.5)	Eq (7.6)	Eq. (7.8)	Eq.(7.7)
CsCl Structure					
CsCl	15.3	81.0	14.8	14.7	1.41
	0.298	0.256	0.307	0.307	10.65
CsBr	21.3	76.5	12.6	12.5	1.41
	0.300	0.267	0.324	0.324	10.49
CsI	25.6	199.0	20.4	20.3	1.27
	0.312	0.263	0.328	0.328	11.07

7.1.5 Parameters of the Huggins-Mayer form

Table 7.4 Values of the parameters r_+ and r_- and ρ in the Huggins-Mayer form (Eq. 7.9); the parameter $b = 0.338 \times 10^{-12}$ erg /molecule

Crystal	r_+ [Å]	r_- [Å]	ρ [Å]	Ref.
NaCl Structure				
LiF	0.816	1.179	0.299	[7.6]
LiCl	0.816	1.585	0.342	[7.6]
LiBr	0.816	1.716	0.353	[7.6]
LiI	0.816	1.907	0.430	[7.6]
NaF	1.170	1.179	0.330	[7.6]
NaCl	1.170	1.585	0.317	[7.6]
NaBr	1.170	1.716	0.340	[7.6]
NaI	1.170	1.907	0.386	[7.6]
KF	1.463	1.179	0.338	[7.6]
KCl	1.463	1.585	0.337	[7.6]
KBr	1.463	1.716	0.335	[7.6]
KI	1.463	1.907	0.355	[7.6]
RbF	1.587	1.179	0.328	[7.6]
RbCl	1.587	1.585	0.318	[7.6]
RbBr	1.587	1.716	0.335	[7.6]
RbI	1.587	1.907	0.337	[7.6]
CsF	1.720	1.179	0.282	[7.6]
CsCl Structure				
CsCl	1.517	1.450	0.318	[7.2]
CsBr	1.517	1.574	0.325	[7.2]
CsI	1.517	1.757	0.332	[7.2]

7.1.6 Lattice Energy (Theoretical)

Table 7.5 Values of lattice energy (U_{calc}) from different expressions; equation numbers for U_{calc} as in Sec. 7.1.1

Crystal	$-U_{calc}$ [kcal mole^{-1}]; [7.1]					
	Eq. (7.4)	Eq. (7.5)	Eq (7.6)	Eq. (7.7)	Eq. (7.8)	Eq. (7.9)
NaCl Structure						
LiF	242.2	245.4	246.1	239.3	246.4	245.3
LiCl	192.9	197.1	198.1	194.0	198.6	198.0
LiBr	181.0	185.7	186.9	183.3	187.3	186.3
LiI	166.1	168.5	169.8	165.7	169.5	168.4
NaF	215.2	214.8	215.5	210.0	214.8	215.7
NaCl	178.6	182.6	183.7	180.9	184.0	182.2
NaBr	169.2	172.2	173.3	170.5	173.4	172.8
NaI	156.6	158.6	159.9	157.1	159.9	160.0
KF	189.1	189.9	191.1	187.1	190.9	190.9
KC	161.6	164.0	165.6	162.9	165.6	165.7
KBr	154.5	157.6	159.3	157.0	159.3	158.8
KI	144.5	147.4	149.2	147.1	149.3	149.0
RbF	180.4	182.6	184.4	181.0	184.4	183.2
RbC	155.4	158.8	160.9	158.8	161.0	161.0
RbBr	148.3	151.5	153.6	151.5	153.7	154.0
RbI	139.6	143.2	145.5	143.8	145.6	145.6
CsF	171.2	175.6	178.6	176.4	178.8	178.3
CsCl Structure						
CsCl	146.6	150.2	153.8	152.0	153.8	-
CsBr	141.1	144.1	147.9	146.1	147.9	-
CsI	132.7	136.1	140.1	138.6	140.1	-

7.1.7 Lattice Energy (Experimental)

Table 7.6 Thermochemical parameters and experimental (Born-Haber) lattice energy U_{exp} at room temperature; for explanation of parameters, see notes and comments

Crystal	Q	S	$D/2$	C	I	E	$-U_{exp}$
	[kcal mole^{-1}]; Ref. [7.1]						
NaCl Structure							
LiF	145.7	38.4	18.9	3.0	124.4	82.1	242.3
LiCl	96.0	38.4	28.9	3.0	124.4	85.8	198.9
LiBr	83.7	38.4	26.8	3.0	124.4	80.5	189.8
LiI	64.8	38.4	25.5	3.0	124.4	72.4	177.7
NaF	136.3	25.9	18.9	3.0	118.4	82.1	214.4
NaCl	98.2	25.9	28.9	3.0	118.4	85.8	182.6
NaBr	86.0	25.9	26.8	3.0	118.4	80.5	173.6
NaI	68.8	25.9	25.5	3.0	118.4	72.4	163.2

Table 7.6 (Continued)

Crystal	Q	S	$D/2$	C	I	E	$-U_{exp}$
	[kcal mole^{-1}]; Ref. [7.1]						
NaCl Structure							
KF	134.5	21.5	18.9	3.0	100.0	82.1	189.8
KCl	104.2	21.5	28.9	3.0	100.0	85.8	165.8
KBr	93.7	21.5	26.8	3.0	100.0	80.5	158.5
KI	78.3	21.5	25.5	3.0	100.0	72.4	149.9
RbF	131.8	19.5	18.9	3.0	96.3	82.1	181.4
RbCl	103.4	19.5	28.9	3.0	96.3	85.8	159.3
RbBr	93.5	19.5	26.8	3.0	96.3	80.5	152.6
RbI	79.0	19.5	25.5	3.0	96.3	72.4	144.9
CsF	130.3	18.7	18.9	3.0	89.7	82.1	172.5
CsCl Structure							
CsCl	106.9	18.7	28.9	3.0	89.7	85.8	155.4
CsBr	97.7	18.7	26.8	3.0	89.7	80.5	149.4
CsI	83.9	18.7	25.5	3.0	89.7	72.4	142.4

Notes and Comments

1. Q = heat of formation of a molecule of the crystal; S = heat of sublimation of the metal; $D/2$ = half the heat of dissociation of the halogen molecule; C = energy for cooling metal and halogen from 298 K to 0 K; I = ionisation energy of the metal; E = electron affinity of halogen and U_e = experimental lattice energy = $\{-Q - S - I - (D/2)\} + C + E$.
2. For electron affinity values from different methods, see Sec. 7.2.
3. Note the definitions of C and D; they are not to be confused with the Van der Waal parameters.
4. Experimental values of lattice energy of alkali halides can also be found in [7.4, 7.7, 7.8].
5. Schlosser [7.12] proposed the following relation between U_{exp} and the interionic distance r

$$U_{exp}\, r = \text{constant} \tag{7.11}$$

He obtained separate constants for the alkali halides with NaCl structure and those with CsCl structure viz. 3.744 and 4.026 respectively in units of 10^{-18} J Å taking U_{exp} in units of 10^{-18} J per pair and r in Å.
6. Reddy et al. [7.13] proposed the following linear relation between U_{exp} and the interionic distance r:

$$U_{exp} = 1387.75 - 208.91\, r \tag{7.12}$$

where r is in Å and U_{exp} in kJ mole^{-1}

7.2 Electron Affinity

Table 7.7 Values of the electron affinities of halogen atoms

	Electron affinities [eV]; Ref. [7.11]			
Halogen→ Method ↓	F	Cl	Br	I
Experimental				
Surface ionisation	3.622	3.757	3.64	3.31
Dissociation of alkali halides	3.62	-	3.83	-
Magnetron	3.567	3.723	3.49	3.141
Spectrophotometry	3.557	-	3.556	-
Electron impact	-	3.005	3.805	3.002
Flame	-	-	3.761	3.557
Space charge	-	-	3.818	3.236
Theoretical				
Lattice energy	3.47	3.68	3.53	3.14
Molecular constants	3.843	3.88	3.717	3.355

7.3 Ionicity

Table 7.8 Ionicity parameters: electronegativity difference ($X_A - X_B$), fractional ionicity (f_i) and effective ionic charge (e^*/e) (for definitions of f_1, f_2, f_3 and f_4, see notes and comments)

		f_i				
Parameter Ref. Crystal ↓	$(X_A\text{-}X_B)$ [7.14]	f_1 [7.15]	f_2 [7.16]	f_3 [7.15]	f_4 [7.15]	e^*/e [7.18]
NaCl Structure						
LiF	3	0.98	0.915	0.735	-	0.81
LiCl	2	0.94	0.903	0.833	-	0.79
LiBr	1.8	0.93	0.899	0.838	-	0.73
LiI	1.5	0.91	0.890	-	-	-
NaF	3.1	0.98	0.946	0.814	-	0.83
NaCl	2.1	0.94	0.935	0.870	0.94	0.77
NaBr	1.9	0.93	0.934	0.880	-	0.74
NaI	1.6	0.91	0.927	0.877	-	0.73
KF	3.2	0.99	0.955	0.893	0.94	0.91
KCl	2.2	0.95	0.953	0.926	0.93	0.81
KBr	2.0	0.91	0.952	0.921	0.93	0.78
KI	1.7	0.92	0.950	0.922	0.93	0.74
RbF	3.2	0.99	0.960	0.916	-	0.95
RbCl	2.2	0.95	0.955	0.937	0.92	0.83
RbBr	2.0	0.94	0.957	0.936	-	0.80

Table 7.8 (Continued)

Parameter	(X_A-X_B)	f_i				e^*/e
		f_1	f_2	f_3	f_4	
Ref.	[7.14]	[7.15]	[7.16]	[7.15]	[7.15]	[7.18]
Crystal ↓						
NaCl Structure						
RbI	1.7	0.92	0.950	0.932	0.92	0.77
CsF	3.3	0.99	-	-	0.96	0.95
CsCl Structure						
CsCl	2.3	0.97	0.963*	-	0.94	0.85
CsBr	2.1	0.96	0.962*	-	0.94	0.82
CsI	1.8	0.94	0.963*	-	0.94	0.78

* Ref. [7.17]

Notes and Comments

1. The fractional ionicities are defined as follows:
   Pauling [7.19] defined the fractional ionicity f_1 as:

$$f_1 = 1 - \left(\frac{N_{el}}{M'}\right) \exp\left[-\left(\frac{X_A - X_B}{4}\right)\right]^2 \tag{7.13}$$

where N_{el} is the number of bonding electrons, M' the coordination number and X_A and X_B the electronegativities of the alkali and halogen atoms respectively. Values calculated from Eq. (7.13) are given in [7.15].
Van Vechten [7.20] and Phillips [7.16] defined fractional ionicities f_2 and f_3 as:

$$f_2 = C^2 / E_G^2 \quad \text{and} \quad f_3 = 1 - \left(E_h^2 / E_G^2\right) \tag{7.14}$$

where C, E_h and E_G are the band gaps due to ionic effects, due to homopolar effects and the total band gap respectively.
Poole et al. [7.15] expressed the Pauling expression for f_1 (Eq. 7.13) in terms of E_S, the energy separation between the outer valence bands due to the cation and anion, and denoted the modified ionicity as f_4.

2. The effective ionic charge e^* is defined by Szigeti [7.21] as

$$(e^*/e)^2 = (9\pi\mu\nu_{TO}^2 / N_v e^2)\,[(\varepsilon_0 - \varepsilon_\infty)/(\varepsilon_\infty + 2)^2] \tag{7.15}$$

where μ is the reduced mass, ν_{TO} the long wavelength TO mode frequency, N_v the number of ion pairs per unit volume and ε_0 and ε_∞ the static and high frequency dielectric constants.

7.4 Electron Density Distribution

The electron density distributions are shown as electron density projection on a plane (Fig. 7.1) and as radial distribution curves (Figs. 7.2).

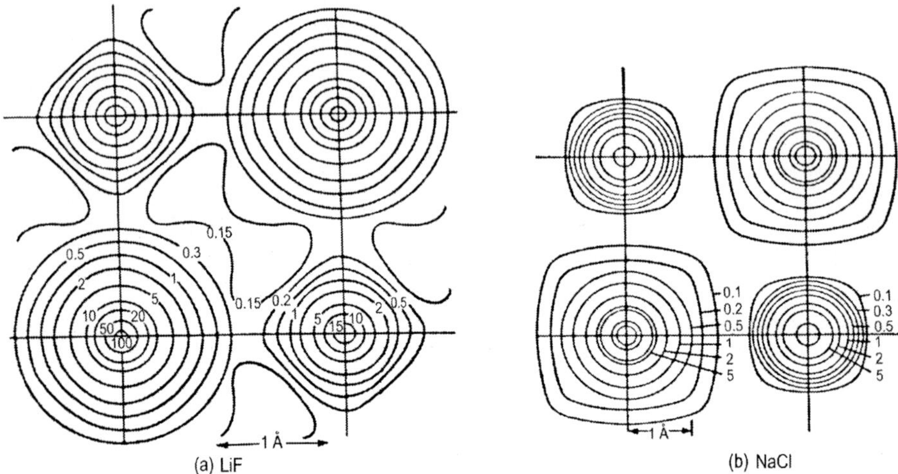

Fig. 7.1 Electron density distribution in (x y 0) plane of **(a)** LiF and **(b)** NaCl; electron density in el Å^{-3} (after [7.22])

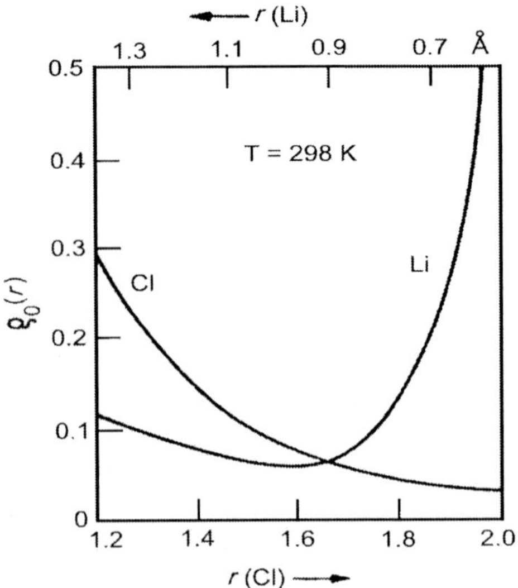

Fig. 7.2 Average electron densities of Li^+ and Cl^- at 298 K (after [7.23])

Notes and Comments

1. The electron density drops to zero along the Na–Cl line whereas the minimum electron density along the Li–F line is 0.19 el Å^{-3} [7.22] and 0.09 el Å^{-3} along the Li–Cl line [7.23].
2. Meisalo and Inkinen [7.24] found from their X-ray study of KBr that the electron density outside the ionic spheres has a low value of about 0.04 el Å^{-3}.

7.5 Force Constant

Table 7.9 Values of the force constant (k_f)

	$k_f\,[10^4\ \text{dyne cm}^{-1}]$	
Ref.	[7.25]	[7.26]
Crystal ↓		
NaCl Structure		
LiF	9.67	8.14
LiCl	5.17	4.64
LiBr	4.24	3.93
LiI	3.36	3.15
NaF	7.31	6.54
NaCl	4.33	4.06
NaBr	3.65	3.53
NaI	2.92	2.93
KF	5.00	4.90
KCl	3.26	3.32
KBr	2.88	2.93
KI	2.38	2.47
RbF	4.35	4.57
RbCl	2.89	3.12
RbBr	2.55	2.77
RbI	2.15	2.33
CsF	3.64	4.23
CsCl Structure		
CsCl	2.36	2.93
CsBr	2.08	2.67
CsI	1.72	2.30

Notes and Comments

1. Jain and Shanker [7.25] calculated k_f from the equation

$$k_f = \frac{M}{3}\left[\frac{1}{\rho^2} - \frac{2}{r\rho}\right] B\,\exp\left(-r/\rho\right) \tag{7.16}$$

where M is the molecular weight and $B\exp(-r/\rho)$ is the repulsive term in the lattice energy.

2. Narain [7.26] calculated k_f from the definition $k_f = 3pr/\psi$ where ψ is the compressibility and p is a constant with values 2 for the NaCl structure and $8/3\sqrt{3}$ for the CsCl structure.
3. Values of force constant for some alkali halides have been given by Waser and Pauling [7.27].
4. The values of k_f [7.26] fit the equation

$$(k_f)^{-1/3} = 0.00788 \ (r + 0.75) \tag{7.17}$$

where r is the interionic distance in Å.
5. The average force constant is useful in correlating and estimating properties and parameters dependent on interatomic forces. For instance, Narain [7.26] estimated the Debye temperatures θ_D from the relation

$$\theta_D = \frac{h}{2\pi \ k_B} \left(\frac{k_f}{\mu} \right)^{1/2} \tag{7.18}$$

where h is Planck's constant, k_B the Boltzmann constant and μ the reduced mass.

References

7.1   M. P. Tosi, Solid State Physics, **16**, 1, 1964.

7.2   D. Cubicciotti, J. Chem. Phys., **31**, 1646, 1959.

7.3   J.E. Mayer, J. Chem. Phys., **1**, 270, 1933.

7.4   M.F.C. Ladd, J. Chem. Phys., **60**, 1954, 1974.

7.5   Y.V.G.S. Murti and T.V. Selvarajan, phys. stat. sol., **(b)108**, 315, 1981.

7.6   M.P. Tosi and F.G. Fumi, J. Phys. Chem. Solids, **25**, 45, 1964.

7.7   *CRC Handbook of Chemistry and Physics*, 76th Ed. (CRC Press, Boca Raton, Florida), 1995–1996.

7.8   S. Nagasaka and Y. Kojima, J. Phys. Soc. Japan, **56**, 408, 1987.

7.9   S. Nagasaka and Y. Kojima, J. Phys. Soc. Japan, **56**, 671, 1987.

7.10   T.L. Bailey, J. Chem. Phys., **28**, 792, 1958.

7.11   S.P. Tandon, M.P. Bhutra and K. Tandon, Ind. J. Phys., **XLI**, 70, 1967.

7.12   H. Schlosser, J. Phys. Chem. Solids **53**, 855, 1992.

7.13   R.R. Reddy, M. Ravi Kumar and T.V.R. Rao, Cryst. Res. Technol., **28**, 973, 1993.

7.14   L. Pauling, *Nature of the Chemical bond*, 3rd Ed. Oxford and IBH Publishing Co., Bombay, 1969.

7.15   R.T. Poole, J.G. Jenkin, R.C.G. Leckey and J. Liesegang, Chem. Phys. Lett., **26**, 514, 1974.

7.16   J.C. Phillips, Rev. Mod. Phys., **42**, 317, 1970.

7.17   P. Lawaetz, Phys. Rev. Lett., **12**, 697, 1971.

7.18   R.P. Lowndes and D. H. Martin, Proc. Roy. Soc., **A308**, 473, 1969.

7.19   L. Pauling, *Nature of the Chemical bond*, 1st Ed. Cornell University Press, Ithaca, 1939.

7.20   J.A. Van Vechten, Phys. Rev., **182**, 891, 1969.

7.21   B. Szigeti, Trans. Farad. Soc., **45**, 155, 1949.

7.22   H. Witte and E. Wolfel, Rev. Modern Phys., **30**, 51, 1958.

7.23   O. Inkinen and M. Jarvinen, Phys. Kondens. Materie, **7**, 372, 1968.

7.24   V. Meisalo and O. Inkinen, Acta Cryst., **22**, 58, 1967.

7.25   V.C. Jain and J. Shanker, phys. stat. sol., (b) 89, 213, 1978.

7.26   S. Narain, phys. stat. sol., **(b)182**, 273,1994.

7.27   J. Waser and L. Pauling, J. Chem. Phys., **18**, 747, 1950.

8 Band-Structure-Related Parameters

8.1 Typical Band Structures

The band structures of LiF, NaCl and KBr calculated by the augmented plane wave (APW) method are shown in Figs. 8.1–8.3.

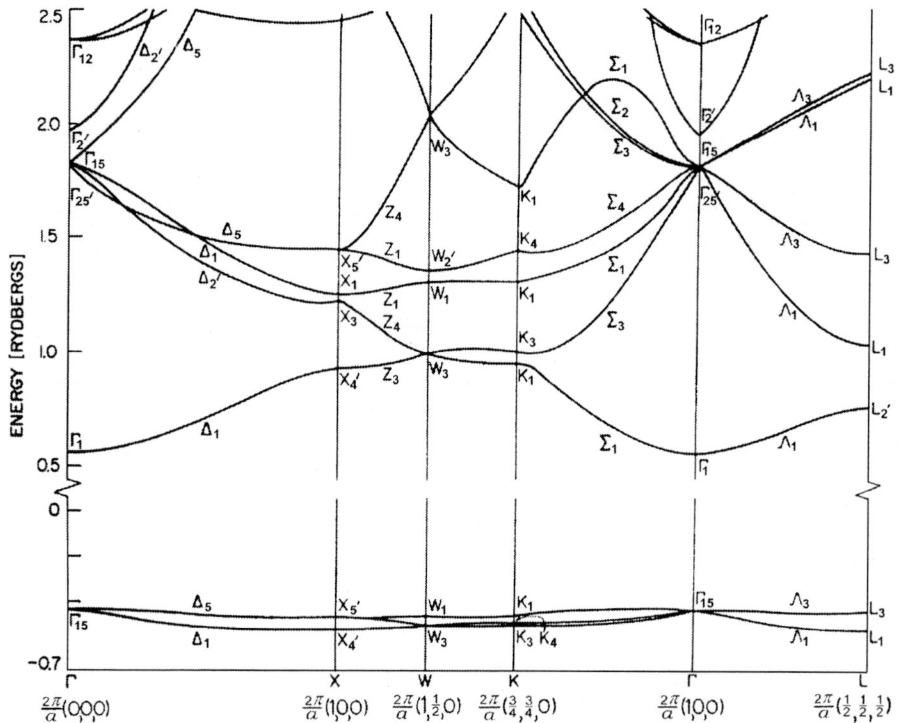

Fig. 8.1 Energy bands of LiF calculated by the APW method. The valence band consists of the $2p^6$ state of F. The band is fairly flat and has a width of 0.88 eV. The conduction band consists of the 2s state of Li, 2p state of Li and 3d state of F. The bands are plane-wave-like. The s-like band does not cross the d-like bands. There are a few Γ_2' higher f states from the F ion around the Γ point (after [8.1])

The band structures are commonly calculated by the augmented-plane-wave (APW) method. This method is concerned with solving the Schrodinger equation for a many-centred potential which exhibits the symmetry of the crystal. As an approximation, a potential is chosen which is spherically symmetric inside spheres centred at each atomic site and constant in the region between the spheres. This is the "muffin-tin" potential. Since the halogen ions are larger than the alkali ions, the APW sphere radii were chosen to be at the point where the potentials of the two ions meet. For construction of the potentials, a superposed free ion potential obtained from relativistic Dirac self-consistent field atomic calculations were used. Very often, the first iteration yielded satisfactory results making self-consistent calculation unnecessary. The values of the APW sphere radii, the Wigner-Seitz radii and the potentials as functions of distance are given in [8.1]

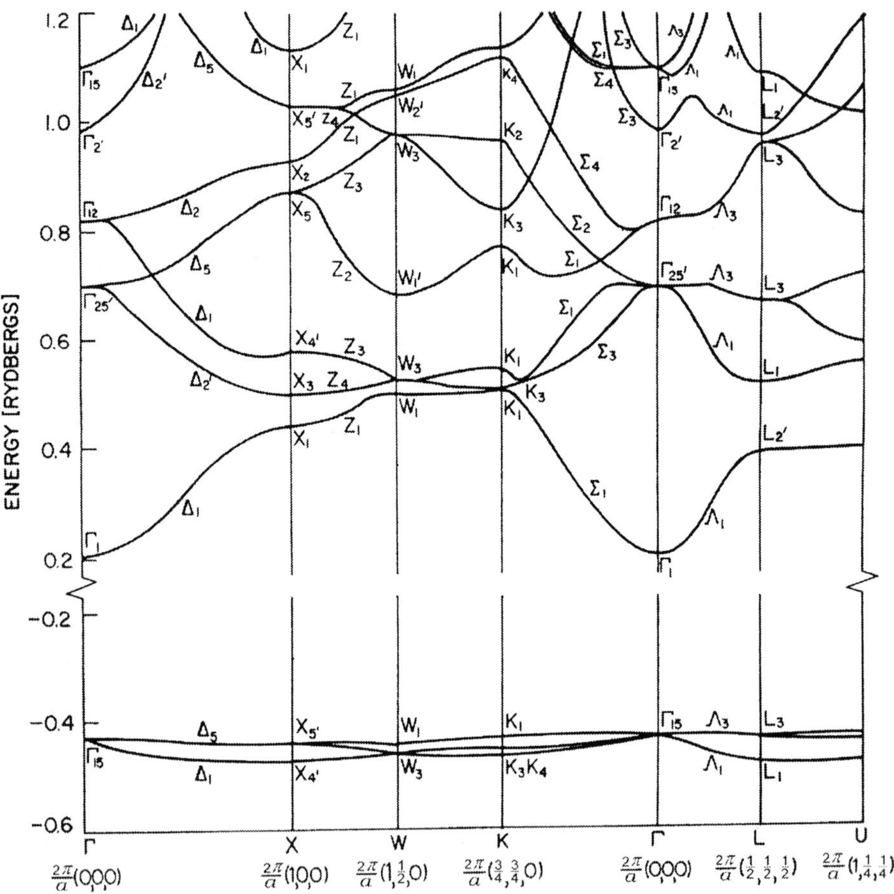

Fig. 8.2 Energy bands of NaCl calculated by the APW method. The valence band of NaCl consists of the $3p^6$ state of Cl. The width of this band is 0.57 eV. The conduction bands of NaCl consist of the 3s band of Na, the 3p band of Na and the 3d band of Cl. There is a low-lying f state which appears around Γ. The bands are not very plane-wave-like (after [8.1])

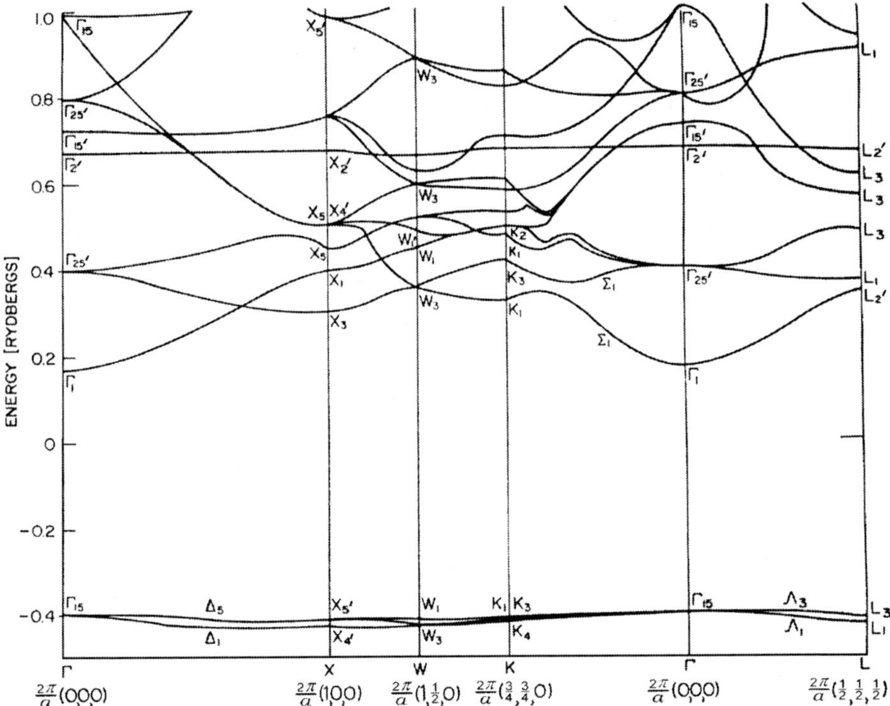

Fig. 8.3 The energy bands of KBr calculated by the APW method. The valence band of KBr consists of the $4p^6$ band of bromine. Its width is 0.31 eV. The conduction band is complicated. It consists of the 4s, 4p and 4d bands of K and 4d and 4f bands of Br (after [8.1])

The energy band calculations for KBr are compared with the optical absorption spectrum in Fig. 8.4.

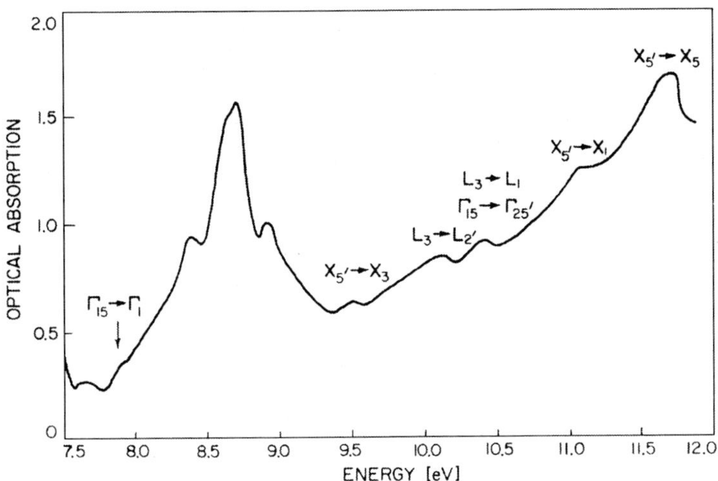

Fig. 8.4 Optical absorption of KBr (optical spectrum after [8.2] and transitions from [8.1])

Notes and Comments

1. Band structure calculations for some select alkali halides are reported in [8.3] for CsI, [8.4] and [8.5] for LiCl, [8.6] for KCl, KBr and KI, [8.7] for NaBr, [8.8] for LiF, NaF, KF, LiCl, NaCl and KCl , [8.9] for LiF and [8.10] for alkali halides (except iodides and cesium halides).
2. Comparison of theoretical [8.1] and experimental [8.2] absorption bands for KBr is shown in Table XV.

Table XV Theoretical and experimental values of transition energies for KBr

Transition	Transition energy [eV]	
	Theory [8.1]	Experiment [8.2]
$\Gamma_{15} \to \Gamma_1$	7.80	7.90
$X_5' \to X_3$	9.50	9.50
$L_3 \to L_2'$	10.30	10.20
$\Gamma_{15} \to \Gamma_{25}'$	10.40	10.40
$X_5' \to X_1$	10.80	11.10
$L_3 \to L_1$	10.60	10.50
$X_5' \to X_5$	11.50	11.80

8.2 Band Structure Parameters

8.2.1 Band Gap Energy

Table 8.1 Values of the experimentally determined band gap energy E_G (energy between outermost valence band and conduction band)

Ref. Crystal ↓	E_G [eV]			
	[8.11]	[8.12]	[8.13]	[8.10]
NaCl Structure				
LiF	-	13.6	13.6	14.2
LiCl	~10	9.4	9.3	9.4
LiBr	~ 8.5	7.6	7.95	7.6
LiI	≥ 5.9	6.1	-	-
NaF	≥10.5	11.5	11.6	11.5
NaCl	8.6	8.75	8.61	9.0
NaBr	7.7	7.1	7.5	7.1
NaI	≥ 5.8	5.9	6.28	-
KF	10.9	10.8	10.72	10.8
KCl	8.5	8.7	8.48	8.7
KBr	7.8	7.4	7.49	7.4
KI	≥ 6.2	6.3	6.07	-

Table 8.1 (Continued)

Ref. Crystal $\downarrow$	E_G [eV]			
	[8.11]	[8.12]	[8.13]	[8.10]
NaCl Structure				
RbF	10.4	10.3	10.35	10.3
RbCl	8.2	8.5	8.35	8.5
RbBr	7.7	7.2	7.51	7.2
RbI	$\geq$ 6.1	6.3	6.15	-
CsF	10.0	9.8	-	-
CsCl Structure				
CsCl	$\geq$ 8.0	8.3	-	-
CsBr	7–8	7.3	-	-
CsI	$\geq$ 6.3	6.1	-	-

Notes and Comments

1. Experimental values of E_G are generally obtained from optical absorption spectra in the ultraviolet region.
2. Considering the differences in values from different sources, the uncertainty in E_G values may be taken as about 5%.
3. Reddy and Ahammed [8.14] reported the following empirical relation between E_G and the refractive index n:

$$(n^2 + 2)^2 (E_G - 0.365) = 154 \tag{8.1}$$

8.2.2 Pressure Derivative of Band Gap

Table 8.2 Values of the pressure derivative of the band gap energy ($\partial E_G / \partial P$)

Crystal	$\partial E_G / \partial P$ [10^{-6} eV bar^{-1}]	
	Experimental [8.15]	Theoretical [8.16]
NaCl Structure		
LiF	-	6.80
LiCl	-	10.5
LiBr	-	10.6
NaF	-	8.4
NaCl	-	11.9
NaBr	13.8	12.9
NaI	14.3	-
KF	-	11.9
KCl	-	16.3
KBr	18.0	17.6
KI	17.1	17.7

Table 8.2 (Continued)

Crystal	$\partial E_G / \partial P \, [10^{-6} \, \text{eV bar}^{-1}]$	
	Experimental [8.15]	Theoretical [8.16]
NaCl Structure		
RbF	-	13.2
RbCl	16.0	17.5
RbBr	17.9	19.7
RbI	19.7	19.2
CsCl Structure		
CsCl	-	15.1
CsBr	15.0	15.4
CsI	18.0	16.7

Notes and Comments

1. The band gap energy E_G is the sum of the exciton energy E_{exc} and the exciton binding energy E_b. The pressure variation of E_b is negligible in comparison with the pressure variation of E_{exc} [8.17]. Hence the experimental values of $\partial E_{exc}/\partial P$ are taken as the experimental values of $\partial E_G /\partial P$.
2. Masunaga and Fujita [8.16] postulated the following relation between the energy gap E_G and the interionic distance r:

$$E_G = (r_0 /r)^n \left(E_G^0 \right) \tag{8.2}$$

where E_G^0 and r_0 are the zero pressure values of r and E_G and n is a constant. From this they derived :

$$\left(\partial E_G /\partial P \right) = n \left(E_G /3B \right) \tag{8.3}$$

where B is the bulk modulus. Using experimental values available for some alkali halides for B, E_G and $(\partial E_G /\partial P)$, they found that $n \approx 1$ for the alkali halides. Hence they calculated $(\partial E_G/\partial P)$ from $(3B/E_G)$ for several alkali halides. These are given as theoretical values.

8.2.3 Valence Band Width

Table 8.3 Theoretical and experimental values of the valence band width

Crystal	Valence band width [eV]	
	Theoretical. [8.10]	Experimental* [8.10]
NaCl Structure		
LiF	3.1	3.5–6.0
LiCl	3.6	5 ± 0.5
LiBr	3.9	-

Table 8.3 (Continued)

Crystal	Valence band width [eV]	
	Theoretical. [8.10]	Experimental* [8.10]
NaCl Structure		
NaF	2.7	-
NaCl	3.0	3.0 ± 0.3
NaBr	3.6	3.5 ± 0.5
KF	2.2	-
KCl	2.4	2.3 ± 0.3
KBr	2.5	-
RbF	1.7	-
RbCl	2.2	1.6 ± 0.2
RbBr	1.3	1.6 ± 0.2

* The experimental values are obtained from photoemission or X-ray emission data.

8.2.4 Effective Mass

Table 8.4 Values of the effective mass m_e* of electrons at the bottom of the conduction band as a fraction of the free electron mass

Crystal	m_e* (Calculated)	
	Ref. [8.18]	Ref. [8.1]
NaCl Structure		
LiF	-	1.2
NaCl	0.6	-
KCl	0.4, 0.496	-
KBr	0.48, 0.428	-
KI	0.47, 0.49, 0.40	-
RbCl	0.66, 0.515	-

8.2.5 Interband Transition Energy

Table 8.5 Values of the interband transition energies

Crystal Ref. Transition ↓	Interband transition energy [eV]								
	LiF [8.19]	LiCl [8.19]	NaCl [8.19]	KCl [8.19]	KBr [8.20]	KI	RbCl	RbI	CsI
$\Gamma_{15} \rightarrow \Gamma_1$	13.5	9.9	8.7	8.6	7.9	-	-	-	-
$L_3 \rightarrow L_2'$	16.0	-	10.2	-	10.2	-	-	-	-
$X_5' \rightarrow X_4'$	18.0	-	-	-	-	-	-	-	-
$L_3 \rightarrow L_1$	21.5	-	-	-	10.5	-	-	-	-
$X_5' \rightarrow X_3$	24.4	-	11.4	12.6	9.5	-	-	-	-

Table 8.5 (Continued)

Crystal Ref. Transition ↓	Interband transition energy [eV]								
	LiF [8.19]	LiCl [8.19]	NaCl [8.19]	KCl [8.19]	KBr [8.20]	KI	RbCl	RbI	CsI
$X_5' \rightarrow X_1$	-	13.3	13.0	13.4	11.1	-	-	-	-
$\Gamma_{15} \rightarrow \Gamma_{25}'$	-	-	15.2	18.0	10.4	-	-	-	-
$L_3 \rightarrow L_3'$	-	-	16.0	-	-	-	-	-	-
$X_5' \rightarrow X_5$	-	-	-	-	11.8	-	-	-	-
The following transition energies are from [8.2]									
$\Gamma_8^- \rightarrow \Gamma_6^+$				8.9	8.0	6.3	8.5	6.3	6.4
$\Gamma_8^- \rightarrow \Gamma_8^+$				10.7	9.5	7.2	9.9	6.9	6.5

8.3 UV Absorption Spectra

8.3.1 UV Absorption Spectra (5–12 eV)

The UV absorption spectra recorded at RT and at 80 K are shown in Fig. 8.5.

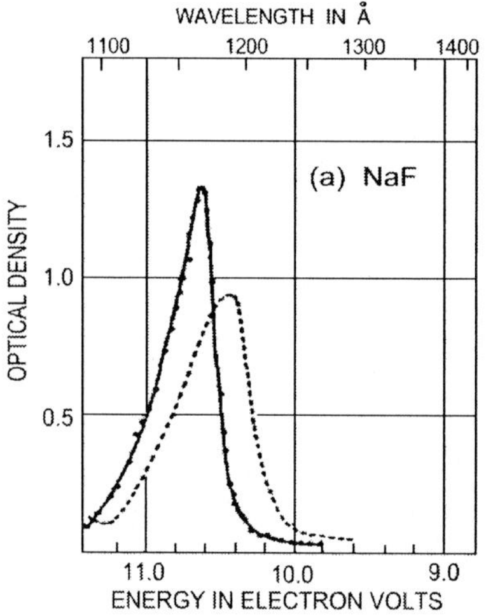

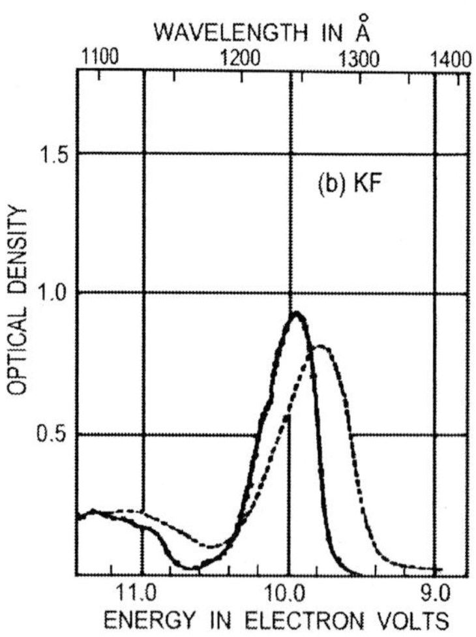

Fig. 8.5 Optical absorption spectra (5–12 eV) of thin films of alkali halides deposited on LiF substrates. The dashed curves represent room temperature data and the points connected by solid curves represent data at 80 K. The curve for CsCl I represents a freshly evaporated film corresponding to the NaCl phase and CsCl II curve represents the same film after annealing at room temperature and corresponds to the CsCl structure. (a) NaF, (b) KF, (c) RbF, (d) CsF, (e) LiCl, (f) NaCl, (g) KCl, (h) RbCl, (i) CsCl (I), (j) CsCl (II), (k) LiBr, (l) NaBr, (m) KBr, (n) RbBr, (o) CsBr, (p) LiI, (q) NaI, (r) KI, (s) RbI, (t) CsI (after [8.21])

Fig. 8.5 (Continued)

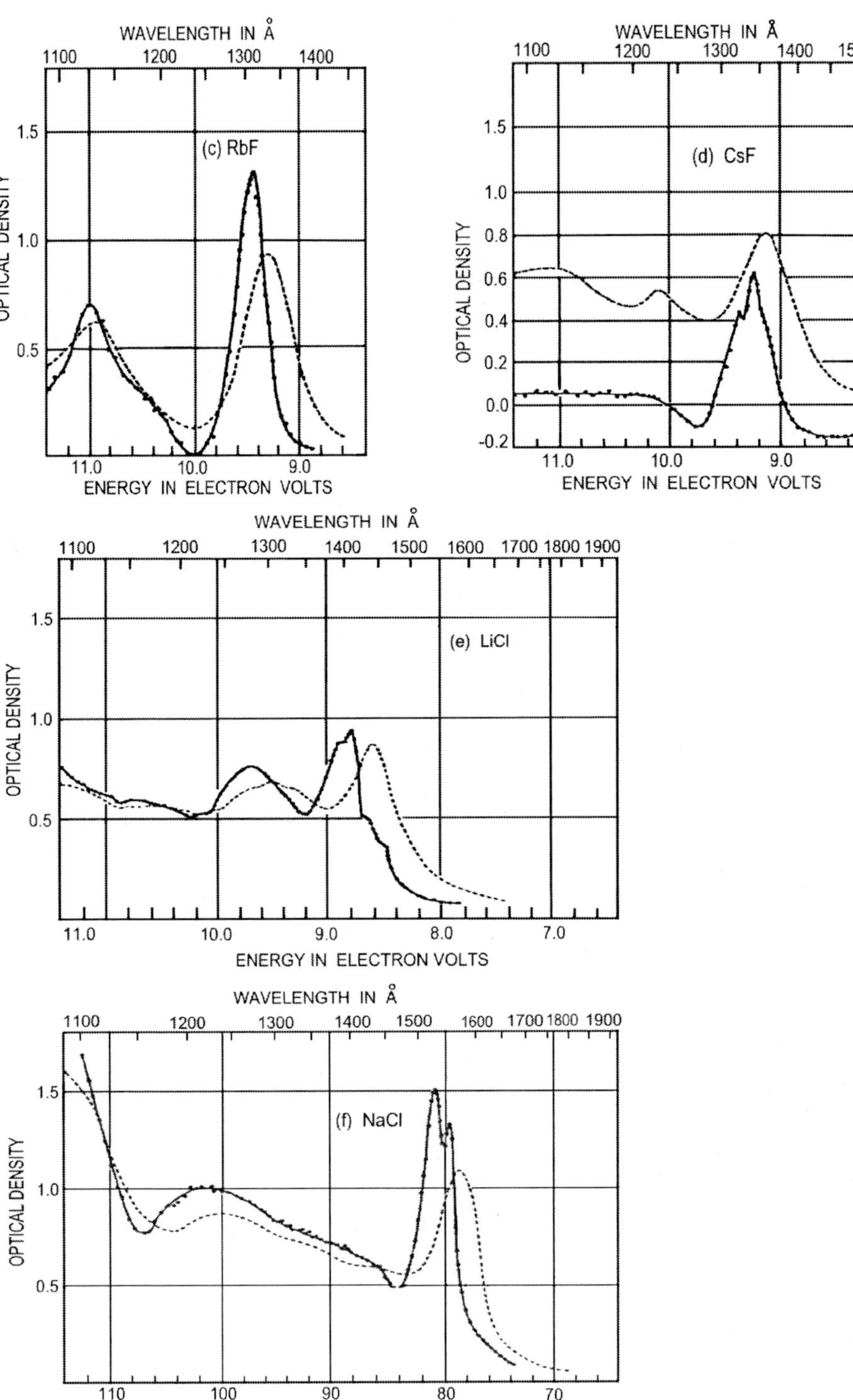

Fig. 8.5 (Continued)

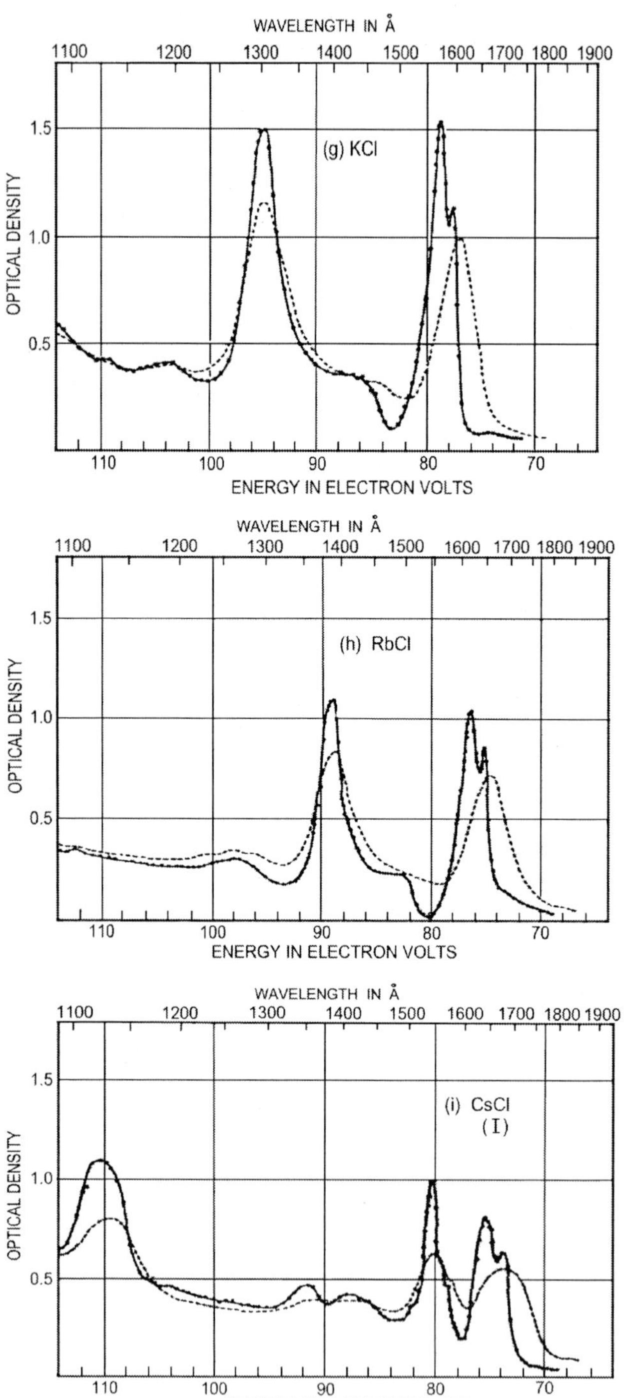

Fig. 8.5 (Continued)

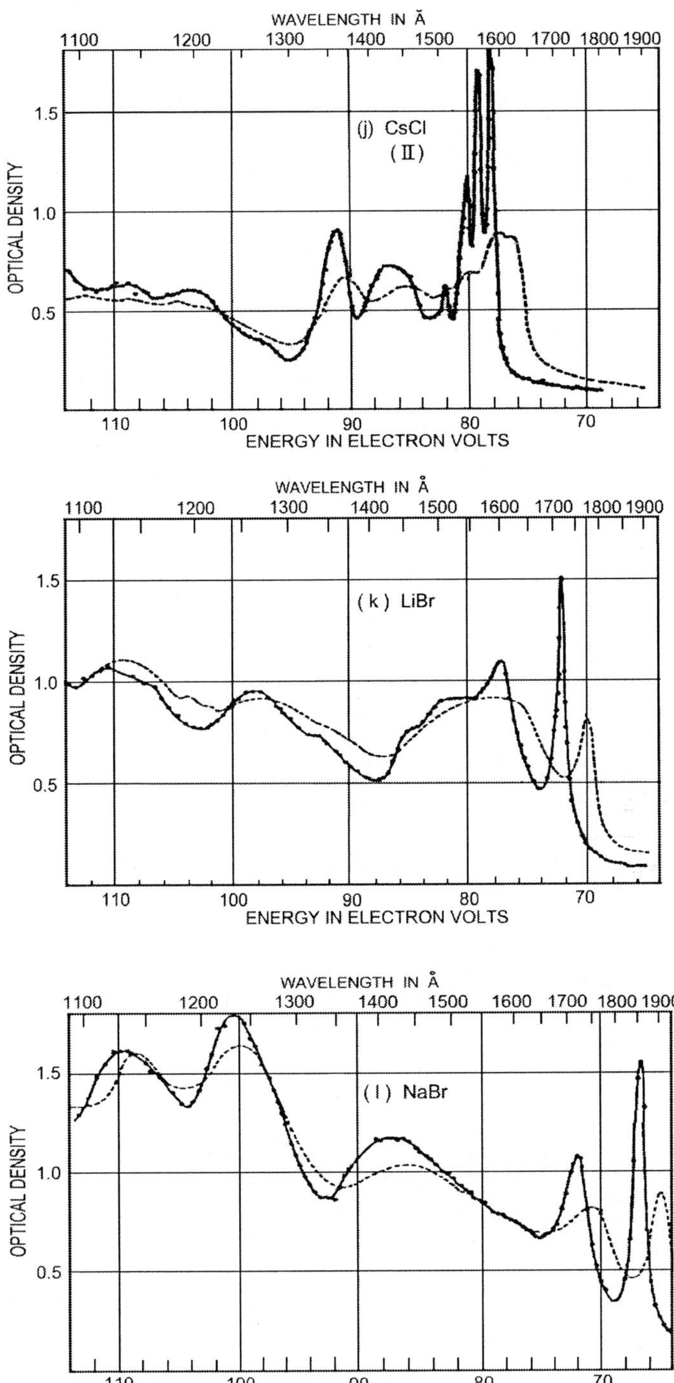

Fig. 8.5 (Continued)

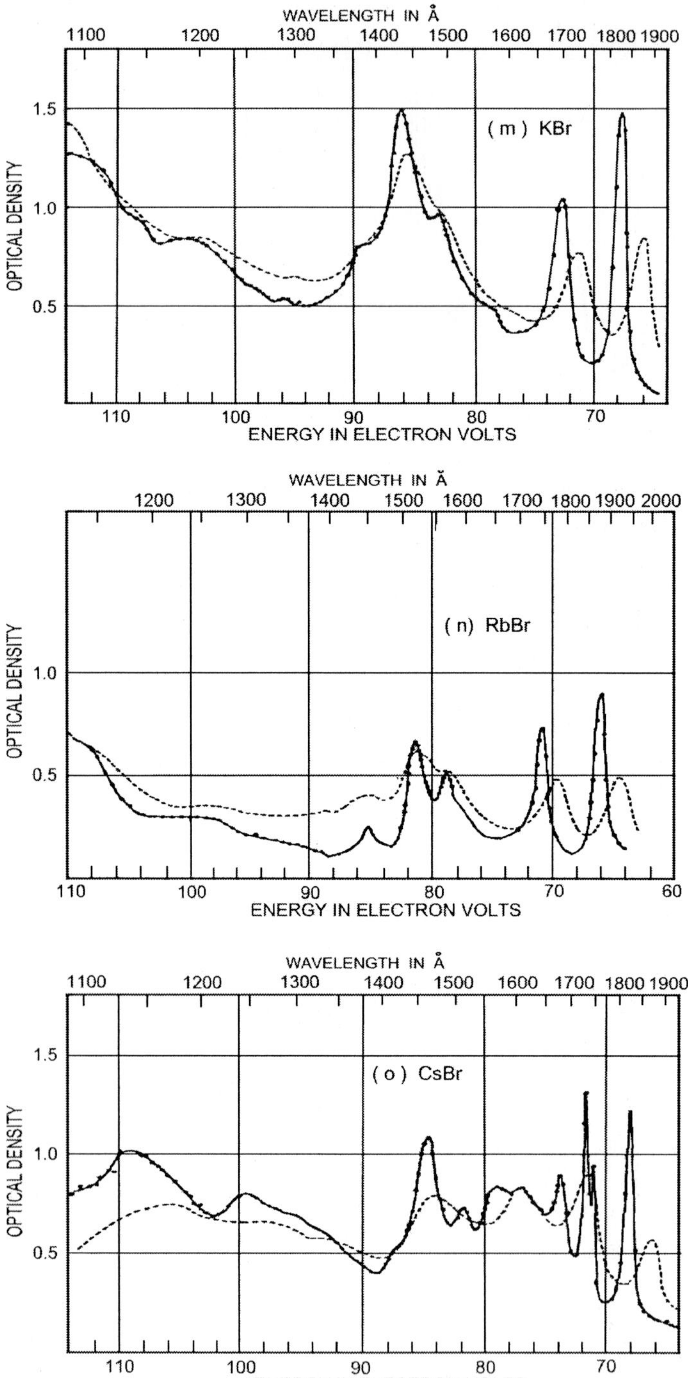

Fig. 8.5 (Continued)

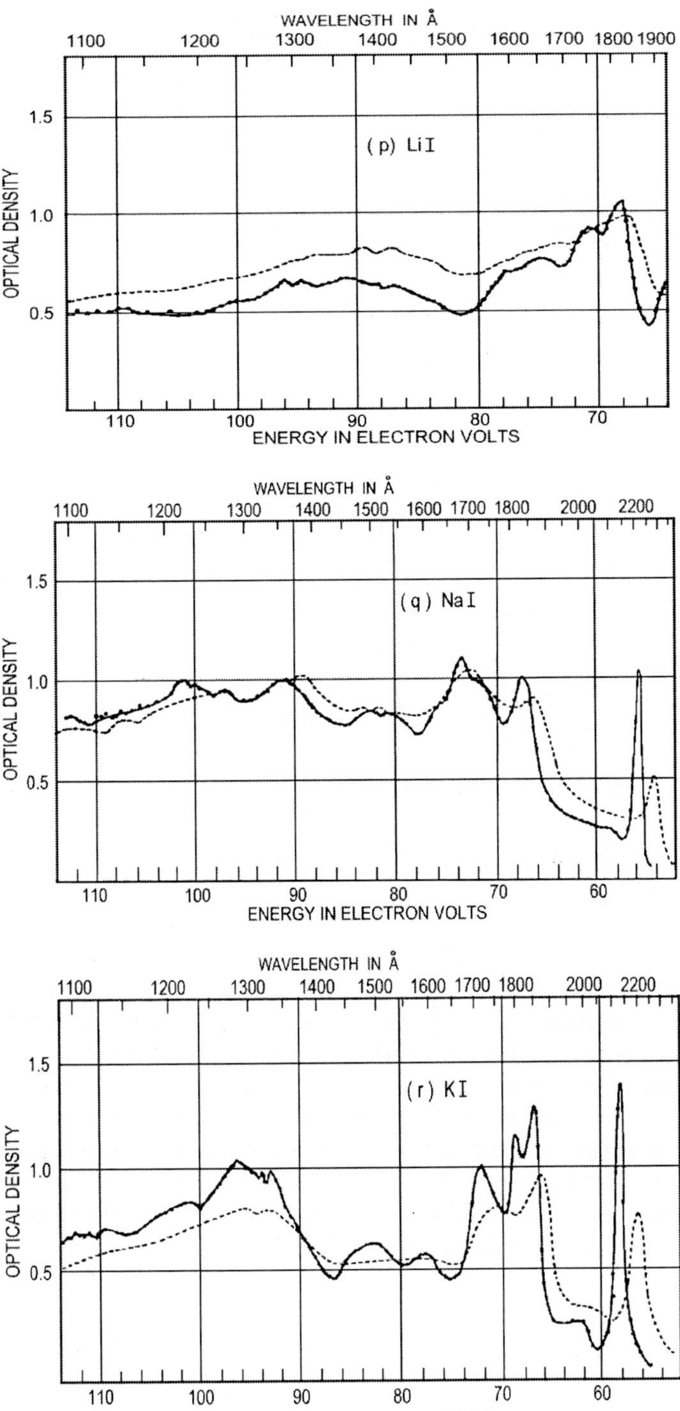

Fig. 8.5 (Continued)

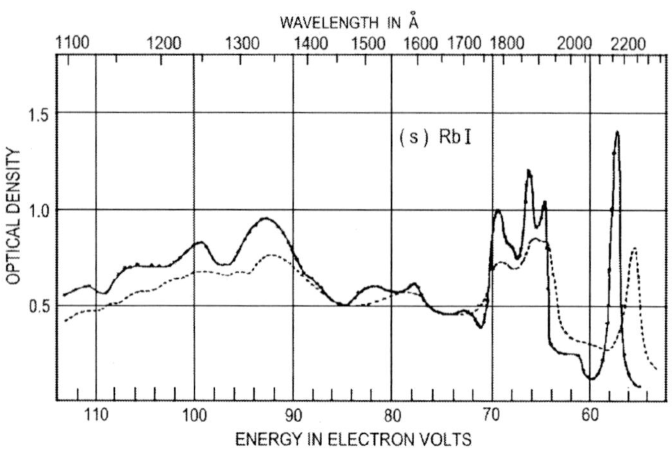

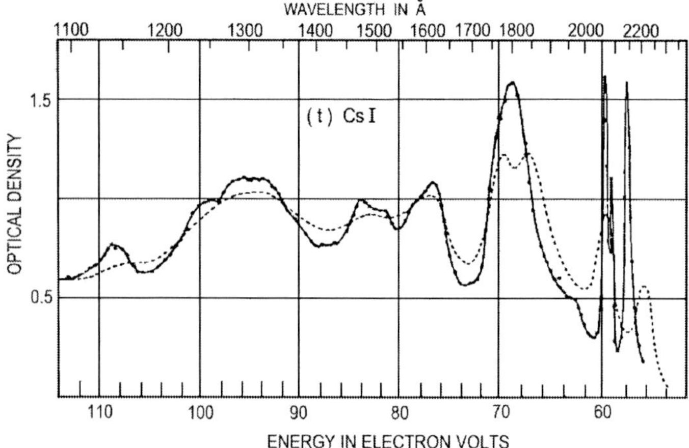

The UV absorption spectra in the range 5-12 eV of thin films of alkali halides on LiF crystals as substrates recorded at 10 K are shown in Fig. 8.6. Although the experimental technique is the same as that used in [8.21] annealing and keeping the sample at 10 K resulted in better resolution in the features. Apart from a better resolution in the shoulder and doublet regions, new features are observed in the iodides which are due to effective mass states associated with the negative ions. These spectra have yielded rich information like interband transition energies, exciton energies and assignment of the spin-orbit doublets (given in relevant sections).

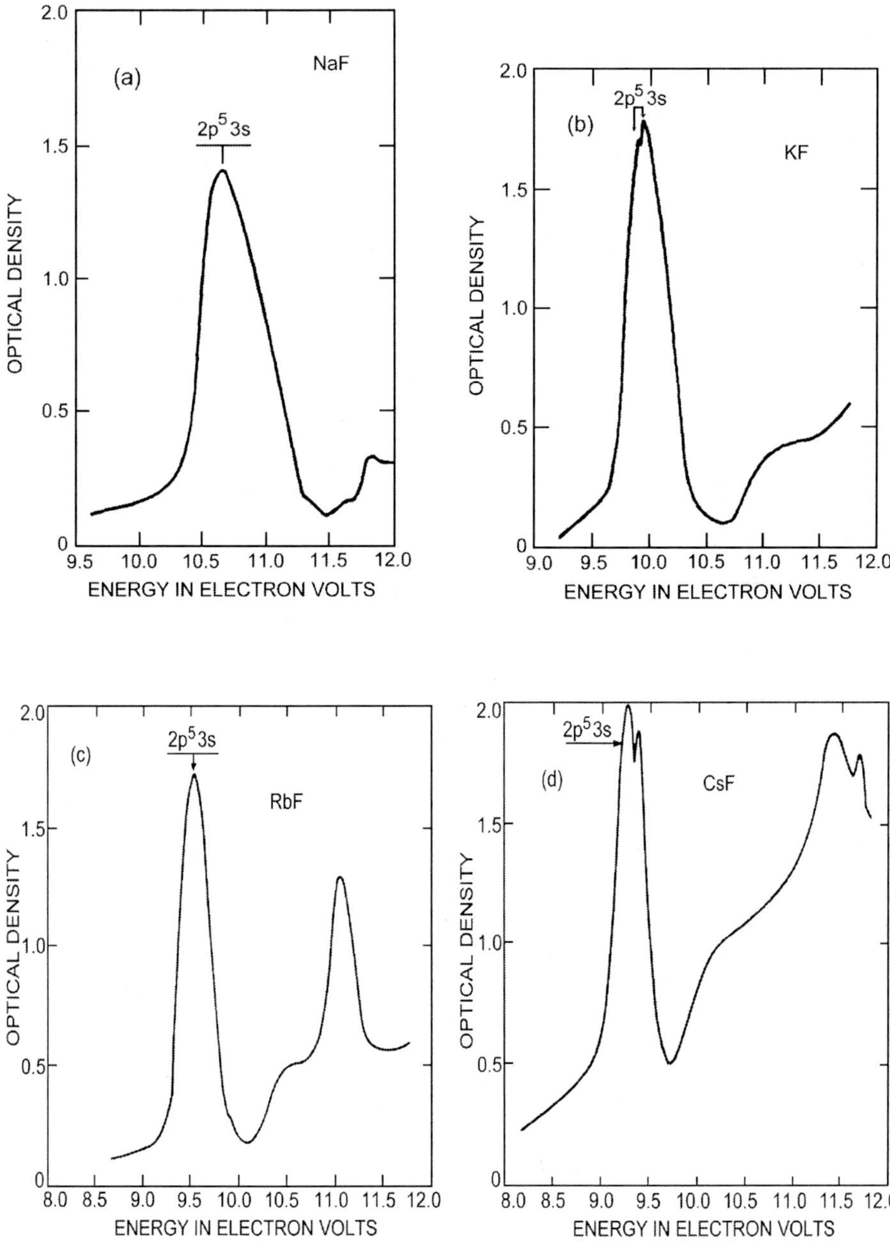

Fig. 8.6 Optical absorption spectra (a–r) of thin films of alkali halides deposited on LiF substrates at 10 K. (a) NaF, (b) KF, (c) RbF, (d) CsF, (e) LiCl (f) NaCl, (g) KCl, (h) RbCl, (i) CsCl, (j) LiBr, (k) NaBr, (l) KBr, (m) RbBr, (n) CsBr, (o) NaI, (p) KI, (q) RbI, (r) CsI (after [8.2])

Fig. 8.6 (Continued)

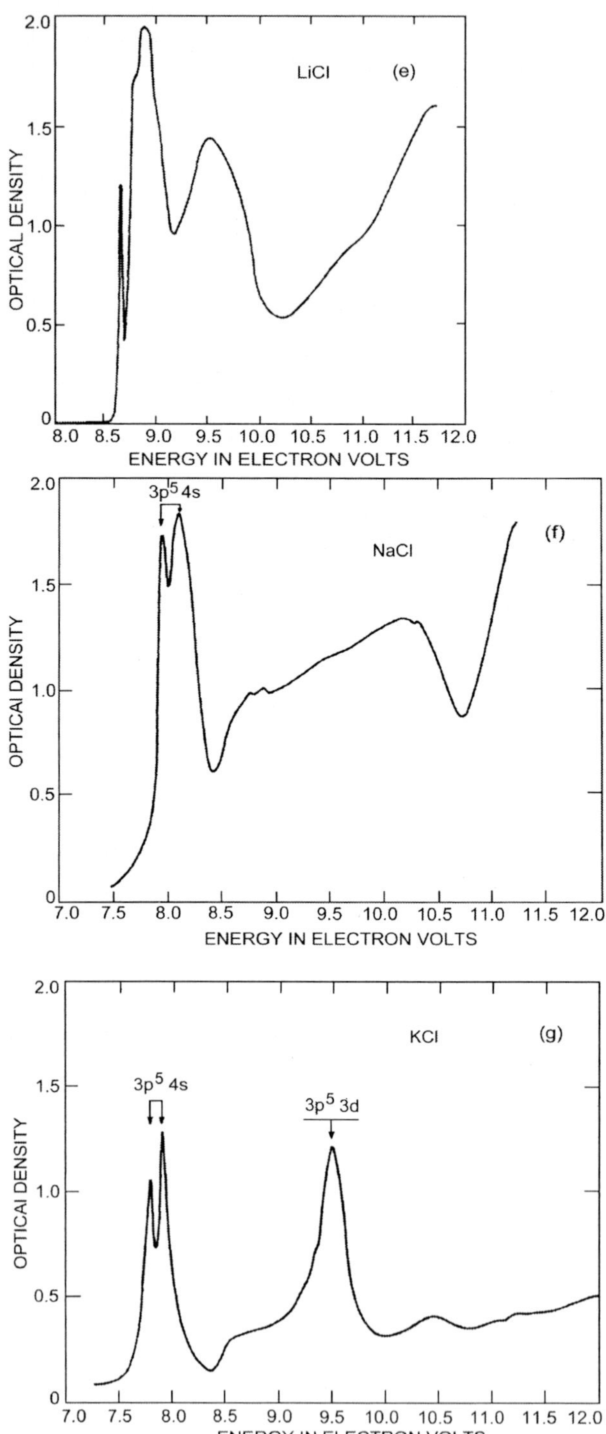

Fig. 8.6 (Continued)

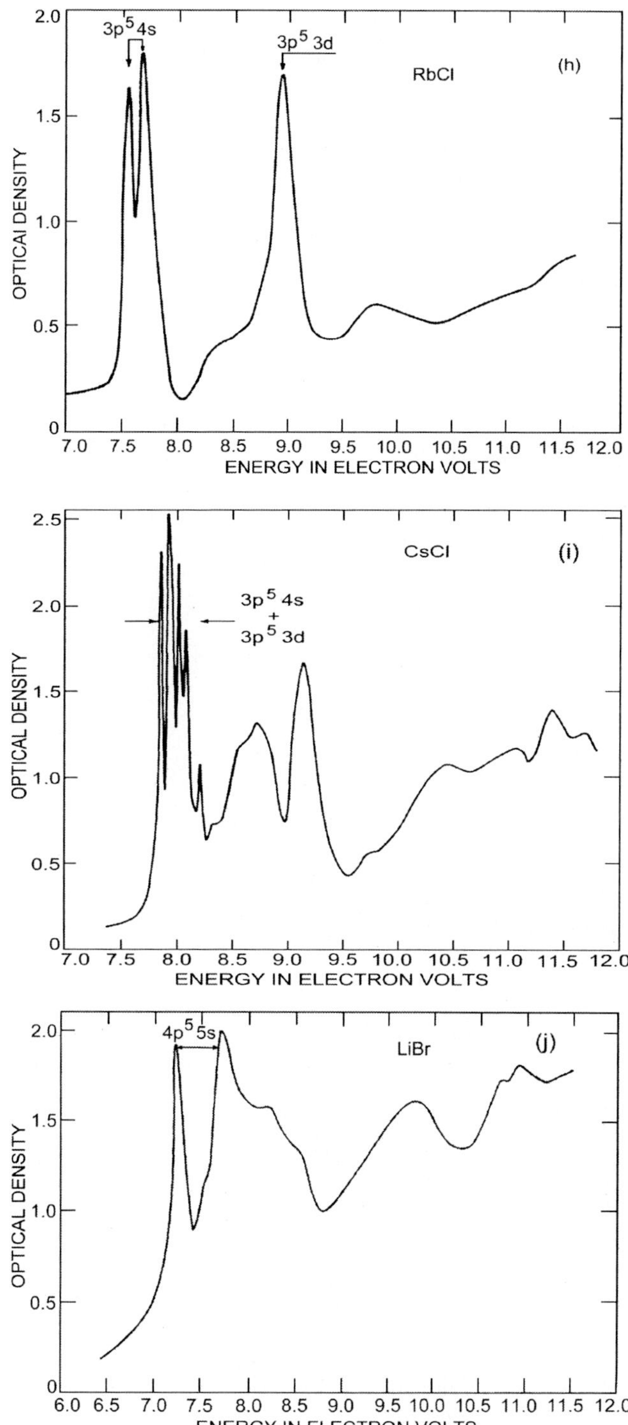

Fig. 8.6 (Continued)

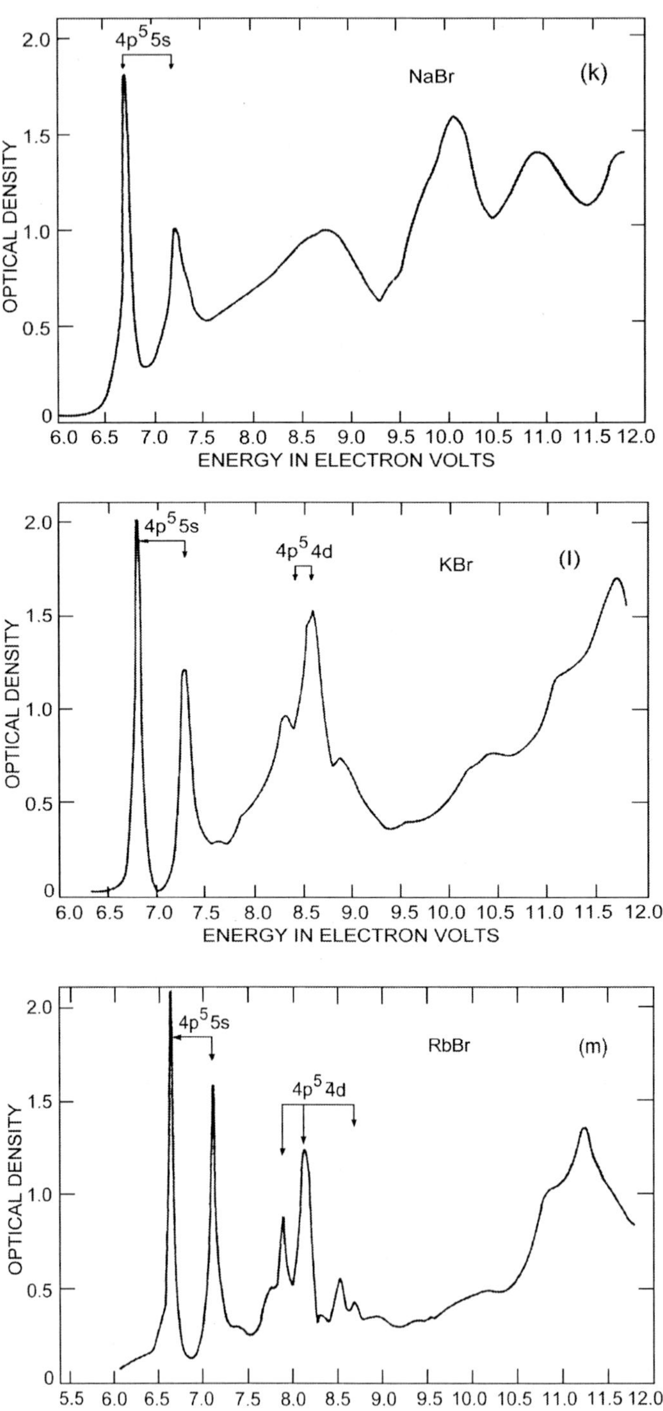

Fig. 8.6 (Continued)

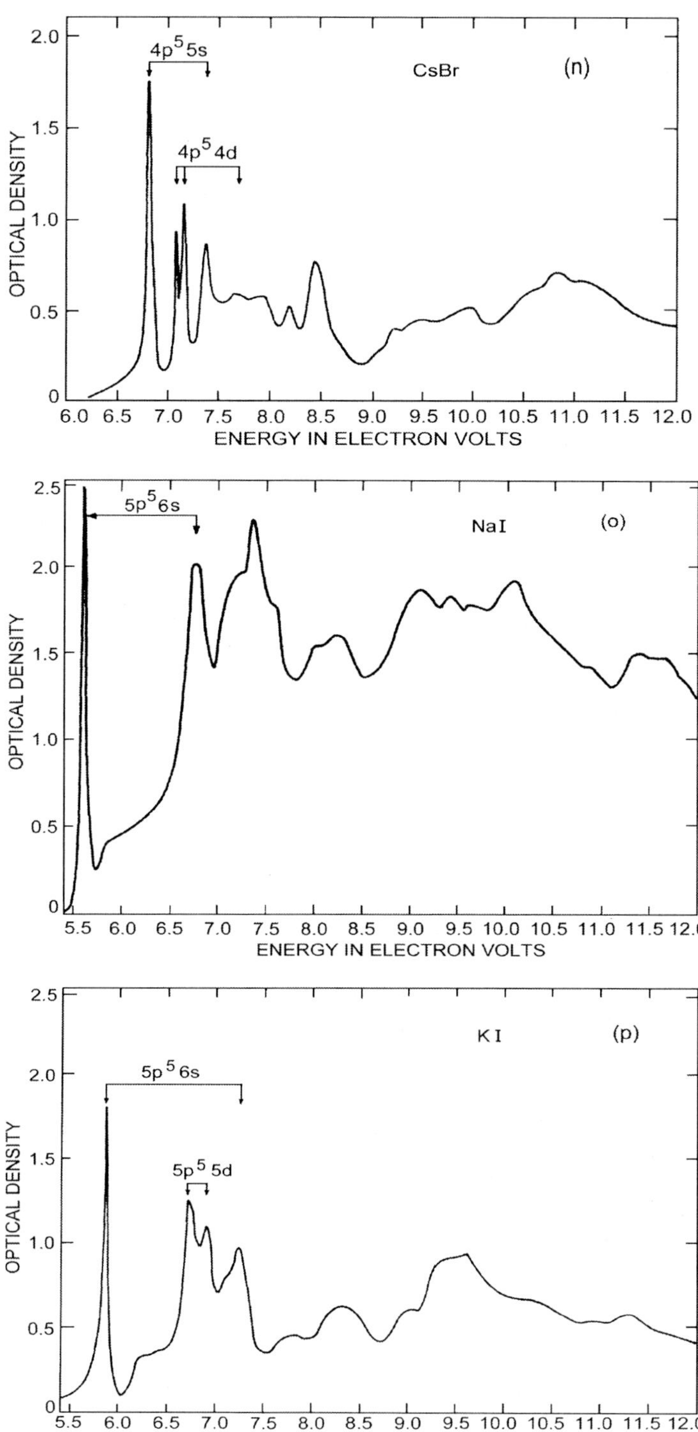

Fig. 8.6 (Continued)

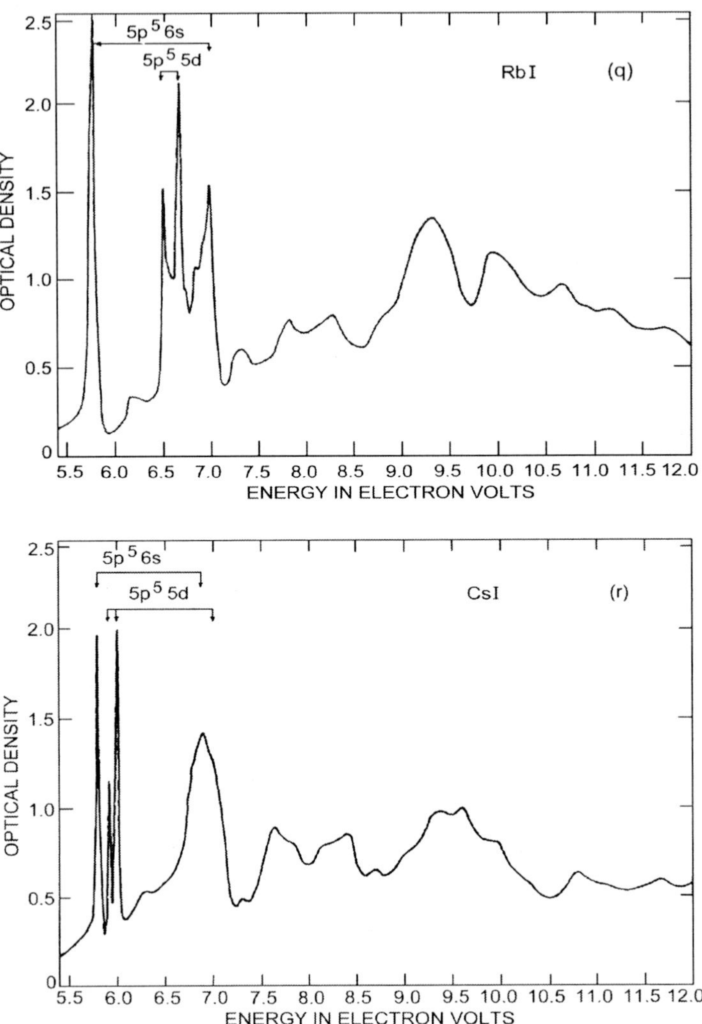

Notes and Comments

1. The lowest energy line is assigned to an exciton associated with the highest valence band Γ_8 (or Γ_{15}) and the lowest conduction level Γ_6 (or Γ_1) at $k = 0$. The energy of this peak is given in Table XVI.

Table XVI Energies of the 1st peak in the absorption spectra [8.21]

Crystal	Energy of 1st peak [eV]	Crystal	Energy of 1st peak [eV]
LiBr	7.20	RbCl	7.51
NaCl	7.96	RbBr	6.60
NaBr	6.68	RbI	5.70
NaI	5.56	CsCl	7.37
KCl	7.76	CsBr	6.80
KBr	6.77	CsI	5.76
KI	5.80		

2. The first peak is due to transition from Γ_{15} (Γ_8^-) to Γ_1 (Γ_6^+). But the Γ_{15} splits into (Γ_8^-) and (Γ_6^-) due to a spin-orbit effect. The predicted energy difference between these split states is 0.047 eV for F^-, 0.103 eV for Cl^-, 0.432 eV for Br^- and 0.889 eV for I^- [8.22]. As a consequence, the 1st line should be a doublet [($\Gamma_8^- - \Gamma_8^+$), ($\Gamma_6^- - \Gamma_6^+$). The lowest energy line is the low energy component of this doublet. The doublet structure is not observed in the RT spectra but shows up in the 80 K spectra [8.21]. The doublet structure is seen with better resolution in the 10 K spectra [8.2]. Assignment of the second component of the halogen spin-orbit doublet is not straightforward due to appearance of several other lines. The assignment in the paper by Eby et al. [8.21] has been revised in the paper by Teegarden and Baldini [8.2].

3. Absorption 'shoulders' have been observed by Eby et al. [8.21] and have been assigned to onset of band-to-band transitions. The shoulders occur at energies given in Table XVII.

Table XVII Energies corresponding to the shoulders in the absorption spectra

Crystal	Shoulder energy [eV]	Crystal	Shoulder energy [eV]
LiI	5.9	KI	6.2
NaCl	8.6	RbF	10.4
NaBr	7.7	RbCl	8.2
NaI	5.8	RbBr	7.7
KF	10.9	RbI	6.1
KCl	8.5	CsF	10.0
KBr	7.8	CsI	6.3

4. The assignment of other features in the spectra is highly involved and is not unambiguous (for details, see, [8.2, 8.21, 8.22]).

8.3.2   Extreme UV Absorption Spectra (50–250 eV)

Optical absorption spectra in the 50–250 eV region are shown in Fig. 8.7.

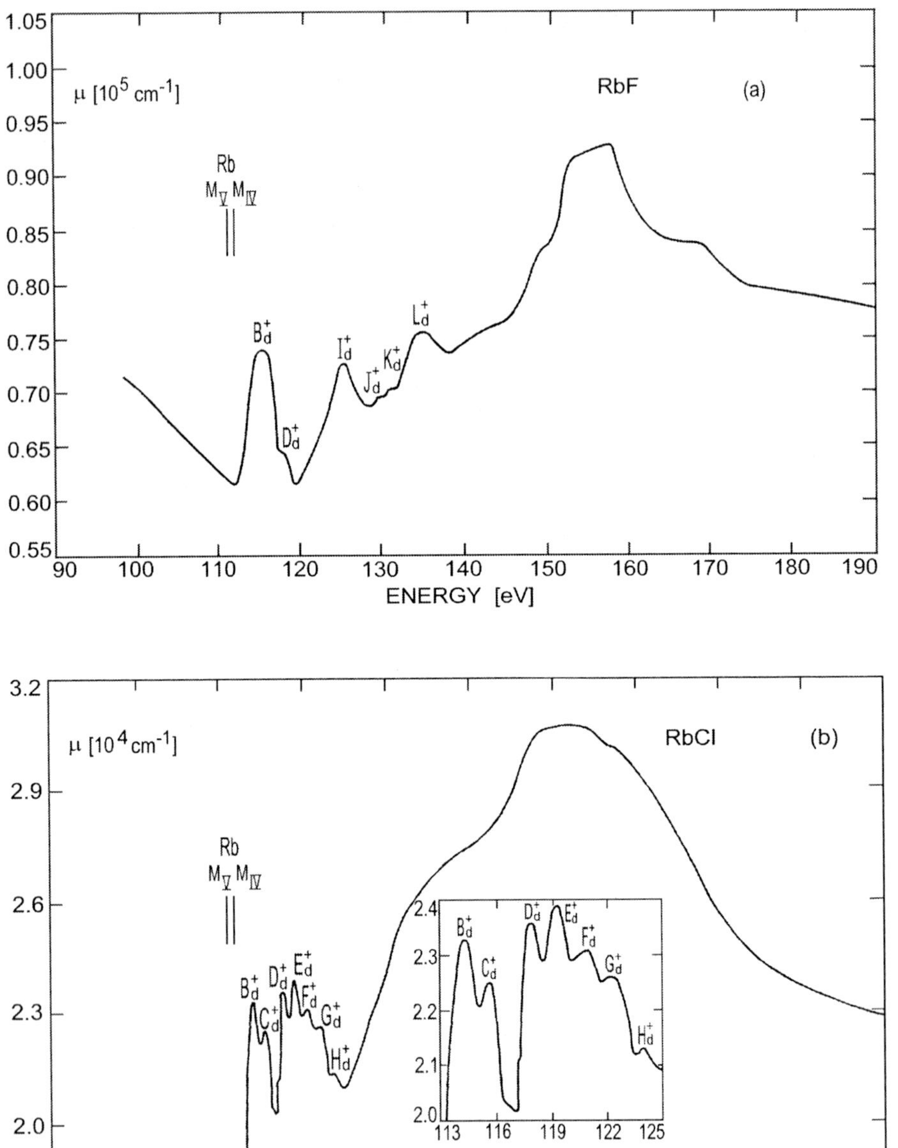

Fig. 8.7. Optical absorption spectra of evaporated thin films of Rb halides (a–d) and Cs halides (e–h) at RT in the extreme UV region (50–250 eV) using 7.5 GeV electron synchrotron radiation (after [8.23])

Fig. 8.7 (Continued)

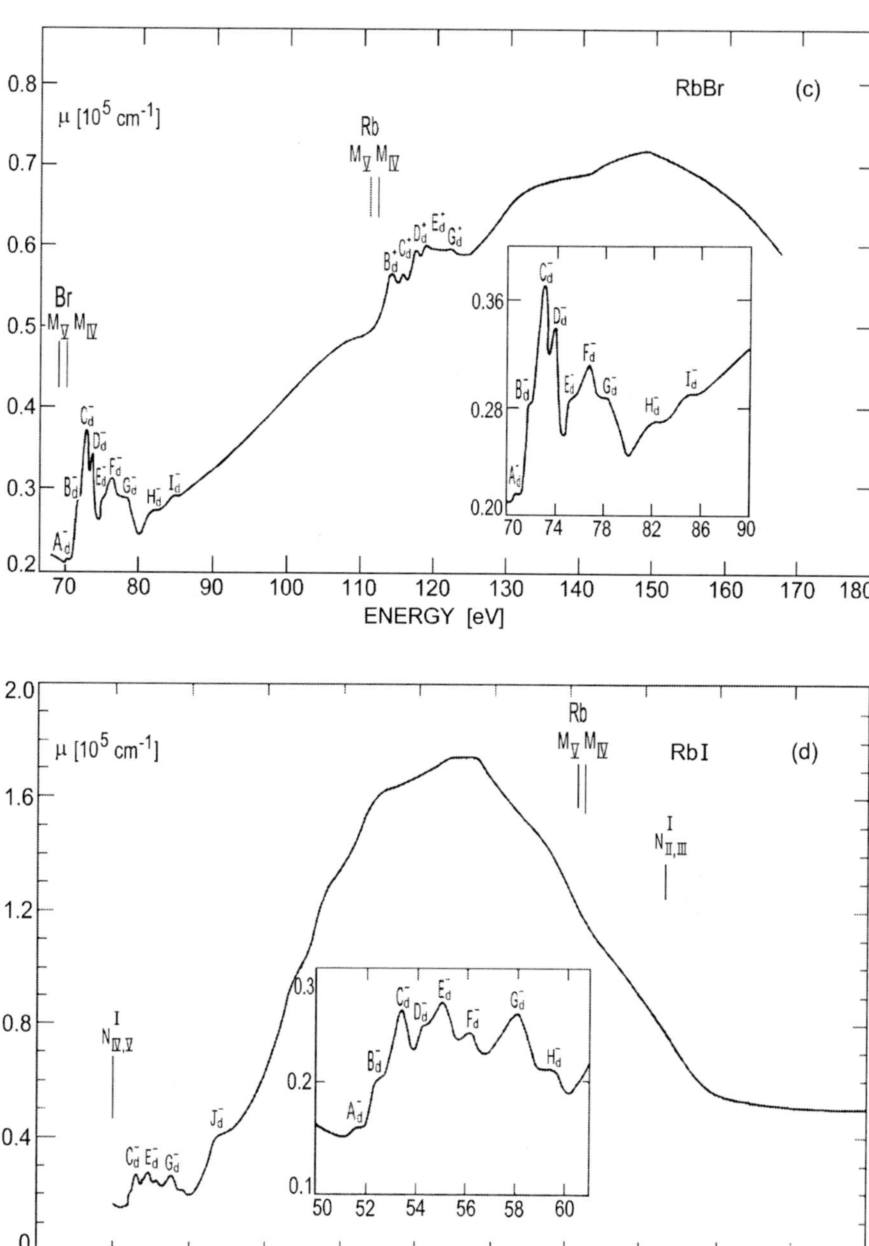

Fig. 8.7 (Continued)

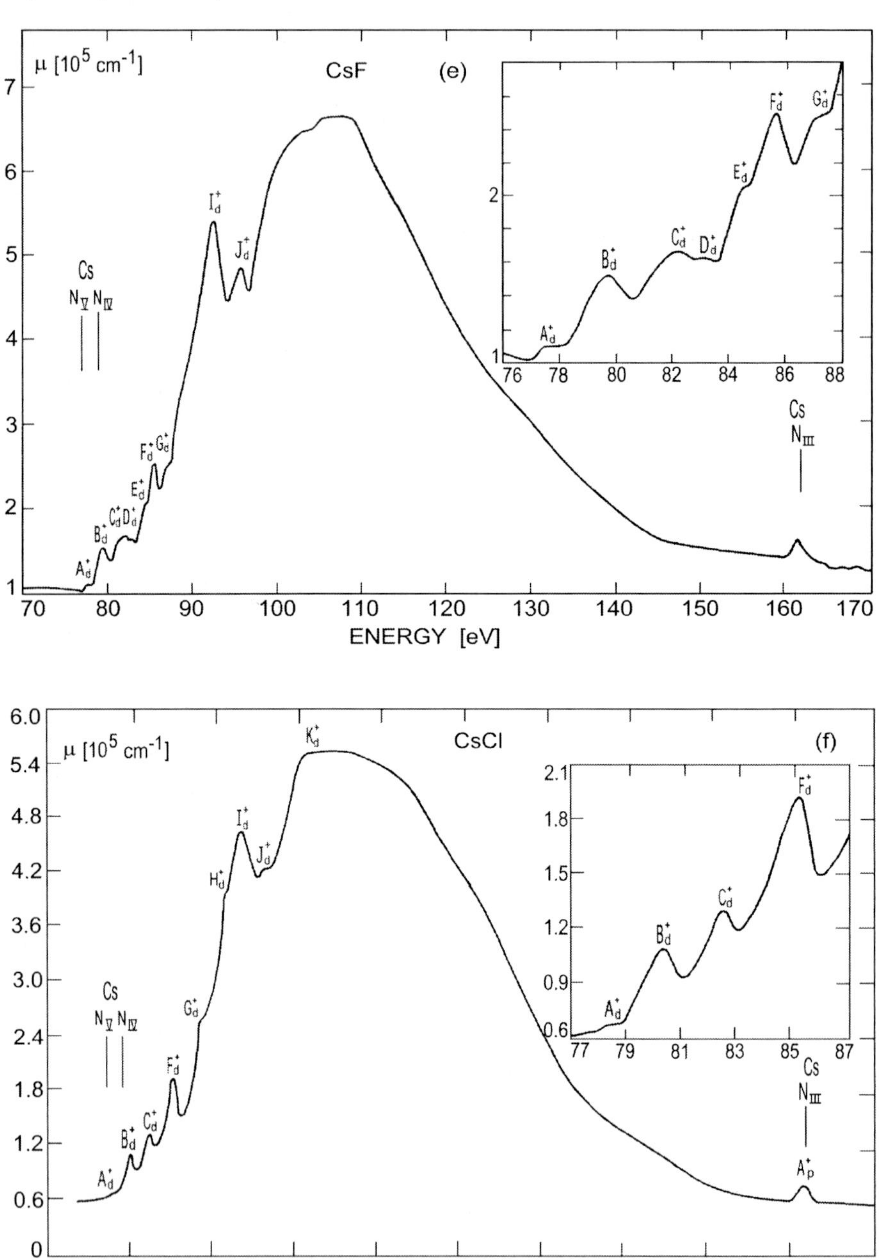

Fig. 8.7 (Continued)

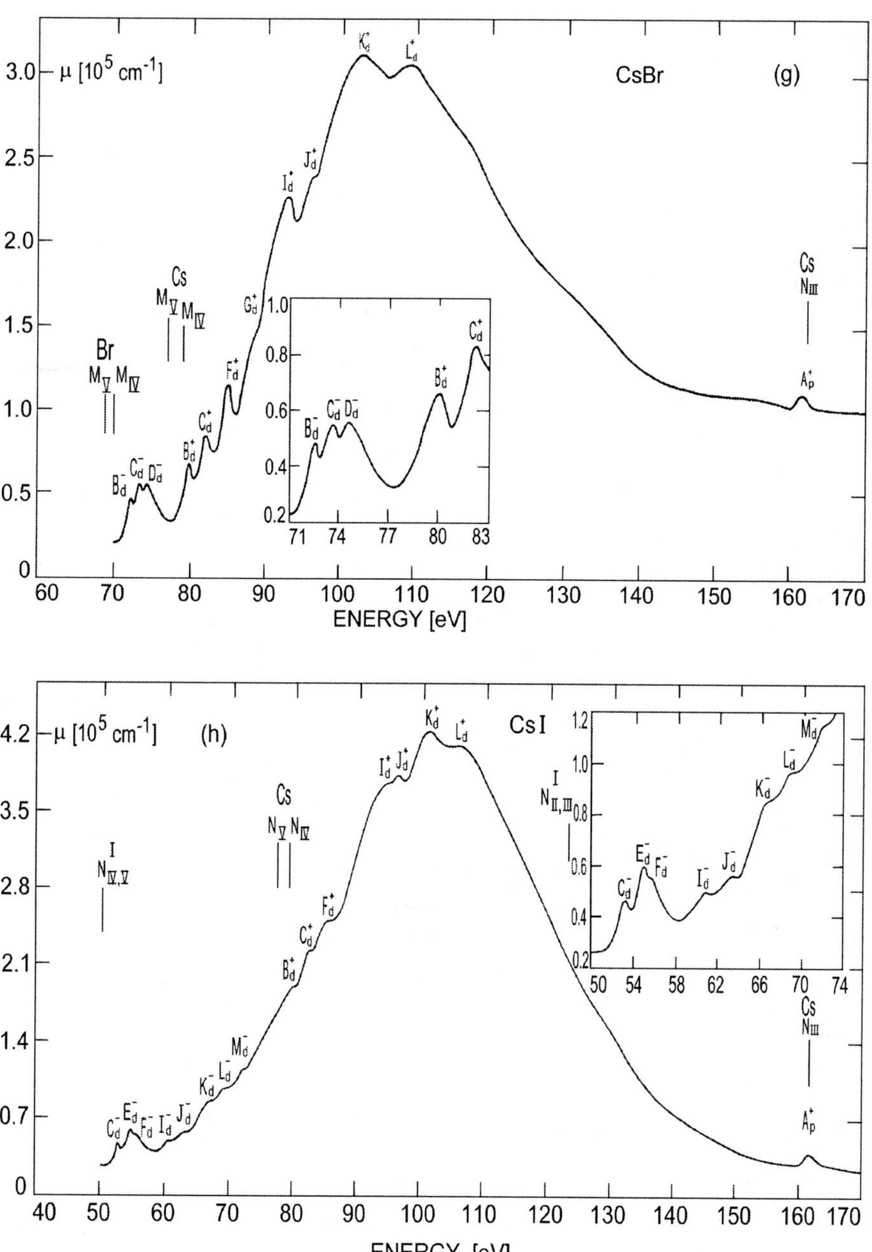

Notes and Comments

1. The energies of peaks in the absorption spectra of rubidium halides are given in Table XVIII.

Table XVIII Energies of peaks (in eV) in the absorption spectra of Rb halides recorded at room temperature; the energy of the corresponding free atom levels referred to vacuum is also given (nomenclature as in [8.23]); uncertainty $\pm$ 0.3 eV

	Free Atom		RbF		RbCl		RbBr		RbI	
I	$N_{IV,V}$	50								
									A_d^-	51.7
									B_d^-	52.7
									C_d^-	53.4
									D_d^-	54.4
									E_d^-	55.0
									F_d^-	56.2
									G_d^-	58.1
									H_d^-	59.6
									I_d^-	64
Br	M_V	69								
	M_{IV}	70								
							A_d^-	71.1		
							B_d^-	72.1		
							C_d^-	73.0		
							D_d^-	73.9		
							E_d^-	75.6		
							F_d^-	76.7		
							G_d^-	78.2		
							H_d^-	82.9		
							I_d^-	85.2		
Rb	M_V	111								
	M_{IV}	112								
					B_d^+	114.2	B_d^+	114.0		
			B_d^+	115.5						
					C_d^+	115.6	C_d^+	115.8		
					D_d^+	117.8	D_d^+	117.4		
			D_d^+	118.4						
					E_d^+	119.2	E_d^+	118.8		
					F_d^+	120.9				
							G_d^+	122.0		
					G_d^+	122.3				
					H_d^+	123.9				
I	$N_{II,III}$	123								
			I_d^+	125.4						
			J_d^+	130.0						
			K_d^+	131.6						
			L_d^+	135.0						

2. The energies of peaks in the absorption spectra of Cs halides are given in Table XIX.

Table XIX Energies of peaks in the absorption spectra of Cs halides recorded at room temperature; the energy of the corresponding free atom levels referred to vacuum is also given (nomenclature as in [8.23]); uncertainty ± 0.3 eV

		Free Atom	CsF		CsCl	CsBr		CsI	
I	$N_{IV,V}$	50							
								C_d^-	53.1
								E_d^-	54.9
								F_d^-	55.7
								I_d^-	60.8
								J_d^-	63.4
								K_d^-	66.4
								L_d^-	68.5
								M_d^-	68.5
Br	M_V	69							
	M_{IV}	70							
							B_d^-	72.6		
							C_d^-	73.7		
							D_d^-	74.6		
Cs	N_V	77							
	N_{IV}	79							
			A_d^+	78.1	(78.7)				
			B_d^+	79.7	80.3	80.1		79.9	
			C_d^+	82.4	82.5	82.2		82.0	
			D_d^+	83.5					
			E_d^+	84.7					
			F_d^+	85.7	85.1	85.1		84.4	
			G_d^+	87.5	88.7	88.2			
			H_d^+		(91.3)				
			I_d^+	92.8	93.2	93.2		(92.5)	
			J_d^+	95.8	96.1	96.2		96.3	
			K_d^+		101.1	102.8		100.7	
			L_d^+			110.0		105.3	
Cs	N_{III}	162							
	N_{II}	172							
			A_P^+	161.6	161.7	161.7		161.7	
			B_P^+	164.3	164.4				
			C_P^+	166.9	167.0			167.0	
			D_P^+	168.5	169.5				
			E_P^+	170.7	170.9			170.8	
			F_P^+	172.7	173.2	172.6		173.2	
			G_P^+		176.7	176			
			H_P^+	183.0	185.8			180.9	
			A_P^-			184.4			
			B_P^-			191.1			

3. Cardona et al.[8.23] assigned several (but not all) peaks in the observed spectra as transitions from the 3d shell of Rb^+ and the 4d and 4p shells of Cs^+ and also transitions from inner shells of the halogen ions. Also some broad absorption structure observed in the spectra has been assigned to d→f continuum transitions.

4. Cardona et al. [8.23] calculated the effective number of electrons N_{eff} from the absorption curves from the formula

$$N_{eff} = 2.3 \times 10^{15}\, a^3 \int_{E_1}^{E_2} \mu(E)\, n(E)\, dE \qquad (8.4)$$

where a is the lattice constant, μ the absorption coefficient and n the real part of the refractive index (~1). The results are given in Table XX.

Table XX Values of N_{eff} calculated from Eq. (8.4); [8.23]

Crystal	E_1 [eV]	E_2 [eV]	$N_{eff.}$
RbF	100	190	2.8
RbCl	100	190	1.3
RbBr	70	170	3.8
RbI	50	150	8.3
CsF	70	190	16.2
CsCl	70	190	15.1
CsBr	70	190	13.1
CsI	50	190	18

5. Additional information on UV absorption spectra of alkali halides is given in [8.24, 8.25]

8.4 Exciton Spectra

8.4.1 Exciton Energy

Table 8.6 Experimental and theoretical values of the exciton energy E_{exc}

	E_{exc} [eV]					
	Experimental				Theoretical	
Ref. Crystal	[8.2]	[8.21]	[8.26]	[8.27]	[8.28]	[8.30]
NaCl Structure						
LiF	-	-	12.9	-	-	-
LiCl	8.67	-	-	-	-	-
LiBr	7.23	7.20	-	-	-	-
LiI	-	-	-	5.94	-	-
NaF	10.66	-	-	-	-	-

Table 8.6 (Continued)

	E_{exc} [eV]					
	Experimental				Theoretical	
Ref.	[8.2]	[8.21]	[8.26]	[8.27]	[8.28]	[8.30]
Crystal						
NaCl Structure						
NaCl	7.96	7.96	-	-	7.81	-
NaBr	6.71	6.68	-	-	6.80	7.0
NaI	5.61	5.56	-	-	5.61	-
KF	9.88	-	-	-	-	-
KCl	7.79	7.76	-	-	7.70	7.5
KBr	6.71	6.77	-	-	6.81	-
KI	5.88	5.80	-	-	5.71	-
RbF	9.54	-	-	-	-	-
RbCl	7.54	7.51	-	-	7.56	-
RbBr	6.64	6.60	-	-	6.71	-
RbI	5.73	5.70	-	-	5.67	5.9
CsF	9.27	-	-	-	-	8.7
CsCl Structure						
CsCl	7.85	7.37	-	-	-	-
CsBr	6.83	6.80	-	-	-	-
CsI	5.30	5.76	-	-	-	-

Notes and Comments

1. The exciton energy is estimated from the lowest energy band in the exciton (UV absorption) spectra. Data from [8.21] are from spectra at 80 K or RT. The data for LiF [8.26] is from a spectrum at 300K, for LiI [8.27] from a spectrum at 14 K and for the others [8.2] from spectra at 80 K or RT.

2. Damm and Chovj [8.28] showed that the exciton energy values ($E_{exc.}$) for the chlorides, bromides and iodides of Na, K and Rb, follow the relation

$$E_{exc} = 44.82 \; (r_+ / r_-)^{2/3} \; r^{-3/2} \tag{8.5}$$

   where r_+ and r_- are ionic radii of cation and anion respectively and r is the interionic distance. The values calculated from this equation are given in Table 8.6 (Ref. [8.28]).

3. Hilsch and Pohl [8.29] proposed the following relation for predicting the exciton energy ($E_{exc.}$)of alkali halides:

$$E_{exc} = E_A - E_I + \alpha_r (e^2 / r) \tag{8.6}$$

   where E_A is the electron affinity of the halogen, E_I the ionisation energy of the alkali atom, α_r the Madelung constant and r the interionic distance. The values calculated from this relation by Knox [8.30] are given in Table 8.6. It is seen that these calculated values are in fair agreement with the experimental values.

8.4.2 Pressure Derivative of Exciton Energy

Table 8.7 Values of the pressure derivative $(\partial E_{exc}/\partial P)$ of the exciton energy

Crystal	$\partial E_{exc}/\partial P$ $[10^{-6}\ eV\ bar^{-1}]$	Ref.
NaCl Structure		
NaBr	13.8	[8.17]
NaI	14.3	[8.15]
KBr	18.0	[8.17]
KI	17.1	[8.15]
RbCl	16.0	[8.15]
RbBr	17.9	[8.17]
RbI	19.7	[8.15]
CsCl Structure		
CsBr	15.0	[8.15]

Notes and Comments

1. The pressure variation of E_{exc} was studied by Fujita and co-workers [8.15, 8.17] by recording and analysing the UV spectrum up to 3000 bars. Over this pressure range, the pressure variation of E_{exc} was found to be linear.
2. Fujita and co-workers [8.15, 8.17] have suggested that the pressure derivative $\partial E_{exc}/\partial P$ may be considered equal to the pressure derivative of the energy gap dE_G/dP.
3. Limited studies of the pressure variation of exciton energies have been reported in [8.31, 8.32].

8.5 UV Photoelectron and X-ray Photon Emission

Table 8.8 Values of E_S, the energy separation between outer valence bands of cation and anion

	E_S [eV]; uncertainty $\pm\,0.2$ eV		
Method	X-ray photon emission		UV photoelectron emission
Ref. Crystal $\downarrow$	[8.33]	[8.34]	[8.13]
NaCl Structure			
LiF	47.7	-	-
LiCl	-	51.3	-
NaF	21.6	-	-
NaCl	-	25.5	-
NaBr	-	26.4	-
NaI	-	27.6	-

Table 8.8 (Continued)

Method	E_S [eV]; uncertainty $\pm$ 0.2 eV		
	X-ray photon emission		UV photoelectron emission
Ref. Crystal $\downarrow$	[8.33]	[8.34]	[8.13]
NaCl Structure			
KF	-	-	10.5
KCl	-	-	12.9
KBr	-	-	13.7
KI	-	-	14.6
RbF	-	-	7.4
RbCl	-	-	9.5
RbBr	-	-	10.2
RbI	-	-	11.3
CsF	-	-	4.8
CsCl Structure			
CsCl	-	-	6.7
CsBr	-	-	7.5
CsI	-	-	8.4

Notes and Comments

1. For a given cation, the E_S values decrease in the sequence F–Cl–Br–I. Also, for a given anion, the E_S values decrease in the sequence Li–Na–K–Rb–Cs.
2. Damm and Chovj [8.28] have shown that the E_S values fit the relation:

$$E_S = 140.6 \left(r_+ / r_- \right)^{-2} r^{-2.2} \tag{8.7}$$

   where r is the interionic distance and r_+ and r_- are the cation and anion radius respectively.
3. Poole et al. [8.13] used the E_S values to estimate the ionicity. The ionicity values thus obtained agree well with the Pauling ionicity values.

8.6 Characteristic Electron Energy Loss Spectra

Table 8.9 Values of Characteristic energy losses (E_L); the identification of the energy losses is: V–Valence electron plasma, X–Unidentified plasma, I–Ionization, T–Interband transition, G–Excited state, U–Unidentified

Crystal	E_L [eV]; Ref. [8.35]
	(uncertainty in last digit indicated in parenthesis)
NaCl Structure	
LiF	14.5(2) I, 17.0(3) T, 24.9(2) V, 43.2(4) T, 50.6(4) V, 62.0(2) T, 68.5(5) T

Table 8.9 (Continued)

Crystal	E_L [eV]; Ref. [8.35]
	(uncertainty in last digit indicated in parenthesis)

NaCl Structure

LiCl	9.9(5) I, 16.1(2) V, 23.1(4) U, 29.5(3) U, 45.2(6) T, 61.0(3) T
NaF	11.6(1) I, 17.3(1) T, 20.9(2) V, 23.9(2) T, 33.9(2) I, 39.3(2) T, 46.8(2) T, 55.7(9) I, 64.5(5) I
NaCl	8.7(1) I, 12.7(2) T, 15.5(2) V, 22.2(2) I, 33.5(2) I, 44.5(5) I
NaBr	7.6(3) I, 13.9(3) V, 20.8(2) I, 28.1(3) V, 33.4(3) I, 44.0(4) I
NaI	12.3(2) V, 17.8(2) I, 25.9(3) V, 33.1(2) I
KF	10.1(2) U, 12.4(2) X, 17.1(3) V, 22.5(3) U, 26.9(4) I, 31.0(2) G, 37.7(3) I, 61.4(3) G
KCl	10.2(2) X, 13.8(2) V, 19.6(2) I, 23.0(4) U, 26.4(3) U, 29.3(2) G, 36.2(3) I
KBr	9.2(2) X, 12.5(1) V, 18.2(3) I, 22.0(2) U, 25.4(2) V, 28.0(2) G, 34.7(3) I
KI	8.0(3) X, 11.2(1) V, 16.0(2) I, 21.9(3) V, 27.1(3) G, 33.0(2) I
RbCl	9.8(2) X, 13.1(2) V, 19.5(2) U, 22.8(2) U, 25.6(2) G, 33.5(5) I, 50.8(4) G

CsCl Structure

CsBr	8.7(2) X, 11.1(2) V, 19.2(3) U, 22.5(3) G, 27.3(4) I
CsI	9.8(1) V, 18.2(2) U, 20.7(3) G, 26.8(2) I, 39.8(8) G

Notes and Comments

1. The peak assigned as V in the energy loss spectrum is attributed to valence electron plasma oscillation. The energy of this peak agrees with the plasma oscillation energy calculated theoretically.
2. The energy of the peak marked X bears a constant ratio of $(1/\sqrt{2})$ with the energy of the V peaks. This peak is attributed to longitudinal plasma oscillations.
3. The electron energy loss spectra correlate with X-ray absorption spectra. As shown in Table XXI, the difference $(E_V - E_X)$ is close to the energy $(E_B - E_A)$ where E_A and E_B are the energies of the first two maxima in X-ray absorption spectra.

Table XXI Values of energy difference $(E_V - E_X)$ and $(E_B - E_A)$; [8.35]

	KF	KCl	KBr	KI
$(E_V - E_X)$ in [eV]	4.7	3.6	3.3	3.2
$(E_B - E_A)$ in [eV]	4.4	3.6	3.1	1.8

4. Data on energy loss spectra is also given in [8.36–8.38].

8.7 Plasma Oscillation Frequency

Table 8.10 Values of the valence electron plasma energy $\hbar\omega_p$ (where ω_p is the plasma oscillation frequency)

Crystal	$\hbar\omega_p$ [eV]	
	Theoretical [8.39]	Experimental [8.35]
NaCl Structure		
LiF	25.96	24.9
LiCl	17.99	16.1
LiBr	10.87	-
LiI	13.16	-
NaF	20.11	20.9
NaCl	15.68	15.5
NaBr	14.37	13.9
NaI	12.74	12.3
KF	16.83	17.1
KCl	13.29	13.8
KBr	12.38	12.5
KI	11.19	11.2
RbF	15.03	-
RbCl	12.40	13.1
RbBr	11.59	-
RbI	10.53	-
CsF	13.41	-
CsCl Structure		
CsCl	12.54	-
CsBr	11.77	11.1
CsI	10.73	9.8

Notes and Comments

1. The experimental values were estimated from electron energy loss spectra; [8.35].
2. The theoretical values given in [8.39] were calculated from the formula

$$\hbar\omega_p = 28.8 \, (Z\rho / M)^{1/2} \tag{8.8}$$

   where Z is total number of valence electrons ($Z_{cation} + Z_{anion}$), ρ the density [g cm^{-3}] and M the molecular weight [g].

3 Reddy et al. [8.39] reported the following correlation between the lattice energy U and the valence electron plasma energy $\hbar\omega_p$ for the alkali halides:

$$U = 25.84 \, \hbar\omega_p + 378 \tag{8.9}$$

4. Reddy et al. [8.39] used the ω_p values to calculate the electronic polarisability α of the alkali halides from the correlation given by Eq. (8.10):

$$\alpha = \left[\frac{(\hbar\omega_{\mathrm{P}})^2 S_0}{(\hbar\omega_{\mathrm{P}})^2 S_0 + 3E_{\mathrm{P}}^2} \right] \times \left(\frac{M}{\rho} \right) \times 0.396 \times 10^{-24} \ \mathrm{cm}^3 \qquad (8.10)$$

where

$$S_0 = 1 - (E_{\mathrm{P}}/4E_{\mathrm{F}}) + (1/3)(E_{\mathrm{P}}/4E_{\mathrm{F}})^2$$

$$E_{\mathrm{P}} = (\hbar\omega_{\mathrm{P}})/(\varepsilon_\infty - 1)^{1/2} \ \mathrm{eV}$$

$$E_{\mathrm{F}} = 0.295 (\hbar\omega_{\mathrm{P}})^{4/3} \ \mathrm{eV}$$

ε_∞ = Electronic dielectric constant

The α values thus calculated agree well with those calculated from the Clausius-Mosotti relation.

8.8 Metallisation and Superconductivity

Table 8.11 Values of the pressure (P_{M}) at which metallisation is theoretically predicted from band structure calculations (BSC) or experimentally observed from optical absorption (OA) are given along with the structure (Str.) at P_{M}; the pressure (P_{S}) at which the superconductivity sets in, calculated from BSC or experimentally observed from electrical resistance measurements (ERM), are also given.

Substance	Metallisation			Superconductivity				Ref.
	P_{M} [GPa]	Str.	Method	P_{S} [GPa]	Str.	T [K]	Method	
RbI	85	CsCl	OA	-	-	-	-	[8.40]
	122	CsCl	BSC	-	-	-	-	[8.41]
KI	115	CsCl	OA	-	-	-	-	[8.40]
	155	CsCl	BSC	-	-	-	-	[8.41]
NaCl	> 135	CsCl	BSC	-	-	-	-	[8.42]
	> 135	CsCl	OA	-	-	-	-	[8.43]
	-	-	-	> 135	CsCl	< 7	ERM	[8.44]
CsI	105	CuAu I	OA	-	-	-	-	[8.45]
	111	CsCl	BSC	> 111	HCP	0.04	BSC	[8.46]
	115	CsCl	ERM	180	HCP	2	ERM	[8.47]

References

8.1   L.J. Page and E.H. Hygh, Phys. Rev., **B1**, 3472, 1970.

8.2   K. Teegarden and G. Baldini, Phys. Rev., **155**, 896, 1967.

8.3   U. Rossler, phys. stat. sol., **34**, 207, 1969.

8.4   A.B. Kunz, J.Phys. C: Solid State Phys., **3**, 1542, 1970.

8.5   A.B. Kunz, Phys. Rev., **B2**, 5015, 1970.

8.6   H. Overhof, phys. stat. sol., **(b)43**, 575, 1971.

8.7   A.B. Kunz and N.O. Lipari, Phys. Rev., **B4**, 1374, 1971.

8.8   F. Perrot, phys. stat. sol., **(b)52**, 163, 1972.

8.9   G.E. Laramore and A.C. Switendick, Phys. Rev., **B7**, 3615, 1973.

8.10   A.B. Kunz, Phys. Rev., **B26**, 2056, 1982 and references therein.

8.11   *American Institute of Physics Handbook*, 2nd Ed., McGraw Hill, New York, 1963.

8.12   F.C. Brown, C. Gahwiller, H. Fujita, A.B. Kunz, W. Sheifley and N. Carrera, Phys. Rev., B2, 2126, 1970 and references therein.

8.13   R.T. Poole, J.G. Jenkin, R.C.G. Leckey and J. Liesegang, Chem. Phys. Lett., **26**, 514, 1974 and references therein.

8.14   R.R. Reddy and Y.N. Ahammed, Infrared Physics and Technology, **36**, 825, 1995.

8.15   H. Fujita, K. Yamauchi, A. Akasaka, H. Irie and S. Masunaga, J. Phys. Soc. Japan, **68**, 1994, 1999.

8.16   S. Masunaga and H. Fujita J. Phys. Soc. Japan, **67**, 2146, 1998.

8.17   H. Fujita and S. Masunaga, J. Phys. Soc. Japan, **66**, 4036, 1997.

8.18   R.S. Knox and K.J. Teegarden, in *Physics of Colour Centres*, Ed. W.B. Fowler, Academic Press, New York, 1968, and references therein.

8.19   C. Gout and F. Pradal, J. Phys. Chem. Solids, **28**, 1507, 1967.

8.20   T.D. Clark and K.L. Kliewer, Phys. Lett., **27A**, 167, 1968.

8.21   J.E. Eby, K.J. Teegarden and D.B. Dutton, Phys. Rev., **116**, 1099, 1959.

8.22   R.S. Knox and N. Inchauspe, Phys. Rev., **116**, 1093, 1959.

8.23   M. Cardona, R. Haensel, D.W. Lynch and D. Sonntag, Phys. Rev., **B2**, 1117, 1970.

8.24   Y. Iguchi, Sci. Light, **19**, 1, 1970.

8.25   G.W. Rubloff, J. Freeouf, H. Fritzsche and K. Murase, Phys.Rev. Lett., **26**, 1317, 1971.

8.26   A. Milgram and M.P. Givens, Phys. Rev., **125**, 1506, 1962.

8.27   F. Fischer and R. Hilsch, Z. Physik, **158**, 553, 1960.

8.28   J.Z. Damm and Z. Chovj, phys. stat. sol., **(b)114**, 413, 1982.

8.29   R. Hilsch and R.W. Pohl, Z. Physik, **48**, 384, 1928; **57**, 145, 1929; **59**, 812, 1930.

8.30   R.S. Knox, *Theory of Excitons*, Academic Press, New York, 1963.

8.31   H. Zhang, W.B. Daniels and R.E. Cohen, Phys. Rev., **B50**, 70, 1994.

8.32   K. Reimann, High Pressure Research, **15**, 73, 1996.

8.33   V.V. Nemoshkalenko, A.I. Senkevich and V.G. Aleshin, Sov. Phys. Doklady, **19**, 936, 1973 quoted in [8.13].

8.34   P.H. Citrin and T.D. Thomas, J. Chem. Phys., **57**, 4446, 1972 quoted in [8.13].

8.35   P.E. Best, Proc. Phys. Soc., **79**, 133, 1962

8.36   L.B. Leder, Phys. Rev., **103**, 1721, 1956; **107**, 1569, 1957.

8.37   L.B. Leder, H. Mendlowitz and L. Marton, Phys. Rev., **101**, 1460, 1956.

8.38   H. Watanabe, J. Electron Microscopy, **4**, 24, 1956.

8.39   R.R. Reddy, M. Ravi Kumar and T.V.R. Rao, Cryst. Res. Technol., **28**, 973, 1993.

8.40   K. Asaumi, T. Suzuki and T. Mori, Phys. Rev., **B28**, 3529, 1983.

8.41   R.M. Amirthakumari, G. Pari, R. Rita and R. Asokamani, phys. stat. sol., **(b)199**, 157, 1997.

8.42 V.A. Zhadnov, V.A. Kuchin and V.V. Polyakov, Izv. vyssh. uch, zav. ser. Fizika, N3, 1973.

8.43 L.F. Vereschagin, E.N. Yakovlev, B.V. Vinogradov and V.P. Sakun, Zh. Eksp. Teor. Fiz., Pis'ma Red., 20, 540, 1974.

8.44 G.N. Stepanov, E.N. Yakovlev and T.V. Valanskaya, Proc. VII Int. Conf. High Pr. Sci. Tech.., 1979.

8.45 E. Knittle and R. Jeanloz, J. Phys. Chem. Solids, 46, 1179, 1985.

8.46 R. Asokamani, G. Subramoniam and R.M. Amirthakumari, Proc. XIII AIRAPT Int. Conf. on High Pr. Sci. and Tech., 1991.

8.47 M.I. Eremets, K. Shimizu, T.C. Kobayashi and K. Amaya, Science, 281, 1333, 1998.

9 Defect State Parameters

9.1 Schottky Defects

9.1.1 Temperature Variation of Ionic Conductivity (Diagrams)

The temperature variation of ionic conductivity is shown in Fig. 9.1:

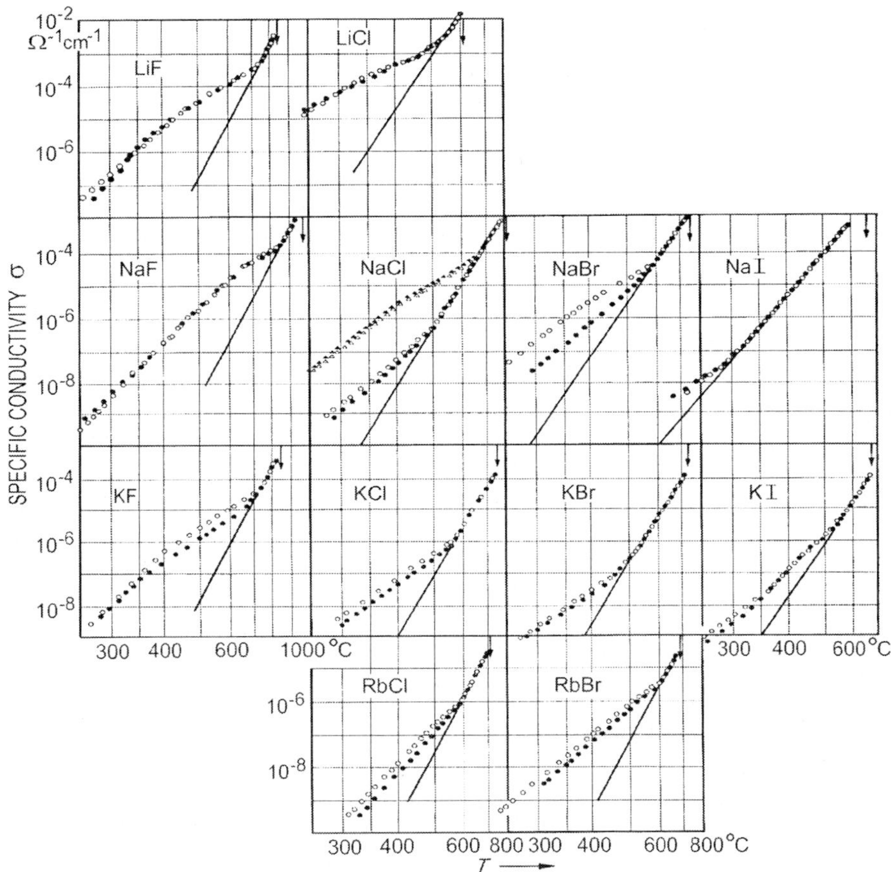

Fig. 9.1 The logarithm of the ionic conductivity plotted against the reciprocal of the absolute temperature for a number of alkali halides. In each case the complete curve divides into an intrinsic high temperature part and a structure-sensitive low temperature part which depends on the particular specimen studied (after [9.1])

9.1.2 Temperature Variation of Ionic Conductivity (2-Parameter Equation)

The temperature variation of ionic conductivity is commonly represented by the 2-parameter equation

$$\sigma(T) = \sigma_0 \exp\left(-W / k_B T\right) \tag{9.1}$$

where $\sigma(T)$ is the ionic conductivity at T K, σ_0 a constant, W the activation energy of conduction and k_B the Boltzmann constant.

Table 9.1 Values of the parameters σ_0 and W occurring in Eq. (9.1); Ref. [9.2]

Crystal	Temp. range [°C]	σ_0 [ohm^{-1} cm^{-1}]	W [eV]
NaCl Structure			
LiF	close to T_m	3.0×10^6	1.99
LiCl	400–550	2.5×10^5	1.42
	30–350	1.2	0.59
LiBr	350–500	4.2×10^5	1.22
	30–300	3.3	0.56
LiI	250–350	1.8×10^5	0.92
	30–150	1.4×10^{-1}	0.36
NaF	close to T_m	1.5×10^6	2.25
	330–980	1.3×10^3	1.42
NaCl	520–720	5.8×10^8 / T	1.89
	350–520	1.0×10^2 / T	0.83
NaBr	490–600	1.5×10^8 / T	1.66
	340–450	2.1×10^2 / T	0.80
NaI	350–600	8.1×10^3	1.23
	170–350	6×10^{-2}	0.60
KF	close to T_m	3×10^7	2.35
KCl	370–740	1.0×10^6	2.05
	100–250	2.1×10^{-5}	0.77
KBr	370–700	3×10^5	1.87
	60–190	8×10^{-6}	0.65
KI	450–675	4×10^4	1.62
	220–400	6×10^{-2}	0.85
RbCl	close to T_m	3×10^6	2.12
RbBr	close to T_m	1.8×10^6	2.03
CsF	550–657	1.6×10^5	1.55
	330–550	2	0.85
High temperature NaCl Phase			
CsCl	469–580	1	0.95
CsCl Structure			
CsCl	330–469	5×10	1.04
CsBr	475–590	2.5×10^5	1.44
	300–475	2.5×10^4	1.28
CsI	480–595	2.2×10^5	1.43
	300–480	1.4×10^4	1.25

Notes and Comments

1. W is referred to as the 'total activation energy'.
2. Mizuta and Yanagida [9.3] found linear relations of the type $W = aU - b$ between W and the lattice energy U where a and b are constants. With W and U both in eV, the constants a and b have values (0.35, 1.53), (0.35, 0.95), (0.35, 0.43) and (0.35, 0.24) for the halides of Li, Na, K and Rb. Mizuta and Yanagida [9.3] also found that the plots of W versus lattice constant are smooth curves.
3. W is actually made up of two components: $W = E_m + (h_S/2)$ where E_m is the activation energy for migration of vacancies and h_S is the energy of formation of a Schottky pair. E_m and h_S can be separately estimated from the activation energies from the intrinsic and extrinsic parts of the log σ-T^{-1} plots.
4. Aduev et al. [9.4] observed that the conductivity of KCl, KBr, NaCl and CsI excited by electron pulses (50 ps, 0.2 MeV) increased linearly with beam current.

9.1.3 Temperature Variation of Ionic Conductivity (4-Parameter Equation)

The temperature variation of ionic conductivity may be more rigorously represented by the double exponential 4-parameter equation:

$$\sigma T = \sigma_a T + \sigma_c T = A_0 \exp(-W_a / k_B T) + C_0 \exp(-W_c / k_B T) \qquad (9.2)$$

where σ is the conductivity at T K and σ_a and σ_c are the anion and cation contributions to σ; and W_a and W_c are anion and cation activation energies for conduction respectively and A_0 and C_0 are constants.

Table 9.2 Parameters pertaining to Eq. (9.2); Ref. [9.5]

Crystal	A_0 [ohm^{-1} cm^{-1} K]	W_a [eV]	C_0 [ohm^{-1} cm^{-1} K]	W_c [eV]
NaCl Structure				
NaCl	1.2×10^9	2.07	4.7×10^8	1.86
KCl	3.85×10^9	2.17	4.63×10^7	1.84
RbCl	8.85×10^{11}	2.55	3.58×10^6	1.58

9.1.4 Temperature Variation of Ionic Conductivity (6-Parameter Equation)

In the most sophisticated analysis, the ionic conductivity σ is represented by a 6-parameter equation:

$$\sigma T = A x_0 [\exp(\Delta S_1 / k_B) \exp(-\Delta h_1 / k_B T) + \exp(\Delta S_2 / k_B) \exp(-\Delta h_2 / k_B T) \qquad (9.3)$$

with $x_0^2 = B \exp(S_S / k_B) \exp(-h_S / k_B T)$ where A and B are constants, k_B the Boltzmann constant, h_S, Δh_1 and Δh_2 the enthalpies of formation of a Schottky pair and of the motion of the cations and anions. S_S, ΔS_1 and ΔS_2 are the entropies associated with the formation of a Schottky pair and of the motion of the cations and anions respectively. The determination of these six parameters needs data on conductivities in the intrinsic region and with divalent anionic and cationic doping as well as diffusion data.

Table 9.3 Parameters of Eq. (9.3)

Crystal	h_S	Δh_1	Δh_2	S_S	ΔS_1	ΔS_2	Ref.
	[eV]			$[10^{-3}\ eV\ (deg)^{-1}]$			
NaCl Structure							
LiF	2.34–2.68	0.65	1.1	0.827	0.086	0.301	[9.6]
KCl	2.59	0.73	0.99	0.828	0.232	0.356	[9.7]
KBr	2.53	0.65	1.22	0.887	0.163	0.629	[9.8]
KI	2.21	0.63	1.29	0.765	0.136	0.805	[9.9]
RbI	2.1	0.60	1.6	0.5	0.14	0.13	[9.10]

Notes and Comments

1. Jacobs and Vernon [9.11] made conductivity measurements on pure RbCl and RbCl doped with Sr and S ions and interpreted the results in terms of a model that included Schottky as well as Frenkel defects on anion and cation sublattices. A set of defect energies and entropies of formation, migration and association (18 parameters) has been established.

9.1.5 Diffusion Parameters

The temperature variation of diffusion is represented by the equation

$$D = D_0 \exp\left(-W / k_B T\right) \tag{9.4}$$

where D is the diffusion coefficient at temperature T K, D_0 a constant, W the activation energy for diffusion and k_B the Boltzmann constant.

Table 9.4 Parameters of the diffusion equation; Ref. [9.2]

Crystal	Isotope	Temp. range [°C]	$D_0\ [cm^2\ s^{-1}]$	W [eV]	Vacancy
NaCl Structure					
NaCl	Na^{24}	550–720	3.1	1.80	Na^+
	Cl^{36}	380–720	5.9×10^3	2.55	Cl^-
NaBr	Na^{24}	425–700	6.7×10^{-1}	1.53	Na^+
	Br^{82}	450–690	5.0×10^{-2}	2.02	Br^-
KF	K^{42}	580–840	2	1.78	K^+
KCl	K^{42}	670–750	1.5	1.74	K^+
	Cl^{36}	540–700	1.0×10^4	2.6	Cl^-

Table 9.4 (Continued)

Crystal	Isotope	Temp. range [°C]	D_0 [cm^2 s^{-1}]	W [eV]	Vacancy
NaCl Structure					
KBr	K^{42}	470–730	1×10^{-2}	1.26	K$^+$
	Br^{82}	490–730	2×10^{-2}	1.43	Br$^-$
KI	K^{42}	430–690	1×10^{-5}	0.64	K$^+$
	I^{131}	430–690	1.2×10^{-3}	1.12	I$^-$
CsF	Cs^{137}	480–640	3.1	1.67	Cs$^+$
CsCl Structure					
CsCl	Cs^{137}	290–465	1×10^{-5}	0.69	Cs$^+$
	Cl^{36}	290–465	1.3×10^{-3}	0.87	Cl$^-$
CsBr	Cs^{134}	320–550	1.5×10	1.54	Cs$^+$
	Br^{82}	415–530	3.9	1.42	Br$^-$
CsI	Cs^{134}	320–550	1.4×10	1.53	Cs$^+$
	I^{131}	410–540	2.1	1.37	I$^-$

9.1.6 Enthalpy of Formation of a Schottky Pair (Different Methods)

Table 9.5 Values of the enthalpy of formation of a Schottky pair (h_S)

Method	h_S [eV]					Theoretical
	Experimental		Empirical relations			
	Ionic conductivity	Thermal Expansion	Melting point	Debye temperature	Compressibility	Mott-Littleton theory
Ref. Crystal ↓	[9.6]	[*]	[9.16]	[**]	[9.16]	[9.19]
NaCl Structure						
LiF	2.34–2.68	2.42	-	2.50	-	2.45
LiCl	2.2	-	-	1.89	-	1.94
LiBr	1.8	-	1.75	1.72	1.98	1.81
LiI	1.34	-	1.54	1.47	1.86	1.62
NaF	-	2.42	2.60	3.12	2.42	2.99
NaCl	2.18–2.38	2.19	-	2.44	-	2.43
NaBr	1.72	-	2.20	2.31	2.15	2.26
NaI	-	-	1.98	2.00	1.97	2.04
KF	-	-	2.47	2.73	2.47	2.65
KCl	2.26	2.24	-	2.36	-	2.48
KBr	2.3–2.5	1.99	-	2.23	-	2.38
KI	1.6	1.83	-	2.20	-	2.22
RbF	-	-	2.21	2.81	2.25	2.38
RbCl	-	-	2.11	2.21	2.24	2.42
RbBr	-	2.14	2.04	2.09	2.11	2.32
RbI	-	1.81	1.96	1.98	2.12	2.21

| CsF | - | | - | | - | | - | | - | | 2.17 |

Table 9.5 (Continued)

Method	h_S [eV]					
	Experimental		Empirical relations			Theoretical
	Ionic conductivity	Thermal Expansion	Melting point	Debye temperature	Compressibility	Mott-Littleton theory
Ref.	[9.6]	[*]	[9.16]	[**]	[9.16]	[9.19]
Crystal ↓						
CsCl Structure						
CsCl	1.86	-	-	-	-	1.97
CsBr	2.0	1.74	-	1.56	-	1.85
CsI	1.9	-	-	1.58	-	1.88

* NaCl, KCl, CsBr [9.12]; LiF [9.13]; NaF, KBr, RbBr [9.14]; KI, RbI [9.15]
** NaCl structure crystals [9.17]; CsCl structure crystals [9.18]

Notes and Comments

1. The terms "Enthalpy of formation h_S" and "Energy of formation E_f" are used synonymously in literature [9.6].
2. Barr and Lidiard [9.6] showed that the h_S values correlate linearly with the melting point T_m according to the equation

$$h_S = 2.14 \times 10^{-3} T_m \quad (eV) \tag{9.5}$$

   Values of h_S have been calculated by Pathak and Vasavada [9.16].
3. Pathak and Trivedi [9.17] calculated h_S for NaCl type crystals from the Debye temperature (θ_M) using the relation

$$\theta_M = 4183 \, (h_S / MV^{2/3})^{1/2} \tag{9.6}$$

   where M is the molecular mass and V the molar volume. Subhadra and Sirdeshmukh [9.18] calculated h_S for CsCl type crystals using a similar relation.
4. Pathak and Vasavada [9.16] calculated h_S from the relation

$$h_S = 3.4 \times 10^{-13} V / \psi \quad (eV) \tag{9.7}$$

   where ψ is the compressibility.

9.1.7 Pressure Variation of Ionic Conductivity (σ-P Plots)

The pressure variation of ionic conductivity is shown in Figs. (9.2–9.6).

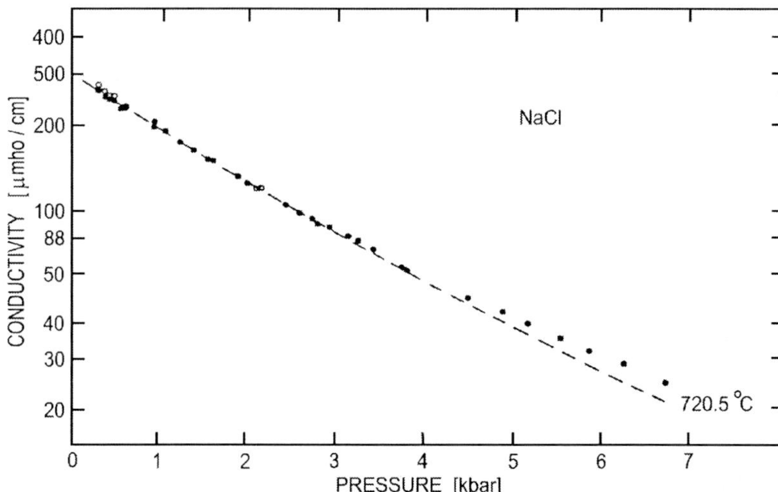

Fig. 9.2 Pressure dependence of conductivity in pure NaCl in the intrinsic temperature range (after [9.20])

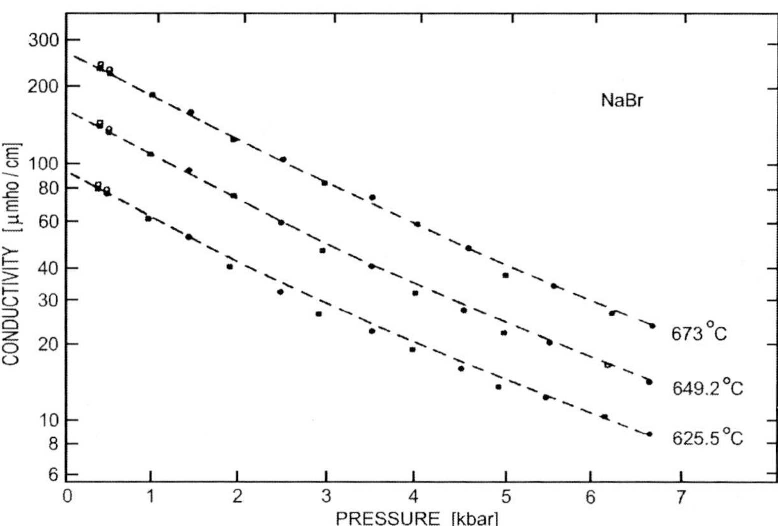

Fig. 9.3 Pressure dependence of conductivity in pure NaBr in the intrinsic temperature range (after [9.20])

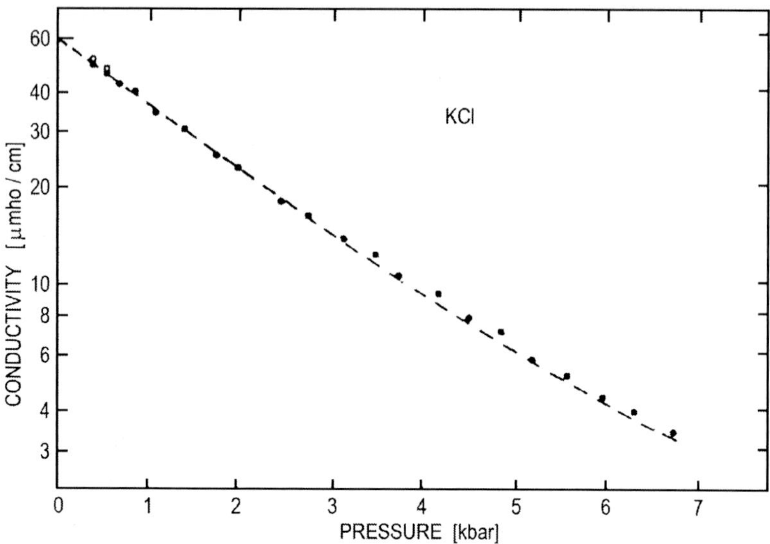

Fig. 9.4 Pressure dependence of conductivity in pure KCl in the intrinsic temperature range at 704 °C (after [9.20])

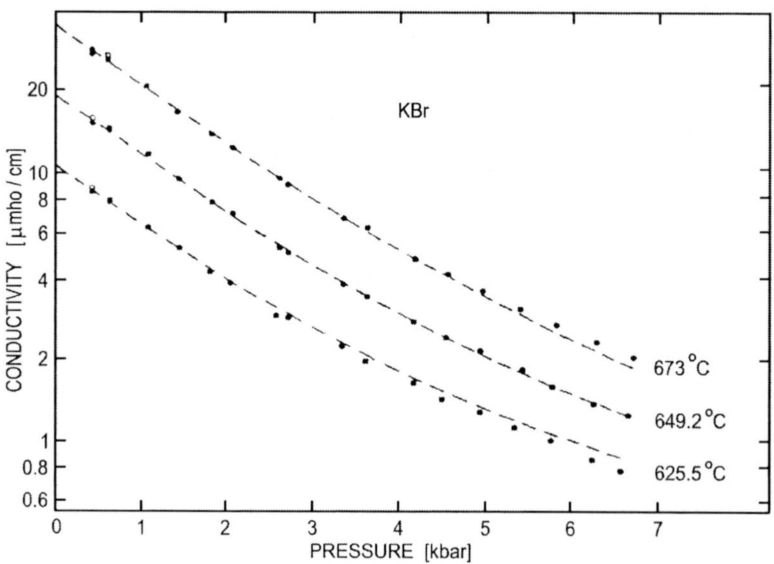

Fig. 9.5 Pressure dependence of conductivity in pure KBr in the intrinsic temperature range (after [9.20])

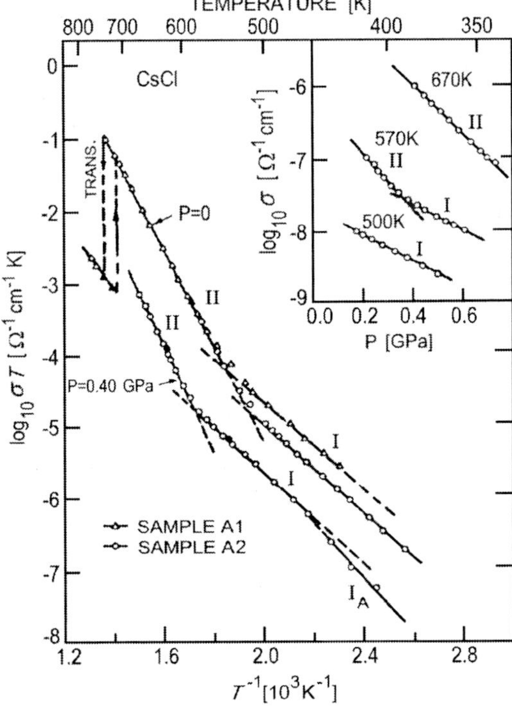

Fig. 9.6 Temperature dependence of the ionic conductivity of CsCl at 0 and 0.40 GPa showing the various conduction regimes. Also shown at atmospheric pressure is the change in conductivity at the structural transition to the NaCl structure. The inset shows the pressure dependence of the ionic conductivity at constant temperature in regimes I and II (after [9.21])

9.1.8 Pressure Variation of Ionic Conductivity (Pressure Coefficient and Activation Volume)

Table 9.6 Values of the pressure coefficient of ionic conductivity (d $\log\sigma$/ d P); migration activation volumes $\Delta V_{m.v}^{+}$, $\Delta V_{m.v}^{-}$ for cation and anion vacancies; molar volume V_M and activation volume ΔV_f associated with formation of Schottky defects; [9.20–9.22]

Crystal	(d $\log\sigma$/ d P) [kbar^{-1}]	$\Delta V_{m.v}^{+}$	$\Delta V_{m.v}^{-}$ [cm^3 mole^{-1}]	ΔV_f	$\Delta V_f/V_M$	Ref.
NaCl Structure						
NaCl	-0.43 ± 0.02	7 ± 1	-	55 ± 9	1.9 ± 0.3	[9.20]
NaBr	-	8 ± 1	-	44 ± 9	1.2 ± 0.3	[9.20]
KCl	-0.49 ± 0.02	8 ± 1	-	61 ± 9	1.5 ± 0.2	[9.20]
KBr	-	11 ± 1	-	54 ± 9	1.1 ± 0.2	[9.20]
CsCl Structure						
CsCl	-	18 ± 2	5.5–9	80–87	1.8	[9.21]

Notes and Comments

1. $\Delta V_{m,v}$ is derived from σ-P data in the intrinsic region while ΔV_f is derived by combining σ-P data in intrinsic as well as extrinsic regions. The relevant equations are given by Samara [9.22].
2. Note that $\Delta V_f / V_M > 1$ in all the cases studied. This means that the lattice volume relaxation associated with the Schottky defects is outward which is in disagreement with theoretical calculations which show that the relaxation is inward [9.23].

9.1.9 Vacancy-Impurity Dipoles

Table 9.7 Values of the activation energy ϕ for orientation of vacancy-impurity dipole

Crystal	Impurity	ϕ [eV]
	[9.24]	
NaCl Structure		
NaCl	Mg^{2+}	0.64
	Ca^{2+}	0.70
	Sr^{2+}	0.73
	Ba^{2+}	0.77
	Mn^{2+}	0.65
	Ni^{2+}	0.65
	Co^{2+}	0.65
	Zn^{2+}	0.65
	Cd^{2+}	0.65
	Pb^{2+}	0.66
KCl	Ca^{2+}	0.64
	Sr^{2+}	0.67
	Ba^{2+}	0.70
	Pb^{2+}	0.68
KBr	Sr^{2+}	0.65
	Ba^{2+}	0.66

Notes and Comments

1. The relaxation frequency ω_m (obtained from measurement of dielectric loss $\tan \delta$) is given by

$$\omega_m = \tau_0^{-1} \exp\left(-\phi / k_B T\right) \tag{9.8}$$

where τ_0 is the relaxation time; the values of ϕ are obtained from plots of $\log \omega_m$ versus T^{-1}.

2. Varotsos and Miliotis [9.24] showed that when the ϕ values are plotted against the radius of the divalent cation impurity, the plot is linear with positive slope for cations with p electrons on their outer subshells (Mg^{2+}, Ca^{2+}, Sr^{2+}, Ba^{2+}).

The ϕ values for cations with d electrons on their outer subshells (Mn^{2+}, Co^{2+}, Ni^{2+}, Zn^{2+}, Cd^{2+}) do not depend upon the ionic radii of the impurities.

9.1.10 Solution Enthalpy of Divalent Defects

Table 9.8 Experimental and theoretical values of the solution enthalpy for divalent defects (ΔH_{dd})

Crystal	Impurity	ΔH_{dd} [eV]		
		Experimental	Theoretical	
		[9.25]	[9.26]	[9.27]
NaCl Structure				
NaF	Ca^{++}	-	-	2.66
NaF	Sr^{++}	-	-	4.61
KF	Ca^{++}	-	-	1.55
KF	Sr^{++}	-	-	2.60
RbF	Ca^{++}	-	-	0.95
RbF	Sr^{++}	-	-	1.81
NaCl	Ca^{++}	0.64	-	0.42
NaCl	Sr^{++}	0.78, 0.72	-	1.52
NaCl	Ba^{++}	0.99	1.18	-
KCl	Ca^{++}	-	-	0.60
KCl	Sr^{++}	1.02, 1.30	-	1.24
KCl	Ba^{++}	-	1.79	-
RbCl	Ca^{++}	-	-	0.44
RbCl	Sr^{++}	1.43	-	0.96
RbCl	Ba^{++}	-	1.88	-

9.2 Polarons

9.2.1 Polaron Coupling Constant

Table 9.9 Values of the polaron coupling constant (α) and effective band mass m_e^* (expressed as fraction of free electron mass m_e)

Parameter	$(m_e^*/m_e)_{exp}$	α_{calc} [9.28]	α_{calc} [9.29]
	[9.28]	using $(m_e^*/m_e)_{exp}$	using $m_e^* = m_e$
Crystal $\downarrow$			
NaCl Structure			
LiF	-	-	6.4
LiCl	-	-	4.8
LiBr	-	-	3.5
LiI	-	-	2.4
NaF	-	-	5.9

Table 9.9 (Continued)

Parameter	$(m_e^* / m_e)_{exp}$ [9.28]	α_{calc} [9.28] using $(m_e^*/m_e)_{exp}$	α_{calc} [9.29] using $m_e^* = m_e$
Crystal $\downarrow$			
NaCl Structure			
NaCl	-	-	5.0
NaBr	-	-	5.0
NaI	-	-	4.8
KF	-	-	5.8
KCl	0.496	3.97	5.8
KBr	0.428	3.52	5.4
KI	0.398	2.60	4.8
RbF	-	-	5.7
RbCl	0.515	4.14	6.3
RbBr	-	-	6.6
RbI	-	-	5.8
CsCl Structure			
CsCl	-	-	6.3
CsBr	-	-	4.1
CsI	-	-	5.5

Notes and Comments

1. α is calculated [9.30] from the formula

$$\alpha = [e^2 / \hbar][m_e^* / 2\hbar\omega_{LO}]^{1/2} [(\varepsilon_0 - \varepsilon_\infty)/\varepsilon_0\varepsilon_\infty] \tag{9.9}$$

where ε_0 is the static dielectric constant, ε_∞ is the optical dielectric constant, ω_{LO} the longitudinal optic frequency and m_e^* the effective band mass of the electron.
2. α values are in the range 2.4–6.4 .
3. α values decrease in the sequence fluoride–iodide (except in Rb and Cs halides) and increase in the sequence Li–Rb (except for the fluorides).

9.2.2 Polaron Mass

Table 9.10 Values of polaron mass m_P expressed as fraction of free electron mass m_e ; [9.28]

Crystal	m_P / m_e
NaCl Structure	
KCl	1.25
KBr	0.93
KI	0.67
RbCl	1.38

Notes and Comments

1. The polaron mass m_P has been calculated from the effective band mass m_e^* of the electron and the coupling constant α using the relation [9.28]:

$$m_P = [1 + (\alpha / 6)] \, m_e^* \tag{9.10}$$

2. The polaron mass is measured from cyclotron resonance from the formula

$$m_P = eH / \omega c \tag{9.11}$$

where H is the magnetic field at maximum of cyclotron response and ω the cyclotron resonance frequency.

9.3 Colour Centres

9.3.1 Glossary of Colour Centres

Table 9.11 Nomenclature and structure of colour centres

Colour Centre	Structure *
F	A single electron trapped by a negative ion vacancy
F$'$	Two anion vacancies on nearest neighbour sites along <110> direction (2 electrons in a −ve ion vacancy)
R (or F_3)	Three F centres with three electrons in a 2-d array forming an equilateral triangle in (111) plane
R_1	F centre and an associated negative ion vacancy
R_2	Two coupled F centres
F_3^+	Ionised R centre
M (or F_2)	F centre joined to a pair of oppositely charged vacancies
F_2^+	Ionised F_2 centre
M_A	F_2 or M centre adjacent to a cation impurity of same valence as host
N_1	Four halogen ion vacancies with four electrons in a 2-d array
N_2	Four halogen ion vacancies with four electrons in a tetrahedral array
$(F_2^+)_H$	A double negatively charged oxygen anion and a charge compensating <110> neighbouring anion vacancy together with an F centre
F_2^-	Two neighbouring anion vacancies binding three electrons
$F_A(Tl)$	F centre attached to a <100> neighbouring Tl^+ ion having large electron affinity
$(F_2^+)^*$	A defect complex formed from a F_2^+ centre, a divalent cation impurity and a charge compensating cation vacancy
$(F_2^+)_A$	F_2^+ centres attached to a nearest neighbour Li^+ or Na^+ impurity
$F_B\,II$	F centre attached to two <100> neighbouring Na^+ cations forming a triangular configuration
$F_A\,II$	F centre attached to a small cationic impurity on a <100> nearest neighbour site

Table 9.11 (Continued)

Colour Centre	Structure *
F_A	F centre adjacent to a cation impurity of same valence as host
$(F_2^+)_{AH}$	A complex centre involving an (F_2^+) centre, an O^{2-} ion and an alkali metal (Na^+) ion
H	Interstitial halide ion bound by a hole to a substitutional halide ion (a $<110>$ (halogen)$_2$ molecule located on a single halogen site)
U	H ion forming solid solution (substituting for the halogen ion) with the host alkali halide
U_1	An interstitial H^- (ion)
U_2	An interstitial H^0 (atom)
V_K	An electron with a self-trapped hole (essentially a (halogen)$_2^-$ molecule-ion and, hence, very similar to H centre)
V_3	Molecular halogen centre
X or C	Colloid centre
Z_1	F centre with cation vacancy as nearest neighbour and divalent impurity at a nearby cation site
Z_2	F$'$ centre in the neighbourhood of divalent impurity
Z_3	Ionised Z_2 centre

* A more detailed description of the structure of various colour centres is given in review articles [9.31–9.35]

9.3.2 F Centre Parameters

Table 9.12 Peak position λ_{max}, half-width W and oscillator strength f; room temperature data

Parameter	λ_{max} [mμ]	W [eV]	f	
Ref.	[9.31]	[9.31]	[9.36]	[9.32]
Crystal ↓				
NaCl Structure				
LiF	249	0.7	0.82	-
LiCl	385	-	-	-
LiBr	(452)	-	-	-
LiI	(531)	-	-	-
NaF	340	0.51	-	-
NaCl	466	0.49	0.71	0.6
NaBr	536	-	-	-
NaI	588	-	-	-
KF	455	-	-	-
KCl	563	0.39	0.66	0.6
KBr	630	0.42	0.71	0.75
KI	689	0.35	0.46	0.83
RbF	525	-	-	-

Table 9.12 (Continued)

Parameter	λ_{max} [µm]	W [eV]	f	
Ref.	[9.31]	[9.31]	[9.36]	[9.32]
Crystal ↓				
NaCl Structure				
RbCl	609	0.31	-	0.85
RbBr	694	0.38	-	0.70
RbI	756	0.36	-	0.81
CsF	(531)	-	-	-
CsCl Structure				
CsCl	603	0.36	-	-
CsBr	675	0.32	0.38	-
CsI	785	0.36	-	-

Notes and Comments

1. Mollwo [9.37] derived the relation

$$\lambda_{max} = A\, r^2 \tag{9.12}$$

   where r is the interionic distance and A is a constant. The values in parenthesis in Table 9.12 are estimated from the Ivey formula (Eq. 9.13).
2. By fitting data for a large number of crystals, Ivey [9.38] showed that a better relation between λ_{max} and r is

$$\lambda_{max} = 703\, r^{1.84} \tag{9.13}$$

   for alkali halides with NaCl structure and

$$\lambda_{max} = 251\, r^{2.5} \tag{9.14}$$

   for alkali halides with CsCl structure. Here λ_{max} and r are in Angstrom units.
3. Bansigir [9.39] reanalysed data on NaCl-type alkali halide crystals and claimed better agreement with the relation

$$\lambda_{max} = 640\, r^{1.91} \tag{9.15}$$

4. The oscillator strength f is an important parameter. It enters into Smakula's formula for the concentration (N) of F centres:

$$N = 1.29 \times 10^{17} \times \frac{n}{(n^2 + 2)^2} \left(\frac{\alpha_{max}\, W}{f} \right) \; cm^{-3} \tag{9.16}$$

   where α_{max} is the absorption coefficient at the maximum of the F centre band, W the half width in eV and n the refractive index [9.32].

9.3.3 F Centre Formation Energy

Table 9.13 Values of E_F the formation energy of F centres

Ref. Crystal	E_F [eV]	
	at 4 K using 40 kV X-rays [9.40]	at 195 K using 500 eV electrons [9.41]
NaCl structure		
LiF	6.2×10^2	1×10^3
LiCl	-	9×10^3
LiBr	-	8×10^5
NaF	1.3×10^3	0.5×10^3
NaCl	1.4×10^4	1×10^3
NaBr	8.3×10^5	1×10^6
NaI	-	2×10^3
KCl	1.3×10^3	-
KBr	1.4×10^3	2×10^3
KI	5.2×10^4	-
RbF	-	9×10^3
RbCl	-	0.3×10^3
RbBr	-	2×10^3
CsF	-	0.5×10^6

Notes and Comments

1. To determine E_F, a crystal is irradiated for a fixed time by radiation of known energy and the resulting concentration of centres is determined by the use of Smakula's formula. E_F is the ratio of the energy of the radiation to the number of F centres produced.

2. Fig. 9.7 shows a plot of the F centre formation energy values against the parameter S/D where S is the space between adjacent halides in <110> direction and D is the diameter of the halogen ion. This ratio represents the space available to a halogen for insertion between two halogens along the face diagonal. As S/D decreases it is expected that it will be more difficult to produce an F centre-H centre pair and total energy for producing an F centre increases. This is seen in the initial part of the curve from NaBr to KBr. Beyond this point i.e. S/D ~ 0.5 the energy required is insensitive to increasing S/D values thus leading to stable H centres.

3. Townsend [9.42] redrew the Rabin-Klick diagram using 400 keV electron data from [9.41]. In the new plot, the initial steeply changing region stops at S/D ~ 1/3 and the plateau region extends to S/D ~ 1. Beyond this, the curve again rises by a thousand-fold (Fig. 9.8).

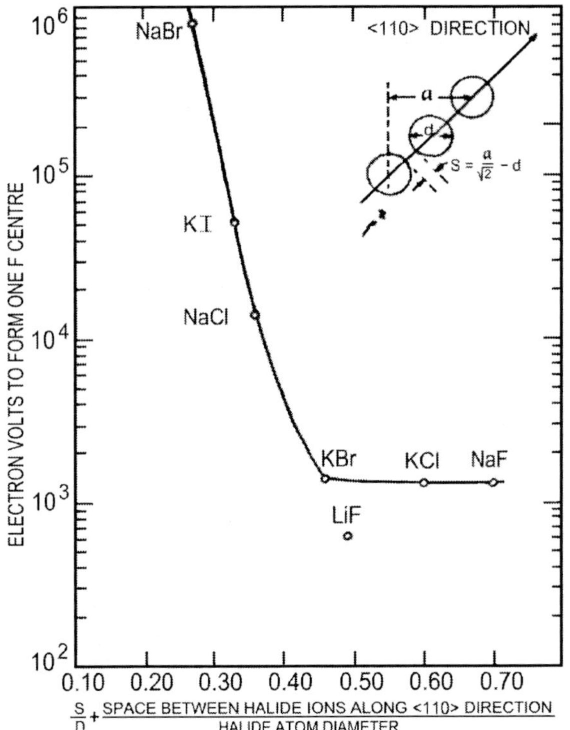

Fig. 9.7 Total x-ray energy required to form one F centre at liquid helium temperature as a function of the ratio S/D (after [9.40])

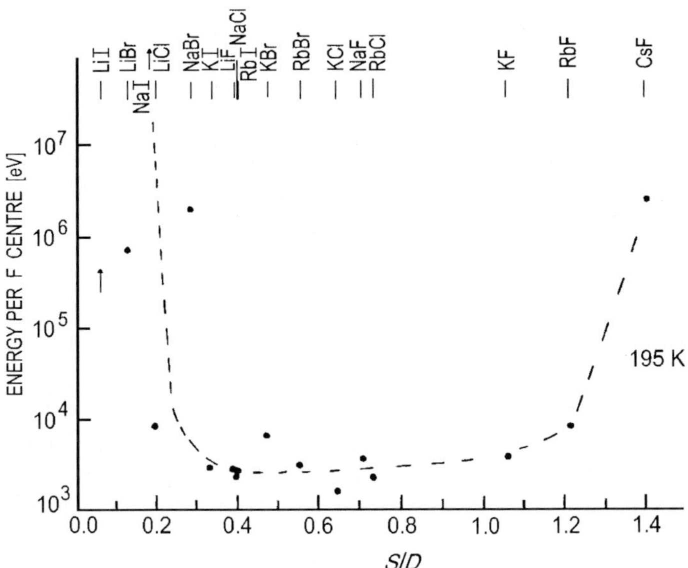

Fig. 9.8 The Rabin and Klick diagram with data measured by Hughes et al. [9.41] at 195 K using 400 keV electrons (after [9.42])

9.3.4 EPR and ENDOR Parameters for F Centres

Table 9.14 g values and half-widths W obtained from electron paramagnetic resonance (EPR) [9.32]; hyperfine interactions $(a/h)_I$, $(a/h)_{II}$, $(b/h)_I$, $(b/h)_{II}$ obtained from electron nuclear double resonance (ENDOR) [9.32]

Crystal	EPR		ENDOR			
	g	W [gauss]	$(a/h)_I$	$(a/h)_{II}$	$(b/h)_I$	$(b/h)_{II}$
			[Mc]			
NaCl Structure						
LiF	2.001	150	39.0	105.9	3.2	14.9
LiCl	2.001	53	19.1	11.2	1.7	0.9
LiBr	1.999	140	-	-	-	-
NaF	2.000	220	107.0	96.8	5.3	9.8
NaCl	1.997	145	62.4	12.5	3.0	1.0
NaBr	1.984	250	-	-	-	-
KF	1.996	91	34.3	35.5	1.6	4.1
KCl	1.995	47	20.7	6.9	0.9	0.5
KBr	1.982	125	18.3	42.8	0.8	2.7
KI	1.964	225	15.1	49.5	0.6	3.0
RbCl	1.980	425	98.0	5.8	4.0	0.4
RbBr	1.967	390	87.7	36.9	4.0	2.0
RbI	1.949	380	-	-	-	-
CsCl Structure						
CsCl	1.968	700	-	-	-	-
CsBr	1.958	700	-	-	-	-
CsI	1.95	730	-	-	-	-

Notes and Comments

1. The EPR spectrum of several alkali halides (LiCl, NaCl, NaBr, K-halides, RbBr and RbI) consists of a single broad Gaussian-shaped line without structure. In some cases (LiF, NaF, RbCl, CsCl), the EPR spectrum is broad with a number of components.
2. The line widths lie between 45 to several hundred Gauss.
3. The g values in all cases are slightly smaller than the g value (2.0023) for a free electron. The values listed in the table have been corrected for 2nd order hyperfine structure effects.
4. The ENDOR spectrum depends on the orientation of the magnetic field with respect to crystallographic directions. The spectrum consists of several lines with frequency ratio equal to ratio of the respective nuclear g factors. Each line is a doublet with a separation of $2\nu_k$, where ν_k is the Larmor frequency of free nuclei.
5. ENDOR data facilitate determination of the wave function of the F centre electron in its ground state.

9.3.5 F Centres (Faraday Rotation and Circular Dichroism)

Table 9.15 Values of the spin orbit splitting Δ of the excited state and the orbital g factor (g_{orb}) of the excited state; [9.43]

Parameters→	Δ [meV]	Δ [meV]	g_{orb}
Method→ Crystal ↓	Faraday rotation	Circular dichroism	Faraday rotation
NaCl Structure			
LiF	-2.5	-	-
NaCl	-5.1 ± 1	-7.7 ± 1.00	0.38
NaBr	-	-28.3 ± 6.0	-
KCl	-11.4 ± 2.3	-10.1 ± 1	0.62
KBr	-19.2 ± 3.8	-29.8 ± 4.0	0.47
KI	-30.0 ± 3.2	-57.0 ± 8	0.83
RbCl	-15.1	-15.1	-
RbBr	-32.4 ± 6.5	-26.6	0.47
RbI	-50	-	-
CsCl Structure			
CsCl	-	-37.2 ± 5.0	-
CsBr	-42.0 ± 8.4	-40.9 ± 5.0	-

Notes and Comments

1. Δ is negative for all alkali halides.
2. Δ increases if the atomic number of either alkali or halogen ion increases.

9.3.6 F Centres (Dissociation Energy)

Table 9.16 Values of the dissociation energy of F centres E_D; [9.44]

Crystal	E_D [eV]
NaCl Structure	
NaCl	2.45
NaBr	2.25
NaI	(1.99)
KCl	2.88
KBr	2.66
KI	2.33

Notes and Comments

1. The values of E_D in Table 9.16 were obtained by Dalal et al. [9.44] from measurement of F centre mobilities at different temperatures.
2. The temperature variation of mobility of F centres (μ_F) follows the equation

$$\mu_F = \mu_0 \exp(-W / k_B T) \tag{9.17}$$

where W is the activation energy of migration. W is related to the dissociation energy E_D of F centres and the enthalpy of formation of a Schottky pair (h_S) through the relation

$$W = E_D - (h_S / 2) \tag{9.18}$$

3. Dalal et al. [9.44] found linear plots between E_D and interionic distance r and also between E_D and the refractive index n. The value of E_D for NaI was estimated from an extrapolation of the E_D versus r plot for NaCl and NaBr.

9.3.7 F Centres (Temperature Variation of Peak Position)

Data on temperature variation of F centre peak position (λ_{max} [μm] or ε_{max} [eV]) are given in Tables 9.17, 9.18 and Fig. 9.9.

Table 9.17 F centre Peak position (λ_{max}) at 5 K; [9.40]

Crystal	LiF	NaF	NaCl	NaBr	KCl	KBr	KI
λ_{max}	243	336	450	526	539	602	666

Table 9.18 F centre λ_{max} at different temperatures for NaCl ([9.45] quoted in [9.46])

T [K]	20	90	300	873	973
λ_{max} [mμ]	454	455	459	525	540

Notes and Comments

1. λ_{max} at 5 K follows the Ivey-type relation

$$\lambda_{max} = 69.6 \, r^{1.79} \tag{9.19}$$

where r is the interionic distance [9.38].

2. Assuming that the F wavelength follows an Ivey-type relation at different temperatures Jacobs [9.48] calculated the parameter

$$\eta_P(T) = -(\partial \log \nu_{max} / \partial \log r)_P \tag{9.20}$$

from Mollwo's [9.37] data (see Table XXII).

Table XXII Values of $\eta_P(T)$ from Eq. (9.20); [9.48]

Crystal	NaCl	NaBr	KCl	KBr	KI	RbCl	CsCl
$\eta_P(T)$	4.0	3.8	5.0	4.5	4.4	5.2	4.9

From high pressure data, Jacobs [9.48] obtained the parameter

$$\eta_T(P) = -(\partial \log \nu_{max} / \partial \log r)_T \tag{9.21}$$

and derived the relation

$$\eta_P(T) - \eta_T(P) = b / (\varepsilon_{max}\,\theta\alpha) \tag{9.22}$$

where ε_{max} is the F centre energy at ν_{max}, θ a characteristic temperature, α the coefficient of linear expansion and b is a constant.

3. Bansigir [9.39] derived the following equations for the temperature variation of λ_{max}. These equations are applicable to alkali halides NaCl as well as KCl, KBr and KI.

$$T < 140\ \text{K} \qquad \lambda_{max} / r^{1.91} = 640 \exp(0.6\times10^{-4}T) \tag{9.23}$$

$$T > 140\ \text{K} \qquad \lambda_{max} / r^{1.91} = 615.5 \exp(0.2292\times10^{-6}T^2 + 0.6\times10^{-4}T) \tag{9.24}$$

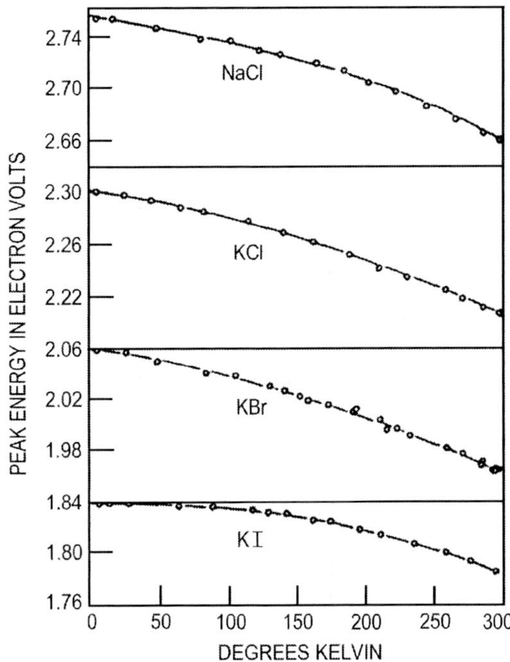

Fig. 9.9 Variation of the peak of the F centre absorption band (ε_{max}) with temperature for NaCl, KCl, KBr and KI (after [9.47])

9.3.8 F Centres (Temperature Variation of Half-Width)

The temperature variation of the half-width (W) of the F-band for LiF, NaCl, KCl, KBr and KI is shown in Fig. 9.10

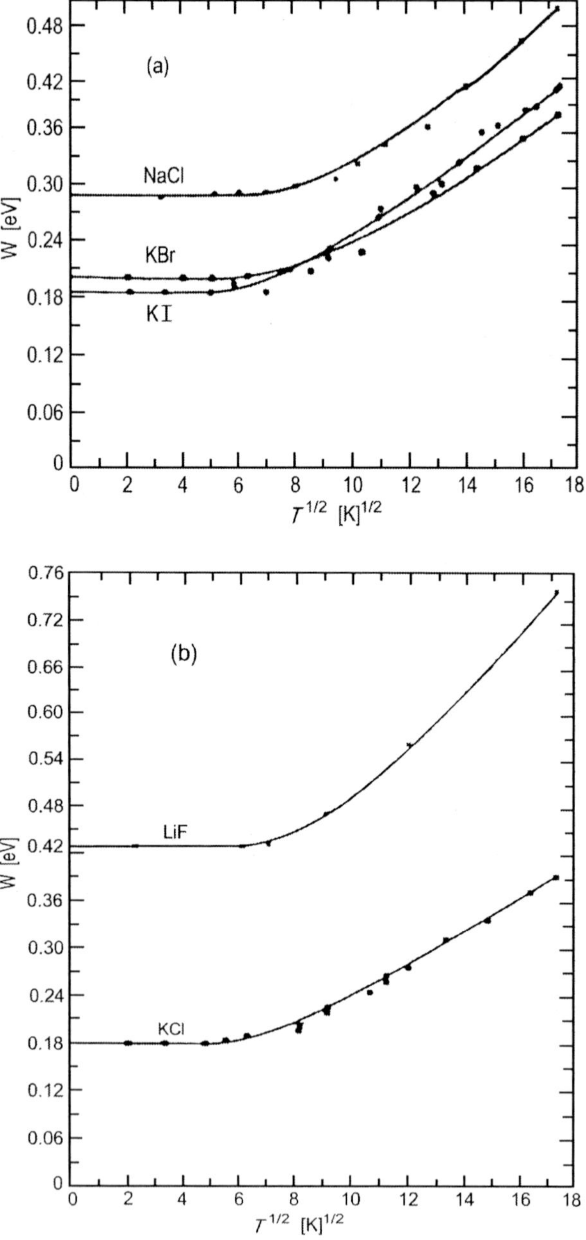

Fig. 9.10 Variation of the width of the F band in (a) NaCl, KBr and KI and (b) LiF and KCl at half maximum as a function of temperature.(after [9.47])

Notes and Comments

1. The W-$T^{1/2}$ plots can be fitted to the equation

$$W = A \left[\coth\left(h\nu_g / 2k_B T\right)\right]^{1/2} \qquad (9.25)$$

where A is a constant and ν_g a characteristic frequency. Russel and Klick [9.47] found the following values for ν_g (10^{12} cps), LiF: 4.1; NaCl: 4.4; KCl: 2.6; KBr: 2.6 and KI: 3.6. The continuous curves are best fits to experimental data using Eq. (9.25).

9.3.9 F Centres (Pressure Variation)

Table 9.19 Values of peak position (ν_{max}) of F centres at various pressures (P); [9.48]

	ν_{max} [10^4 cm^{-1}]								
P [bar] Crystal $\downarrow$	1	1000	2000	3000	4000	5000	6000	7000	8000
NaCl Structure									
NaCl	2.147	2.161	2.175	2.190	2.203	2.215	2.226	2.238	2.250
NaBr	1.865	1.876	1.886	1.897	1.909	-	-	-	-
KCl	1.787	1.798	1.810	1.821	1.832	1.842	-	-	-
KBr	1.593	1.604	1.616	1.628	1.640	1.652	-	-	-
KI	1.446	1.460	1.474	1.489	1.503	1.517	-	-	-
RbCl	1.578	1.591	1.604	1.615	1.624	1.634	-	-	-
CsCl	1.656	1.666	1.676	1.685	1.693	1.701	-	-	-

Notes and Comments

1. Assuming that the Ivey relation $\lambda_{max} = A\,r^n$ is valid at high pressure, we get the parameter η as

$$\eta_T(P) = -(\partial \log \nu_{max} / \partial \log r)_T \qquad (9.21)$$

The values of $\eta_T(P)$ at 1 atm. evaluated by Jacobs [9.48] are given in Table XXIII.

Table XXIII Values of $\eta_T(P)$ from Eq. (9.21)

Crystal$\rightarrow$	NaCl	NaBr	KCl	KBr	KI	RbCl	CsCl
$\eta_T(P) \rightarrow$	4.4	3.6	3.5	3.5	3.7	3.4	2.8

2. A similar parameter $\eta_P(T)$ may be defined from data on temperature variation of λ_{max} or ν_{max} as

$$\eta_P(T) = -(\partial \log \nu_{max} / \partial \log r)_P \qquad (9.20)$$

3. Jacobs showed that

$$\eta_P(T) - \eta_T(P) = b/(\varepsilon_{max}\,\theta\alpha) \tag{9.22}$$

where ε_{max} is the F centre energy at λ_{max}, θ a characteristic temperature (taken to be the Debye temperature), α the coefficient of linear expansion and b a constant. Values of b/θ are given by Jacobs [9.48].

4. Measurements were extended by Drickamer [9.49] to 600 kbars. He observed that:

i)   the log ν_{max} versus log r plots are linear at low pressures but show a non-linearity

ii) the rate of change of ν_{max} with pressure shows a discontinuous change at the NaCl-CsCl transition in KCl.

9.3.10 Deformation Bleaching and Mechanoluminescence Parameters

Table 9.20 Values of E_a (band gap between bottom of dislocation band and the ground state energy level of the interacting F centres lying in an edge dislocation), D_B (coefficient of deformation bleaching), ψ (coefficient of deformation generated compatible traps), r_F (distance up to which a dislocation can interact with an F centre) and b (Buergers vector)

Crystal	E_a [eV]	D_B	ψ	r_F / b
	[9.50, 9.51]*	[9.52]	[9.53]	[9.52]
NaCl Structure				
NaCl	0.095	1.54	-	51.3
NaBr	0.08	1.96	-	42.6
KCl	0.08	1.93	0.67	42.4
KBr	0.09	2.00	1.7	66.7

* Values quoted in [9.52]

Notes and Comments

1. The F centre density n_F for a strained crystal is given by

$$n_F = n_{F0}\,\exp\left(-D_B\,\varepsilon\right) \tag{9.26}$$

where ε is the strain and n_{F0} is the value of n_F for zero strain. The deformation bleaching coefficient D_B is obtained from the slope of the linear plot between log n_F and ε.

2. The maximum intensity I_m of mechanoluminescence is given by

$$I_m = \text{constant}\ \exp\left(-E_a/k_BT\right) \tag{9.27}$$

The value of E_a is obtained from the slope of the plot between log I_m and (T^{-1}).

3. The strain dependence of the total intensity I_T of mechanoluminescence is given by

$$I_T = I_{T0}[1 - \exp\{-(D_B + \psi)\varepsilon\}] \tag{9.28}$$

where ε is the strain and I_{T0} the value of I_T for zero strain. The value of $(D_B + \psi)$ is obtained from the slope of the linear plot between $\log [(I_T / I_{T0}) - 1]$ and ε. Since D_B is known as discussed above, ψ can be estimated.

4. r_F is estimated from the interrelation between r_F and D_B.

9.3.11 F Aggregate Centres (R$_1$, R$_2$, M, N$_1$, N$_2$, V$_3$ and F$_3^+$ Centres)

Table 9.21 Peak positions λ_{max} [μm] and half-widths W [eV]; room temperature data

Colour centre	R$_1$	R$_2$	M		N$_1$	N$_2$	V$_3$		F$_3^+$	
Parameter	λ_{max}	λ_{max}	λ_{max}	W	λ_{max}	λ_{max}	λ_{max}	Ref.	λ_{max}	Ref.
Crystal ↓	[9.31]									
NaCl Structure										
LiF	310	380	447	-	520	540	113	[9.54]	-	
NaF	-	415	505	0.16	-	-	-	-	700	[9.56]
NaCl	-	-	725	-	-	-	-	-	-	
KF	-	-	570	-	-	-	-	-	-	
KCl	680	740	820	0.12	955	1080	215	[9.55]	960	[9 57]
KBr	-	-	917	0.12	1080	1080	255	[9.55]	1020	[9 57]
KI	-	-	-	-	-	-	310	[9.55]	-	-
RbCl	-	-	-	-	-	-	205	[9.55]	-	-
RbBr	805	859	957	-	-	-	260	[9.55]	-	-
RbI	-	-	-	-	-	-	345	[9.55]	-	-
CsCl Structure										
CsCl	-	-	1000	0.07	-	-	-	-	-	-
CsBr	-	-	1070	-	-	-	-	-	-	-
CsI	-	-	1220	0.10	-	-	-	-	-	-

Notes and Comments

1. Ivey [9.38] showed that the absorption peak wavelengths for the R$_1$, R$_2$ and M centres for the NaCl type alkali halides correlate with the interionic distance r through the following equations:

$$\text{R}_1 \text{ band: } \lambda_{max} = 816\, r^{1.84} \tag{9.29}$$

$$\text{R}_2 \text{ band: } \lambda_{max} = 884\, r^{1.84} \tag{9.30}$$

$$\text{M band: } \lambda_{max} = 1400\, r^{1.56} \quad \text{for NaCl structure} \tag{9.31}$$

$$\lambda_{max} = 571\, r^{2.23} \quad \text{for CsCl structure} \tag{9.32}$$

2. M centre absorption is anisotropic; this property finds application in information storage.

3. In the case of NaCl and KBr, the N$_1$, N$_2$ bands are not resolved.

4. The V_3 bands are not dependent on r directly. Instead they correlate with the anion radius (r_-) as follows:

$$\lambda_{max} = 813 \ r_-^{1.89} \tag{9.33}$$

5. The Ivey-type relation for the F_3^+ bands is

$$\lambda_{max} = 2835 \ r^{1.07} \tag{9.34}$$

9.3.12 Temperature Variation of M Band Half-Width

The temperature variation of the M band half-width in KCl is shown in Fig. 9.11.

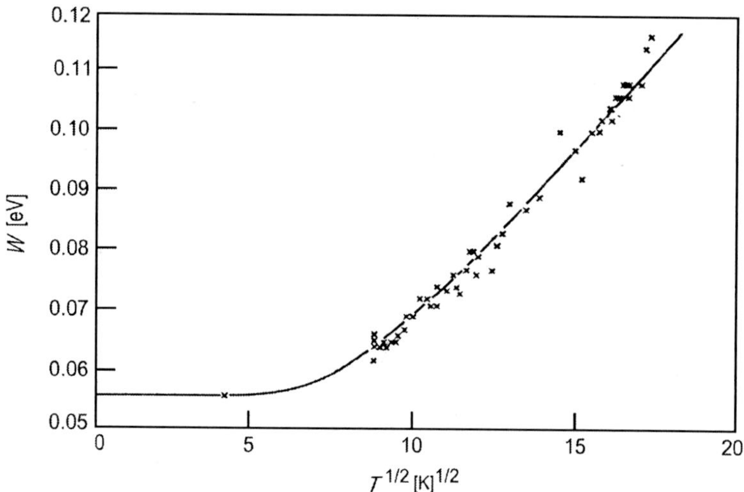

Fig. 9.11 Plot of M band half-width W versus $T^{1/2}$ for KCl (after [9.58])

Notes and Comments

1. The data on half-width fits Eq. (9.35)

$$W = A\,[\coth\,(h\nu_g\,/\,2k_BT)]^{1/2} \tag{9.35}$$

with $A = 0.056$ eV and $\nu_g = 3.2 \times 10^{12}$ cps [9.58].

9.3.13 Pressure Variation of M, R_2 and N Frequencies

Table 9.22 Values of the coefficient $x = \partial \log(v / v_0) / \partial \log(\rho / \rho_0)$ where v_0 and ρ_0 are frequency and density at zero pressure

Crystal	X; [9.59]		
	M	R_2	N
NaCl Structure			
NaCl	0.633	-	-
KCl	0.813	0.633	1.102
KBr	0.850	-	-
KI	0.850	-	-
High Pressure CsCl Phase			
KCl	0.543	-	0.872
KBr	0.543	-	-
KI	0.543	-	-

9.3.14 U Centre (Main UV Bands)

Table 9.23 Values of the peak positions (λ_{max} or ε_{max})

Crystal	λ_{max} [Å]	ε_{max} [eV]	Ref.
NaCl Structure			
NaCl	1920	-	[9.38]
NaBr	2100	-	[9.38]
KCl	2140	-	[9.38]
KBr	2280	-	[9.38]
KI	2440	-	[9.38]
RbCl	2300	-	[9.38]
RbBr	2426	-	[9.38]
CsCl Structure			
CsCl	-	5.43	[9.60]
CsBr	-	4.99	[9.61]
CsI	-	4.60	[9.62]

Notes and Comments

1. The U centre is an H^- ion that substitutes for a halide ion in alkali halides. It produces an absorption band in the ultraviolet, data on which are given in Table 9.23.
2. Mitra et al. [9.61] showed that the plot of $\varepsilon_{max}(F)$ vs $\varepsilon_{max}(U)$ is a straight line with a slope of about 2.5 which is due to the similarity in the structures of the F centre and U centre (F centre: electron in an octahedral box; U centre: electron in an octahedral box of six nearest neighbour alkali ions, in the centre of which is the hydrogen ion).

3. Ivey [9.38] showed that the λ_{max} for U centres and the interionic distance r follow the relation

$$\lambda_{max} = 615\ r^{1.10} \tag{9.36}$$

in crystals with NaCl structure.

4. The U band peaks in the CsCl structure crystals follow a similar relation with 1.54 as the exponent of r [9.60].

9.3.15 U Centre (Localised IR Bands)

Table 9.24 Wavelength, frequency and energy ε(U, IR) of the localised IR modes at liquid nitrogen temperature (data for CsCl, CsBr from [9.63], rest from [9.64])

Crystal	Wavelength [µm]	Frequency [cm^{-1}]	Energy ε(U, IR); [meV]
NaCl Structure			
LiF: H$^-$	9.76	1025	127.0
NaF: H$^-$	11.65	859	106.5
NaF: D$^-$	16.26	615	76.2
NaCl: H$^-$	17.78	563	69.8
NaCl: D$^-$	24.5	408	50.3
NaBr: H$^-$	20.2	496	61.4
NaBr: D$^-$	27.7	361	44.7
NaI: H$^-$	23.4	427	52.9
NaI: D$^-$	31.5	318	39.4
KCl: H$^-$	20.0	499	61.8
KCl: D$^-$	27.9	359	44.5
KBr: H$^-$	22.5	445	55.0
KBr: D$^-$	31.5	319	39.6
KI: H$^-$	26.4	378	46.8
RbCl: H$^-$	21.1	473	58.6
RbCl: D$^-$	29.5	339	42.0
RbBr: H$^-$	23.7	423	52.3
RbI: H$^-$	27.75	360	44.6
CsCl Structure			
CsCl: H$^-$	-	424	-
CsBr: H$^-$	-	367	-

Notes and Comments

1. Apart from giving rise to an UV absorption band, the substitutional H-ion also causes localised vibrational modes which result in IR bands.

2. Damm and Chvoj [9.65] have shown that the energy ε(U, IR) associated with the U centre local modes follows the relation

$$\varepsilon\,(U, IR) = 497.4\ r^{-1.8}\ (r_+ / r_-)^{1/3} \tag{9.37}$$

where r_+, r_- and r are the cation radius, the anion radius and interionic distance (all in Å) respectively.

9.3.16 U_2 Centres

Table 9.25 Values of the peak position ε_{max} of the U_2 centres

Crystal	ε_{max} [eV]; [9.66]
NaCl Structure	
NaCl	5.63
NaBr	4.78
NaI	3.78
KCl	5.27
KBr	4.56
KI	3.68
RbCl	5.02
RbBr	4.42
RbI	3.59

Notes and Comments

1. For a given cation, ε_{max} decreases in the sequence chloride–bromide–iodide. For a given anion, the value of ε_{max} decreases in the sequence Na–K–Rb.
2. Damm and Chvoj [9.65] have shown that the values of ε_{max} fit the relation

$$\varepsilon_{max} = 54.37 \, (r_+ / r_-)^{2/3} \, r^{-2} \tag{9.38}$$

   where r_+, r_- and r are the cation radius, anion radius and interionic distance respectively.
3. Weaker absorption peaks U_2' and U_2'' have also been recorded by Fischer [9.66].

9.3.17 F_A Centres

Table 9.26 Peak position ε_{max} and half-width W of $F_{A\,I}$ and $F_{A\,II}$ bands and characteristic lattice frequency v_g; [9.67]

Parameters System ↓	ε_{max} [eV]		W [eV]		v_g [10^{12} s^{-1}]
	$F_{A\,I}$	$F_{A\,II}$	$F_{A\,I}$	$F_{A\,II}$	
NaCl Structure					
KCl: Na	2.35	2.12	0.19	0.11	3.02
KCl: Li	2.25	1.98	0.19	0.12	3.00
KBr: Na	2.07	1.90	-	-	-
KBr: Li	2.00	1.82	0.235	0.128	-
RbCl: Na	2.09	1.85	0.17	0.10	2.24
RbCl: Li	1.95	1.72	0.15	0.11	2.24
RbBr: K	1.85	1.67	0.20	0.15	-
RbBr: Li	1.78	1.57	0.14	0.11	-

Notes and Comments

1. An F centre in an alkali halide crystal associated with an alkali ion impurity of smaller size (e.g. Na in KCl) constitutes an F_A centre. The introduction of an alkali ion of smaller size reduces the cubic symmetry of the F centre to tetragonal symmetry. This in turn splits the 3-fold degenerate level associated with the F centre into three components, two of which are again degenerate. Thus, the F_A centre has two components : $F_{A\,I}$ and the 2-fold degenerate $F_{A\,II}$.

2. The excited state splitting (energy difference between $F_{A\,I}$ and $F_{A\,II}$) is larger for the chlorides (~ 0.24 eV) than for the bromides (~ 0.19 eV) but is independent of the variation of the host ion (K^+, Rb^+) and impurity ion (Li^+, Na^+, K^+).

3. The temperature variation of the half-width follows the relation

$$W^2(T)/W^2(0) = \coth(h\nu_g/2k_B T) \tag{9.39}$$

This has been verified for the K and Rb chlorides by Fritz et al. [9.68].

9.3.18 Z Centres

Table 9.27 Peak position λ_{max} [nm] and half width W [eV] of Z centres of different types; [9.69]

Crystal	Impurity	Z_1		Z_2		Z_3	Z_4	Z_5
		λ_{max}	W	λ_{max}	W	λ_{max}	λ_{max}	λ_{max}
NaCl Structure								
NaCl	Ca	-	-	471	-	400	-	-
	Sr	506	-	516	-	400	-	-
	Eu	502	0.27	516	0.35	402	-	-
	Sm	505	0.26	517	0.38	-	-	-
	Yb	503	0.30	517	0.30	405	-	-
NaBr	Sr	550	-	560	-	-	-	-
KCl	Ca	590	0.25	598	-	490	-	-
	Sr	590	0.27	635	-	496	837	559
	Ba	590	0.27	639	-	-	-	555
	Eu	595	0.28	675	0.30	512	-	-
	Sm	592	0.28	676	0.32	520	-	-
	Yb	593	0.31	678	0.30	520	-	-
KBr	Ca	-	-	700	-	556	-	-
	Sr	658	-	716	-	558	-	-
	Ba	-	-	716	-	—	-	-
	Eu	657	0.30	720	0.35	549	-	-
	Sm	656	0.35	721	0.35	549	-	-
	Yb	657	0.35	720	0.35	549	-	-
KI	Sr	716	-	840	-	-	-	-
RbCl	Ca	680	-	-	-	-	-	-
	Sr	677	-	-	-	-	-	-
RbBr	Sr	758	-	-	-	-	-	-
RbI	Sr	810	-	-	-	-	-	-

Notes and Comments

1. Z centres occur in alkali halides doped with Ca^{++}, Sr^{++}, Ba^{++}, Eu^{++}, Sm^{++} and Yb^{++} (having low 2nd ionisation potential) but not in alkali halides doped with Pb^{++}, Cd^{++}, Zn^{++} and Bi^{++} (having high 2nd ionisation potential).
2. When an alkali halide crystal containing the divalent impurities mentioned above is coloured and then bleached at room temperature, the Z_1 centres are produced. When the crystal containing Z_1 centres is warmed to $\sim 100°C$, the Z_1 band disappears and a new band (Z_2) is observed. If the bleaching is done at liquid nitrogen temperature instead of at room temperature, another band is observed on the short wavelength side of the F band, this is the Z_3 band. Prolonged bleaching with F light at liquid nitrogen temperature causes another band close to M band; this is the Z_4 band. In KCl, with relatively high concentration of Sr^{++} ions a band is observed at 555 nm which is designated the Z_5 band. The Z_2 band can also be formed in additively coloured crystals quenched from 400 °C in the dark.
3. The peak position of the Z_1 centre depends mainly on the host lattice and is independent of the impurity. This also applies to the Z_2 centres but to a lesser extent; a difference is seen in the Z_2 bands in the alkaline-earth doped crystals and the rare-earth doped crystals.
4. The peak positions (ε_{max} in eV) of the Z_1, Z_2 and Z_3 centres are related to the interionic distance r (in Å) through Ivey-Mollwo type relations: $\varepsilon_{max} = k\,r^n$. The values of the parameters in these equations are $k = 15.6$, 22.35 and 23.82 and $n = -1.79$, -2.16 and -2.00 for the Z_1, Z_2 and Z_3 centres respectively [9.33].
5. Electron paramagnetic resonance studies have shown that the Z_1 and Z_3 centres are paramagnetic whereas the Z_2 centre is diamagnetic [9.69].
6. A number of models have been proposed for the Z centres. Radhakrishna and Chowdari [9.33, 9.69] have critically examined evidence from optical absorption, EPR and ENDOR studies and concluded that the best models are: (a) the model proposed by Rosenberger and Luty [9.70] according to which a Z_1 centre consists of an F centre with a cation vacancy as nearest neighbour and the divalent impurity at a nearby cation site, (b) the model proposed by Okhura and Murase [9.71] according to which a Z_2 centre consists of an F' centre in the vicinity of the divalent impurity and (c) model of Z_3 centre as an ionised Z_2 centre proposed by Okhura [9.72].
7. The formation of Z_1 and Z_2 centres in KCl, KBr and NaCl crystals is useful in optical information storage [9.73].

9.3.19 Colour Centre Information Storage

The properties of photochromism (reversible colour change) and photodichroism (dependence of absorption on the polarisation direction of incident light) displayed by alkali halide colour centres (M and Z_2 centres in particular) has led to their use in optical information storage. The alkali halide systems and the processes responsible for their application in information storage are listed in Table 9.28.

Table 9.28 Alkali halide systems useful in information storage

System	Colour centre	Principle	Ref.
KCl: Na	M	Photodichroism	[9.34]
NaF	M	Photodichroism	[9.34]
KBr	M	Photodichroism	[9.34]
KCl–KBr	M	Photodichroism	[9.74]
KCl: Gd^{2+}	Z_2	Photochromism	[9.75]
NaCl	C	Photothermal	[9.76]

9.3.20 Colour Centre Lasers

The discovery of laser action with some of the colour centres in alkali halides has led to the development of a variety of alkali halide colour centre lasers. A list of alkali halide lasers of various types, based on a combination of host material, impurity and colour centre, is given in Table 9.29.

Table 9.29 Alkali halide colour centre laser systems; [9.35]

Type of laser	Colour centre	Host crystal and impurity
Continuous wave colour centre laser		
	F_A	KCl: Li^+, RbCl: Li^+
	F_B	KCl: Na^+
	F_2^+	NaF, NaCl, LiF, KF, KCl, KBr
	$(F_2^+)^*$	NaF
	$(F_2^+)_A$	KCl: Li^+, KCl: Na^+
	$(F_2^+)_H$	NaCl, KCl, KBr
	$(F_2^+)_{AH}$	KCl: (Na^+, O_2^-), KBr: (Na^+, O_2^-)
	$F_A(Tl)$	KCl (Tl^+), KBr (Tl^+), KF (Tl^+)
Short pulsed mode-locked laser		
	F_2^+	LiF, KF, NaCl
	F_2^-	LiF
	F_A	KCl: Li^+, RbCl: Li^+, KCl: Tl^+
	$(F_2^+)_H$	NaCl
F_H (CN^-) lasers		
	F_H	CsBr: CN^{-1}, CsCl: CN^{-1}

Notes and Comments

1. The colour centre lasers are tunable and, together , they cover the near-infrared region from about 0.8 to 4 μm.
2. The output power offered by colour centre lasers is generally of the order of a few hundred mW. But the KCl: Li^+ $(F_2^+)_A$ laser is capable of a continuous wave power of 1.2 W [9.77].

3. In single-mode operation, the laser line width is as narrow as a few KHz.
4. In mode-locked condition, the pulses have temporal widths of the order of 100 femtosecond.
5. The thermal and photostability of F_2^+ colour centres has been studied in LiF crystals doped with OH^- and Mg^{2+} ions by Ter-Mikirtychev [9.78].

9.4 Luminescence

9.4.1 Intrinsic Luminescence

Table 9.30 Energies f_σ and f_π of σ-polarised and π-polarised emissions

Crystal	f_σ [eV]			f_π [eV]		
	Expt	Ref.	Calc.	Expt.	Ref.	Calc.
NaCl Structure						
LiF	-	-	7.43	-	-	-
LiCl	-	-	5.74	-	-	-
LiBr	5.30	[9.79]	5.34	-	-	-
LiI	-	-	4.86	-	-	-
NaF	-	-	6.40	-	-	-
NaCl	5.28	[9.80]	5.20	3.36	[9.84]	3.26
NaBr	-	-	4.89	4.60	[9.81]	3.64
NaI	-	-	4.49	4.20	[9.81]	4.21
KF	-	-	5.50	-	-	-
KCl	-	-	4.63	2.31	[9.81]	2.45
KBr	4.41	[9.81]	4.40	2.42	[9.80]	2.76
KI	4.12	[9.81]	4.09	3.33	[9.81]	3.25
RbF	5.32	[9.82]	5.19	-	-	-
RbCl	4.45	[9.83]	4.41	2.27	[9.81]	2.21
RbBr	4.20	[9.81]	4.21	2.10	[9.81]	2.50
RbI	3.95	[9.81]	3.93	3.09	[9.85]	2.96

Notes and Comments

1. Cywinski and Damm [9.86] have shown that f_σ and f_π correlate with the interionic distance r and the anion radius r_- as:

$$f_\sigma = 15.6\, r^{-1.06} \tag{9.40}$$

$$f_\pi = 14.2\, r_-^3\, r^{-2.8} \tag{9.41}$$

The values of f_σ and f_π calculated from these equations are given in Table 9.30.
2. Kabler and Patterson [9.81] interpret the emission in terms of two self-trapped exciton states.

9.4.2 Auger-Free Luminescence

Apart from the normal intrinsic luminescence due to self-trapped excitons (STE), the alkali halides show an additional luminescence due to interatomic radiative transition between the halogen valence band and the alkali core band. In this, a photon is emitted without ejecting an Auger electron from the valence band.This luminescence is termed Auger-free luminescence (AFL). The AFL consists of one or more bands. As the AFL overlaps with the normal STE luminescence, it can be observed only with time resolved measurements.The time-resolved AFL spectra and the temperature variation of the half-width for RbF, CsF, CsCl and CsBr are shown in Figs. 9.12, 9.13.

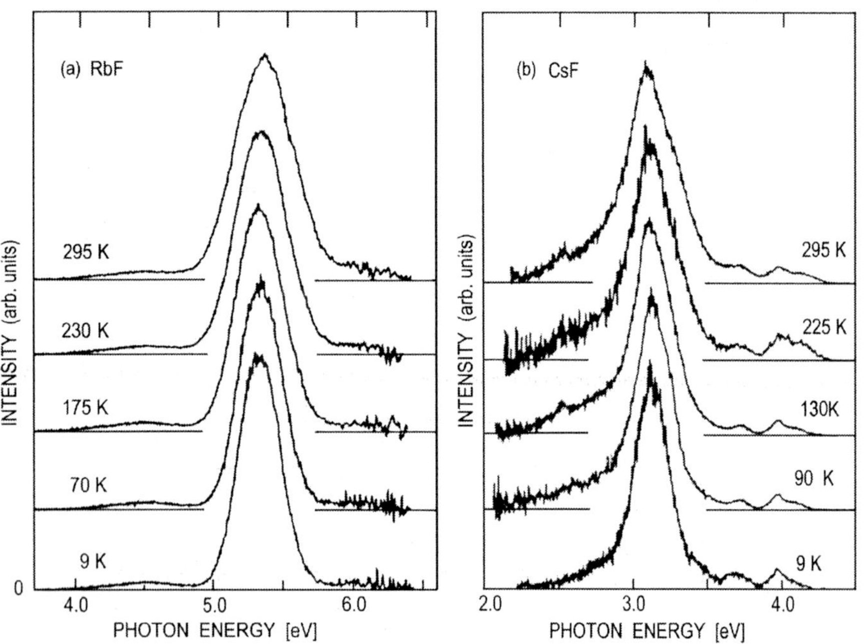

Fig. 9.12 Time resolved AFL spectra of (a) RbF (b) CsF (c) CsCl and (d) CsBr at different temperatures (after [9.87])

Fig. 9.12 (Continued)

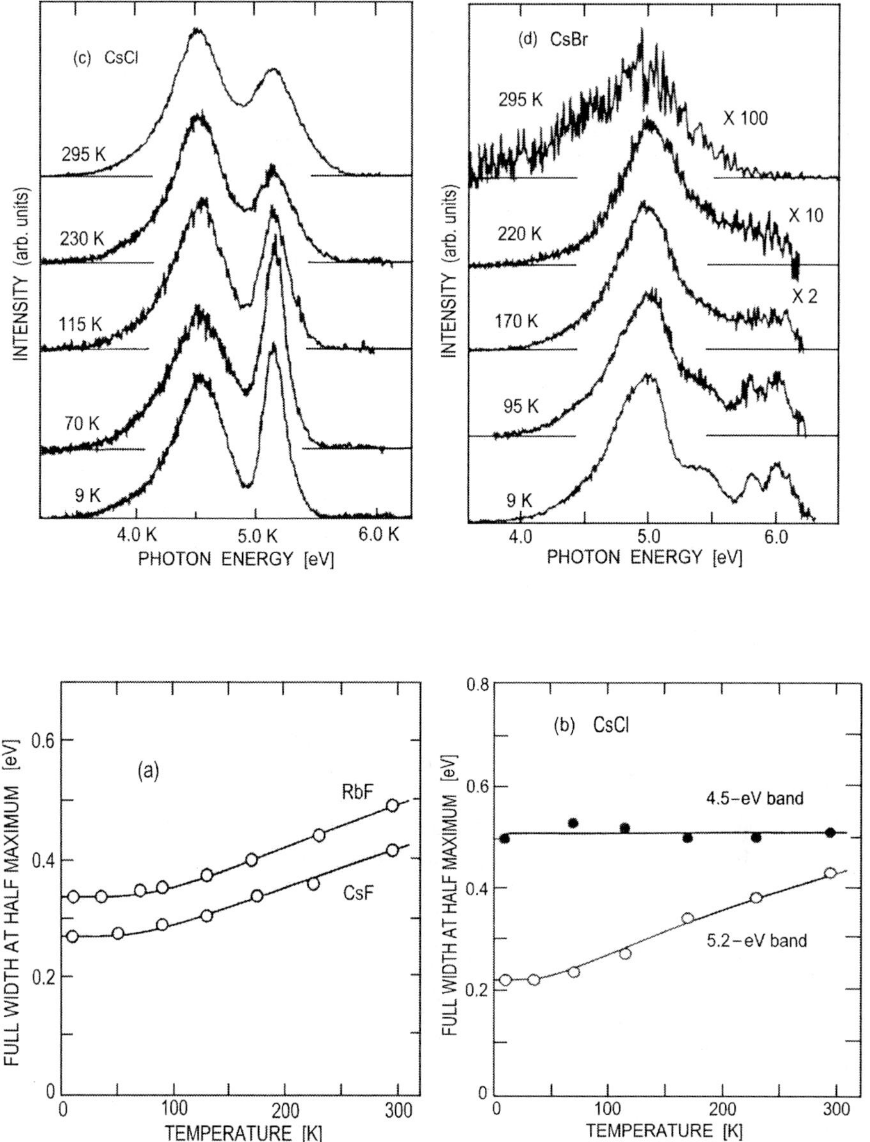

Fig. 9.13. Temperature variation of the half-width of the AFL bands of (**a**) RbF and CsF (**b**) CsCl and (**c**) CsBr (after [9.87])

Fig. 9.13 (Continued)

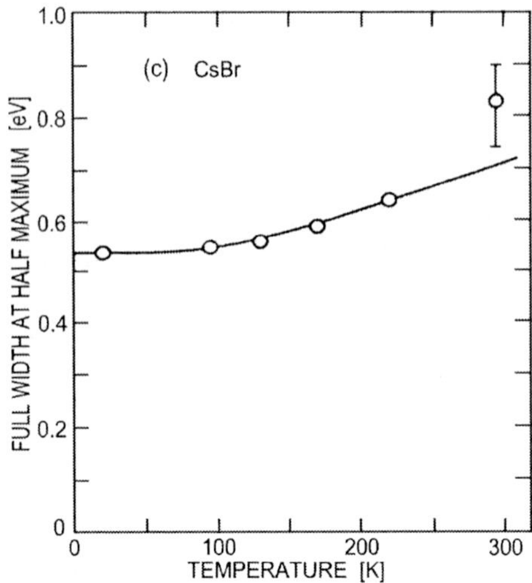

Notes and Comments

1. The temperature variation of the half-width of the AFL bands follows the equations:

$$W(T) = W(0)[\coth{(h\nu / 2k_B T)}]^{1/2} \tag{9.42}$$

and

$$W(0) = (8\log 2)\ (h\nu)^2\ S \tag{9.43}$$

where $h\nu$ is the effective phonon energy and S is the Huang-Rhys factor. The values of the parameters in these equations for some alkali halides are quoted in Table XXIV.

Table XXIV Values of parameters in Eq. (9.43); [9.87]

Crystal	$W(0)$ [eV]	$h\nu$ [meV]	S
NaCl Structure			
RbF	0.34	27	29
CsF	0.27	23	25
CsCl Structure			
CsCl	0.22	14	44
CsBr	0.54	34	45

9.4.3 Thermoluminescence

Table 9.31 Thermoluminescence glow peaks for various pure and doped alkali halides

Crystal	Impurity	Thermoluminescence glow peaks [K]	Ref.
LiF	Pure	26, 29.5, 35.5, 46, 54, 147, 225, 255, 275	[9.88]
	Mg	115, 133, 194, 225, 255, 275, 294	[9.88]
	Mn	115, 133, 147, 158, 194, 255	[9.88]
	(Mg, Ti)	(~)336, 378, 407, 442, 467 (read off from glow curve)	[9.89]
LiCl	Pure	378, 463	[9.89]
	Ca	340, 358	[9.89]
	Sr	348	[9.89]
NaF	Pure	174, 316, 370, 463, 540, 673	[9.89]
	Ca	348, 387, 401, 433, 513, 563	[9.89]
NaCl	Sr	340, 410, 440, 460, 500	[9.89]
	Mg	348, 388, 443, 513, 548	[9.89]
NaBr	Pure	133, 194, 255, 300, 480	[9.90]
NaI	Pure	140, 170	[9.90]
KCl	Pure	408, 463	[9.89]
	Ca	383, 453	[9.89]
	Sr	363, 415	[9.89]
	Ba	362, 383, 453	[9.89]
	Pb	383, 413, 513	[9.89]
	(Eu, Yb)	368, 413	[9.89]
KBr	Pure	358, 377, 383, 459	[9.89]
	Ca	338, 418	[9.89]
	Tl	176, 235, 281, 330, 415, 525	[9.90]
KI	Pure	112, 118, 120, 158, 176, 235	[9.90]
	Tl	106, 140, 190	[9.90]
RbCl	Pure	40, 45, 55, 78, 208, 253	[9.91]
RbBr	Pure	41, 47, 211, 255	[9.91]

Notes and Comments

1. Thermoluminescence of a crystal is a highly structure sensitive property. This is best described by Townsend et al. [9.88] in the following words: "The glow peaks are sensitive to the radiation dose, impurities, manufacturer and, even, vary for crystals cut from a single block. For LiF, as many as 28 glow peaks occur. Not even one of the peaks is common to all the samples".

2. The trap depth E (in eV) can be estimated by the formula given by Randall and Wilkins [9.92]:

$$E = 21\, k_B T_G \tag{9.44}$$

   where T_G is the glow peak temperature.

3. Thermoluminescence is due to release of electrons from traps. The correlation of observed glow peaks with various electron centres has to be done carefully. For a discussion of several examples, see review by Deshmukh and Moharil [9.89].

4. Thermoluminescence is useful in radiation dosimetry. LiF doped with (Mg, Ti) has been found to be an efficient device material as dosimeter. The peak at ~ 470 K is used to monitor dosage. The device is known as LiF_{-TLD} 100 (for details, see [9.93, 9.94]).

9.5 Dislocations

9.5.1 Slip (Glide) Systems

Table 9.32 Primary and secondary glide systems in alkali halides [9.95]

Crystal	Primary glide system		Secondary glide system	
	Direction	Plane	Direction	Plane
NaCl Structure				
LiF	$<110>$	$\{110\}$	$<1\bar{1}0>$	$\{001\}$
NaCl	$<1\bar{1}0>$	$\{110\}$	$<1\bar{1}0>$	$\{001\}$
	$<1\bar{1}0>$	$\{110\}$	$<1\bar{1}0>$	$\{111\}$
NaF	$<1\bar{1}0>$	$\{110\}$		
NaBr	$<1\bar{1}0>$	$\{110\}$		
NaI	$<1\bar{1}0>$	$\{110\}$		
KCl	$<1\bar{1}0>$	$\{110\}$		
KBr	$<1\bar{1}0>$	$\{110\}$		
KI	$<1\bar{1}0>$	$\{110\}$		
RbCl	$<1\bar{1}0>$	$\{110\}$		
CsCl Structure				
CsCl	$<001>$	$\{110\}$		
CsBr	$<001>$	$\{110\}$		
CsI	$<001>$	$\{110\}$		

9.5.2 Stacking Fault Energy

Table 9.33 Values of the stacking fault energy (W_{SF})

Crystal	W_{SF} (110) [J m^{-2}]; [9.95]	
	Expt.	Theor.
NaCl Structure		
LiF	0.390	0.330
NaF	-	0.220
NaCl	0.235	0.195
NaBr	-	0.180
NaI	-	0.157

Table 9.33 (Continued)

Crystal	W_{SF} (110) [J m^{-2}]; [9.95]	
	Expt.	Theor.
NaCl Structure		
KCl	0.135	0.161
KBr	-	0.149
KI	-	0.128

9.5.3 Dislocation Mobility Parameters

Table 9.34 Values of the velocity v of mobile dislocations (for definition, see notes and comments); [9.95]

Crystal	Impurity	Velocity range [ms^{-1}]	$m*$
NaCl Structure			
LiF	Mg 80 ppm	10^{-9}–10^{-3}	17
	Mg 3×10^{-4} %	-	1.2
NaCl	Ca 10^{-2} %	10^{-8}–1	17
	Sr 10^{-2} %	10^{-9}–10^{-1}	29.5
	Ca, Mg, Cu < 1 ppm	10^{-7}–10^{-2}	3.9
KBr	Na and Ca 2×10^{-2} %	10^{-9}–10^{-4}	40
	Mg 4×10^{-5} %, Na 7×10^{-3} % and Ca $< 3 \times 10^{-4}$ %	10^{-7}–10^{-2}	16

Notes and Comments

1. The velocity v of mobile dislocations is given by the equation

$$v = B \left(\tau*\right)^{m*} \tag{9.45}$$

where $\tau*$ is the effective stress, B the velocity produced by unit effective stress and $m*$ is a constant [9.96].

9.5.4 Etchants for Observation of Dislocations

Table 9.35 Composition of etchants used for observation of dislocation etch pits along with time of etching; [9.96]

Crystal	Plane	Composition of etchant* (time of etching, wherever reported)
CsI	(100)	1. 100 ml MeOH + 25 mg CuCl$_2$ 2. 5–40 mg CuCl$_2$. 2H$_2$0 in one litre of 96% EtOH 3. 5–100 mg CuCl$_2$. 2H$_2$0 in one litre MeOH

Table 9.35 (Continued)

Crystal	Plane	Composition of etchant* (time of etching, wherever reported)
KBr	(100)	1. HAc (3 sec, rinse CCl_4)
		2. Abs. EtOH (10–15 sec), 50°C
		3. Saturated solution of $PbCl_2$ in abs. EtOH at 50°C (10–15 s) rinse butanol and ether
		4. HPr + 1.75 % $BaCO_3$(10 sec–1 min), rinse petroleum ether
		5. HxOH saturated with CdI_2
	(110)	1. HxOH saturated with CdI_2, (3–5 s)
	(111)	1. HxOH saturated with CdI_2 (5 –10 s)
		2. 85 % HFr saturated with CdI_2
	(112)	1. 85 % HFr saturated with CdI_2 (3 –5 s)
KCl	(100)	1. EtOH 1/3 –1/4 saturated with $BaBr_2$
		2. HAc saturated with Zn ions
		3. MeOH saturated with CdI_2
		4. HxOH saturated with CdI_2
		5. HPr + 1.75 % $BaCO_3$(10 s –1 min)
	(110)	1. HxOH saturated with CdI_2 (1–2 s)
	(111)	1. MeOH saturated with CdI_2
		2. 85 % HFr saturated with CdI_2
	(112)	1. 85 % HFr saturated with CdI_2
KI	(100)	1. PrOH, rinse CCl_4
		2. Pyridine
		3. HxOH saturated with CdI_2
	(110)	1. 85 % HFr
		2. HxOH saturated with CdI_2(1 –2 s)
	(111)	1. HxOH saturated with CdI_2(3 –5 s)
		2. 85 % HFr
	(112)	1. 85 % HFr(3–5 s)
LiF	(100)	1. Solution of HF + HAc (1:1) + 1 % HF saturated with FeF_3, (30–60 s), 25 °C, rinse EtOH, anhydrous ether. Reveals both fresh and aged dislocations
		2. 2×10^{-6} mole fraction FeF_3 in H_2O (1 min). Etches only fresh dislocations
		3. 0.27 g anhydrous FeF_3 + 50 ml HF (10 s)
	(110)	1. 50 % HFr saturated with $FeCl_3$+ 50 % Fluoboric acid, (100–200 s) Reveals aged dislocations
	(111)	1. 85 % HFr
		2. 50 % HFr saturated with $FeCl_3$ + 50 % Fluoboric acid, (15–30 s)
	(112)	1. 50 % HFr saturated with $FeCl_3$ + 50 % Fluoboric acid, (15–30 s)
NaCl	(100)	1. HAc (0.1 s)
		2. 0.1mg $FeCl_3$ in 1 litre Hac (30–120 s), 25 °C
		3. 3 g $HgCl_2$ in 1 litre EtOH (30 s), 25 °C
		4. $CdCl_2$ in EtOH
		5. 85 % HFr (5–10 s)
		6. (1–10) mg $CdCl_2$ in 1 litre MeOH (15 min)

Table 9.35 (Continued)

Crystal	Plane	Composition of etchant* (time of etching, wherever reported)
NaCl	(110)	1. 85 % HFr (5–10 s)
		2. MeOH saturated with CdI_2
		3. 98–100 % HFr (30 s)
	(111)	1. 80 % HFr (1 s)
		2. MeOH saturated with CdI_2
		3. 98–100 % HFr (30 s)
		4. MeOH
	(120)	1. 85 % HFr saturated with CdI_2
	(112)	1. 85 % HFr saturated with CdI_2
	(775)	1. MeOH
		2. 200 ml HAc + 60 ml HNO_3 + 720 mg CdO
NaF	(100)	1. HPr + 1.75 % $BaCO_3$ (30 s), rinse ether
		2. 10 gm $MgCl_2$ or 3g $Mg(NO_3)_2$ in 1 litre H_2O
		3. 4 g $FeCl_3$ in 1 litre HAc (30 s) with stirring, rinse acetone
NaI	(100)	1. HAc saturated with ZnO and $CuSO_4$
RbI	(100)	1. PrOH saturated with BiI_3 or CdI_2 (30 s)
	(110)	1. 85 % HFr (3–5 s)
	(111)	1. 85 % HFr (3–5 s)
	(112)	1. 85 % HFr (3–5 s)

* Abbreviations: MeOH – Methanol; EtOH – Ethanol; PrOH – Propanol, BuOH – Butanol; HxOH – Hexyl alcohol; HFr – Formic acid; HAc – Acetic acid; HPr – Propionic acid. Composition of Liquid Reagents: CH_3COOH 99.9 %; HNO_3 70 %; HF 48 %; HBF_4 50 %.

References

9.1   W. Lehfeldt, Z. Phys. **85**, 717, 1933.

9.2   *Handbook of the American Institute of Physics*, 2nd Ed., McGraw Hill, N. Y.,1963 and references therein

9.3   S. Mizuta and H. Yanagida, Denki Kagaku, **40**, 616, 1972.

9.4   B.P. Aduev, E.D. Aluker, G.M. Belokurov and V.N. Shvayko, phys. stat. sol., **(b) 208**, 137, 1998.

9.5   R.G. Fuller and M.H. Reilly, Phys. Rev. Lett., **19**, 113, 1967.

9.6   L.W. Barr and A.B. Lidiard, *Physical Chemistry – An Advanced Treatise*, Vol. **10** (Acad. Press, New York, 1970).

9.7   S. Chandra and J. Rolfe, Can. J. Phys., **48**, 412, 1970.

9.8   S. Chandra and J. Rolfe, Can. J. Phys., **49**, 2096, 1971.

9.9   S. Chandra and J. Rolfe, Can. J. Phys., **48**, 397, 1970.

9.10   S. Chandra and J. Rolfe, Can. J. Phys., **51**, 236, 1973.

9.11   P.W.M. Jacobs and M.L. Vernon, J. Phys. Chem. Solids, **58**, 1007, 1997.

9.12   P.D. Pathak and N.G. Vasavada, Acta Cryst., **A26**, 655, 1970.

9.13   P.D. Pathak and N.G. Vasavada, Acta Cryst., **A28**, 30, 1972.

9.14   P.D. Pathak, J.M. Trivedi and N.G. Vasavada, Acta Cryst., **A29**, 477, 1973.

9.15   P.D. Pathak and N.M. Pandya, Acta Cryst., **A31**, 155, 1975.

9.16   P.D. Pathak and N.G. Vasavada, J. Phys., **D3**, 1767, 1970.

9.17   P.D. Pathak and J.M. Trivedi, Acta Cryst.,**A30**, 231, 1974 for NaCl type crystals.

9.18   K.G. Subhadra and D.B. Sirdeshmukh, Pramana, **10**, 597, 1978 for CsCl type crystals

9.19   K. Tharmalingam, Phil. Mag., **24**, 359, 1971.

9.20   D.N. Yoon and D. Lazarus, Phys. Rev., **B5**, 4935, 1972.

9.21   G.A. Samara, Phys. Rev., **B22**, 6476, 1980.

9.22   G.A. Samara, *Solid State Physics*, Vol. **38**, Acad. Press, New York, 1984.

9.23   I.D. Faux and A.B. Lidiard, Z. Natururforsch., **26a**, 62, 1971.

9.24   P. Varotsos and D. Miliotis, J. Phys. Chem. Solids, **35**, 927, 1974.

9.25   R.C. Bowman, Jr., Phys. Rev. Lett., **26.**, 1239, 1971 and references therein.

9.26   R.C. Bowman, Jr., J. Chem. Phys., **59**, 2215, 1973.

9.27   S. Bandopadhyay and S.K. Deb, phys. stat. sol., **(b)124**, 545, 1984.

9.28   J. W. Hodby, J.A. Borders, F.C. Brown and S. Foner, Phys. Rev. Lett., 19, 952, 1967.

9.29   J. Appel, *Solid State Physics*, Vol. **21**, Acad. Press, New York, 1968

9.30   F.C. Brown in *Polarons and Excitons*, Ed. C.G. Kuper and G.D. Whitfield, Oliver and Boyd, London, 1963.

9.31   W.D. Compton and H. Rabin in *Solid State Physics*, Vol.**16**, Acad. Press, New York, 1964 and references therein.

9.32   W.B. Fowler in *Physics of Colour Centres*, Acad. Press, New York, 1968.

9.33   S. Radhakrishna and B.V.R. Chowdari, phys. stat. sol., **(a) 12**, 557, 1972.

9.34   S.V. Pappu, Contemp. Phys., **26**, 479, 1985.

9.35   W. Gellermann, J. Phys. Chem. Solids, **52**, 249, 1991.

9.36   J.J. Markham, *F Centres in Alkali Halides*, Acad. Press., New York, 1966.

9.37   E. Mollwo, Z. Physik, **85**, 56, 1933.

9.38   H.F. Ivey, Phys. Rev., **72**, 341, 1947.

9.39   K.G. Bansigir, Ind. J. Pure and Appl. Phys., **5**, 439, 1967.

9.40   H. Rabin and C.C. Klick, Phys. Rev., **117**, 1005, 1960.

9.41   A.E. Hughes, D. Pooley, H.U. Rahman and W.A. Runciman, AERE Report R5604, 1967, data quoted in [9.42]

9.42   P.D. Townsend, J. Phys. C. Solid State Physics, **6**, 961, 1983.

9.43   C.H. Henry and C.P. Slichter in *Physics of Colour Centres*, Ed. W.B. Fowler, Acad. Press, New York, 1968 and references therein.

9.44   M.L. Dalal, S. Sivaraman and Y.V.G.S. Murti, phys. stat. sol., **(b)110**, 349, 1982.

9.45   E. Mollwo, Nachr. Ges. Wissen., Gottingen, **97**, 1931 quoted by [9.46].

9.46   C.C. Klick, *Amer. Inst. Physics Handbook*, 2nd Ed., McGraw Hill Book Co., London, 1963; also [9.39].

9.47   G.A. Russel and C.C Klick, Phys. Rev., **101**, 1473, 1956.

9.48   I.S. Jacobs, Phys. Rev., **93**, 993, 1954.

9.49   H.G. Drickamer, Solid State Physics, **17**, 1, 1965.

9.50   A.Yu Ossipyan and S.Z.Shmurak, *Defects in Insulating Crystals*, Springer-Verlag, Berlin, 1981.

9.51   M.I. Molotskii, Soviet Sci. Rev. B Chem., **13**, 1, 1989.

9.52   B.P. Chandra, H.L. Vishvakarma and P.K. Khare, phys. stat. sol., **(b) 204**, 625, 1997.

9.53   B.P. Chandra, M. Ramrakhiani, P. Sahu and A.M. Rastogi, Pramana, 54, 287, 2000.

9.54   LiF data from M.R. Mayhugh and R.W. Christy, Phys. Rev., **B2**, 3330, 1970; for other crystals from [9.55].

9.55   E.M. Winter, D.R. Wolfe and R.W. Christy, Phys. Rev., **186**, 949, 1969.

9.56   NaF data from K.R. Murali and Y.V.G.S. Murti, Phys. Lett., **A151**, 425, 1990.

9.57   Data on KCl and KBr from I. Schneider and H. Rabin, Phys. Rev., **140**, 1983, 1965.

9.58   H. Rabin, Phys. Rev., **129**, 129, 1963.

9.59   S. Minomura and H.G. Drickamer, J. Chem. Phys., **23**, 290, 1960.

9.60   S.S. Mitra (unpublished).

9.61   S.S. Mitra, R.S. Singh and Y. Brada, Phys. Rev., **182**, 953, 1969.

9.62   G.A. Tauton, R.A. Shatas, R.S. Singh and S.S. Mitra, J. Chem. Phys., **52**, 538, 1970.

9.63   A.A. Maradudin in *Solid State Physics*, Vol. **19**, Acad. Press, New York, 1966 and references therein.

9.64   M.V. Klein, in [9.32].

9.65   J.Z. Damm and Z. Chvoj, phys. stat. sol., **(b)114**, 413, 1982.

9.66   F. Fischer, Zeit. fur Physik, 204, 351, 1967.

9.67   F. Luty in [9.32].

9.68   B. Fritz, F. Luty and G.Rausch, phys. stat. sol., **11**, 635, 1965.

9.69   S. Radhakrishna and B.V.R. Chowdari, phys. stat. sol., **(a)14**, 11, 1972 and references therein.

9.70   F. Rosenberger and F. Luty, Solid State Commun., **7**, 249, 1969.

9.71   H. Okhura and K. Murase, J. Phys. Soc. Japan, **18** (supplement II), 255, 1963.

9.72   H. Okhura, Phys. Rev., **136**, 446, 1964.

9.73   C. Vijayan and Y.V.G.S. Murti, Optics and Laser Technology, **23**, 353, 1991.

9.74   K.L.N. Prasad, Y.V.G.S. Murti, S.V. Rao and S.V. Pappu, Optics and Laser Technology, **14**, 27, 1982.

9.75   C. Vijayan and Y.V.G.S. Murti, Optics and Laser Technology, **23**, 353, 1991.

9.76   K.R. Murali and Y.V.G.S. Murti, Optics and Laser Technology, **22**, 340, 1990.

9.77   R.S. Afzal, I. Schneider, J. Appl. Phys., **69**, 4178, 1991.

9.78   V.V. Ter-Mikirtychev, J. Phys. Chem. Solids, **58**, 365,1997; **58**, 893, 1997.

9.79   I.M. Blair, D. Pooley and D. Smith, J. Phys., **C5**, 1537, 1972.

9.80   M. Aguilar, F. Jaque and J.M. Cabrera, J. Lum., **17**, 217, 1978.

9.81   M.N. Kabler and D.A. Patterson, Phys. Rev. Lett., **19**, 652, 1967.

9.82   D. Pooley and W.A. Runciman, J. Phys., **C3**, 1815, 1970.

9.83   V.T. Skatov, D.I. Vaisburd and L.A. Ploom, Fiz. tverd . Tela, **16**, 3722, 1974.

9.84   K.S. Song, A.M. Stoneham and A.H. Harker, J. Lum., **12/13**, 303, 1976.

9.85   I.L. Kuusmann, D.H. Liblik, G.G. Lijdia, N.F. Lushchik, Ch.B. Lushchik and T.A. Soovik, Fiz. tverd. Tela, **17**, 3546, 1975

9.86   R. Cywinski and J.Z. Damm, phys. stat. sol., **(b)170**, K 61, 1992.
        M. Itoh, M. Kamada and N. Ohno, J. Phys. Soc. Japan, **66**, 2507, 1997).

9.88   P.D. Townsend, C.D. Clark and P.W. Levy, Phys. Rev., **155**, 908, 1967.

9.89   B.T. Deshmukh and S.V. Moharil, Bull. Mat. Sci., **7**, 427, 1985.

9.90   J. Sharma, Phys. Rev., **101**, 1295, 1956.

9.91   M. Riggin, S. Radhakrishna and P.W. Whippey, phys. stat. sol., **(a) 32**, 711, 1975.

9.92   J.T. Randall and M.H.F. Wilkins, Proc. Roy. Soc., Lond., **A184**, 365, 1945.

9.93   S.M.D. Rao, phys. stat. sol., **22**, 337, 1974.

9.94   M.H. Bradburry and E. Lilley, J. Phys., **D9**, 1009, 1976.

9.95   M.T. Sprackling, *Plastic Deformation in Simple Ionic Crystals*, Acad. Press, New York, 1976.

9.96   Compilation by K. Sangwal and A.A. Urusovskaya, Prog. Crystal Growth and Characterisation, **8**, 327, 1984 and references therein.

10　Miscellaneous Properties

10.1　Mass-Related Parameters

Table 10.1 Values of the masses m_A, m_B of atoms in crystal AB, molar mass M, reduced mass μ and mass ratio

Crystal	m_A	m_B	M (m_A+m_B)	μ $m_A m_B / (m_A+m_B)$	mass ratio (m_A / m_B)
	Masses in atomic mass units relative to (1/12) of the atomic mass of ^{12}C				
LiF	6.941	18.998	25.939	5.084	0.365
LiCl	6.941	35.453	42.394	5.805	0.196
LiBr	6.941	79.904	86.845	6.383	0.087
LiI	6.941	126.904	133.845	6.581	0.055
NaF	22.990	18.998	41.988	10.402	1.210
NaCl	22.990	35.453	58.443	13.946	0.648
NaBr	22.990	79.904	102.894	17.861	0.288
NaI	22.990	126.904	149.894	19.464	0.181
KF	39.098	18.998	58.096	12.785	2.058
KCl	39.098	35.453	74.551	18.980	1.103
KBr	39.098	79.904	119.002	26.252	0.489
KI	39.098	126.904	166.002	29.889	0.308
RbF	85.468	18.998	104.466	15.543	4.499
RbCl	85.468	35.453	120.921	25.058	2.411
RbBr	85.468	79.904	165.372	41.296	1.070
RbI	85.468	126.904	212.372	51.072	0.673
CsF	132.905	18.998	151.903	16.622	6.996
CsCl	132.905	35.453	168.358	27.987	3.749
CsBr	132.905	79.904	212.809	49.902	1.663
CsI	132.905	126.904	259.809	64.918	1.047

10.2 Polishing Agents

Table 10.2 Composition of polishing solutions; [10.1]

Crystal	Plane	Composition of polishing solution* (polishing time, polishing rate wherever available)
CsI	(100)	1. MeOH or EtOH + FeCl$_3$
KBr	(100)	1. 73 % MeOH + 3 % EtOH + 22 % glycerol + 2 % NH$_4$OH; (5–10 min)
LiF	(100)	1. 1.5 % NH$_4$OH in H$_2$O; 25 °C; (1–1.5 µm/min) 2. 20 % H$_2$SO$_4$ + saturated boric acid; (1 µm/sec) 3. HCl, 7 sec; 50 °C; (1 µm/sec)
NaCl	(100)	1. EtOH 2. 95 % EtOH + H$_2$O 3. A solution of 90 % MeOH and 10 % EtOH + 1 % H$_2$O; rinse butanol 4. H$_2$O followed by rinsing in MeOH 5. 80 % MeOH + 4 % EtOH + 16 % H$_2$O; rinse butanol 6. MeOH; 5–20 sec; rinse butanol for 30 sec 7. HCl + H$_2$O 8. MeOH + EtOH (1:2); (0.7 µm/min)
	(110)	1. EtOH 2. 50 ml MeOH + 10 ml H$_2$O + 4 g CaCl$_2$
	(111)	1. EtOH
	(775)	1. 50 ml MeOH + 10 ml H$_2$O + 4 g CaCl$_2$
NaF	(100)	1. 15 ml of a solution of 150 mg Mn^{2+} ions/H$_2$O + 2 drops of HF; (3 sec)

*Abbreviations: MeOH – Methanol; EtOH – Ethanol. Percentage composition of liquid reagents used in preparing polishing solutions: H$_2$SO$_4$ 98 %; HCl 38 %; HF 48–49 %.

10.3   Solubility

Table 10.3 Solubility S in water, CH$_3$OH and C$_2$H$_5$OH at temperature T [°C]; [10.2]

Solvent Crystal ↓		S [g per 100 g of solvent]					CH$_3$OH	C$_2$H$_5$OH		
		Water								
NaCl Structure										
LiF	T	18								
	S	0.27								
LiCl	T	0	10	30	80	100	25	25		
	S	63.7	72	85	115	130	42.4	2.5		
LiBr	T	30						0	30	80
	S	59.62						32.6	72.5	99.1

Table 10.3 (Continued)

S [g per 100 g of solvent]

Solvent / Crystal ↓		Water						CH₃OH	C₂H₅OH		
NaCl Structure											
LiI	T	0	19	40	75	99	130	25	25		
	S	151	164	179	263	476	835	343	250.8		
NaF	T	15	25								
	S	4	4.03								
NaCl	T	0	10	30	100	160	215	20	20		
	S	35.57	35.68	36.05	39.22	43.6	46.2	1.41	0.1		
NaBr	T	0	50	100	210				0	20	50
	S	89.5	116	121	150				2.45	2.32	2.26
NaI	T	0	20	40	60	100	140	25	25		
	S	61.3	64.1	67.6	71.9	75.8	76.9	90.35	46		
KF	T	25	100					20			
	S	98	150					4.79			
KCl	T	0	20	50	100	200	300				
	S	28.0	34.37	43.0	56.0	81.4	109.4				
KBr	T	0	20	30	50	100		20			
	S	53.48	65.9	73.3	81.3	105.0		2.08			
KI	T	0	20	40	60	100		25	25		
	S	127.6	144.8	161.1	175.8	206.3		18. 04	2.16		
RbF	T	18									
	S	130.5									
RbCl	T	1	18.7	44.7	114			25	25		
	S	76.4	90.3	106.24	146.5			1.41	0.078		
RbBr	T	0.5	5	39.7	113.5						
	S	89.6	98.0	131.85	205.2						
RbI	T	6.9	17.4								
	S	137	152								
CsF	T	18						15			
	S	356						192			
CsCl Structure											
CsCl	T	0.7	29.9	60.2	119.4			25			
	S	162.3	197.2	229.4	290.0			0.2			
CsBr	T	18									
	S	106.4									
CsI	T	0	14	61				25			
	S	44	66	150				3.25			

Notes and Comments

1. Solubilities at room temperature, generally increase in the sequence fluoride–chloride–bromide–iodide. Such a regularity is not seen in the cation sequence.
2. Sangwal [10.1] pointed out that the surface energy σ and solubility S of alkali halides are related as follows:

$$\sigma = 280\ S^{-1.5} \tag{10.1}$$

10.4 Magnetic Susceptibility

Table 10.4 Values of the observed susceptibility ψ_{obs}, diamagnetic susceptibility ψ_{dia} and paramagnetic susceptibility ψ_{para} at RT

Crystal	Observed [10.3]	Theoretical [10.4]	
	ψ_{obs}	ψ_{dia}	ψ_{para}
	$[10^{-6}$ emu per mole]; uncertainty in $\psi_{obs} \sim$ 2–3 %		
NaCl Structure			
LiF	−10.1	−11.1	1.0
LiCl	−24.3	−27.3	3.0
LiBr	−34.7	−38.3	3.6
LiI	−50.0	−55.6	5.6
NaF	−16.4	−16.4	0.0
NaCl	−30.3	−34.1	3.8
NaBr	−41.0	−44.1	3.1
NaI	−57.0	−60.0	3.0
KF	−23.6	−26.5	2.9
KCl	−39.0	−44.4	5.4
KBr	−49.1	−54.7	5.6
KI	−63.8	−72.0	8.2
RbF	−31.9	−34.4	2.5
RbCl	−46.0	−53.1	7.1
RbBr	−56.4	−63.2	6.8
RbI	−72.2	−81.1	8.9
CsF	−44.5	−46.4	1.9
CsCl Structure			
CsCl	−56.7	−66.1	9.4
CsBr	−67.2	−75.9	8.7
CsI	−82.6	−93.4	10.8

Notes and Comments

1. Values of ψ are listed in [10.5].
2. The observed (ψ_{obs}) values are the resultant of a diamagnetic part (ψ_{dia}) and a paramagnetic part (ψ_{para}). Ruffa [10.4] has theoretically evaluated ψ_{dia} and esti-

mated ψ_{para} from $\psi_{para} = |\psi_{dia}| - |\psi_{obs}|$. The ψ_{dia} and ψ_{para} values given by Ruffa [10.4] are included in Table 10.4.

2. The ψ_{para} values are proportional to the ψ_{dia} values and follow the relation

$$(\psi_{para})_{calc} = -0.11\,(\psi_{dia})_{calc} \tag{10.2}$$

3. Reddy et al. [10.6, 10.7] have shown that a linear relationship exists between the molecular electronic polarisability (α) and the observed magnetic susceptibility (ψ_{obs}); the relationship is :

$$\psi_{obs} = -(8.82 \times 10^{18}\,\alpha + 5.020 \times 10^{-6}) \tag{10.3}$$

10.5 X-ray Monochromator Parameters

Table 10.5 Miller indices (h,k,l) of reflecting planes and corresponding interplanar spacings (d) of alkali halides commonly used as monochromators; [10.8]

Crystal	Reflecting plane			d [Å]
	h	k	l	
NaCl	2	0	0	2.820
KCl	2	0	0	3.140
KBr	2	0	0	3.290
LiF	2	0	0	2.014
	2	2	0	1.424

10.6   Positron Annihilation

10.6.1 Two-Photon Angular Correlation Cut-off Angle

Table 10.6 Values of the cut-off angle θ_c

Halide ion	Ionic radius [Å]	θ_c [10^{-3} radian]	
		Expt. [10.9]	Theory [10.10]
F$^-$	1.33	10.6 (LiF)	11.68
Cl$^-$	1.81	8.0 (NaCl)	8.58
		7.5 (KCl)	-
Br$^-$	1.96	-	7.92
I$^-$	2.20	6.7 (KI)	7.06

Notes and Comments

1. The cut-off angle θ_c is the intercept on the ζ axis of the linear portion of the angular correlation curve p versus ζ, p being the momentum and

$\zeta = (2\pi r_- m_e c / h\sqrt{5})\theta$ with θ as the projected angle between the directions of the two annihilating photons, r_- the halogen ion radius and m_e the electron mass.

10.6.2 Momentum Distribution of Annihilating Electron-Positron Pairs

The momentum distribution of electron-positron pairs (which is also the momentum distribution of two gamma rays) obtained from experiments is shown in Fig. 10.1 and Fig. 10.2 either as $p\,\rho\,(p)$ vs p plots or $N\,(p)$ vs p plots. Here p is the momentum, $N\,(p)$ the momentum distribution and $\rho\,(p)$ the momentum density.

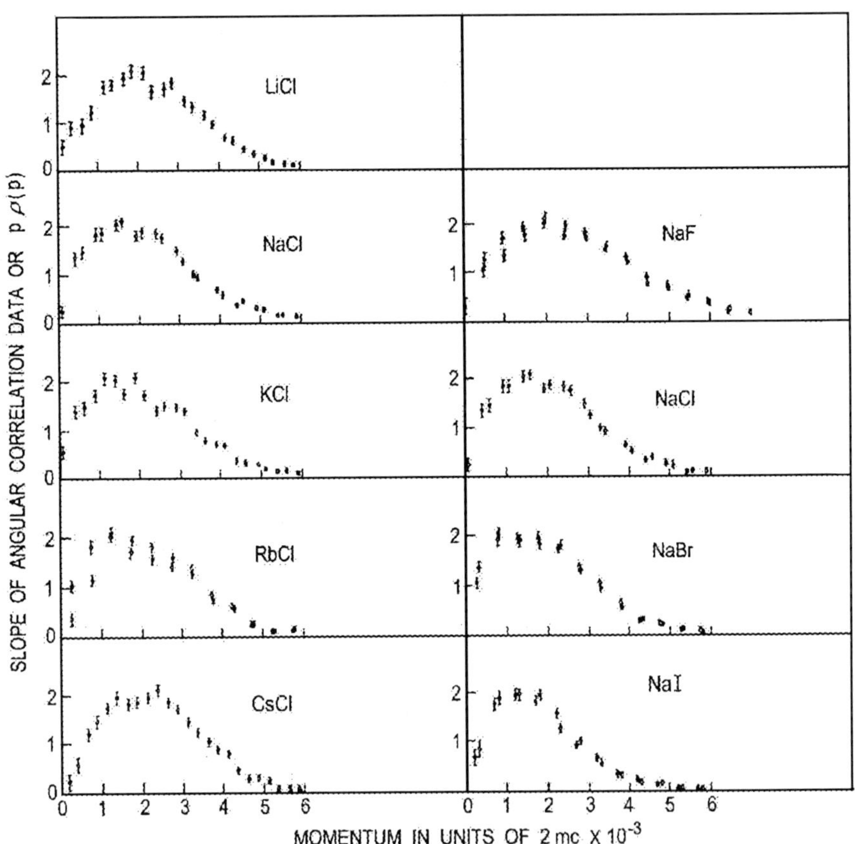

Fig. 10.1 $p\,\rho\,(p)$ versus p plots for alkali chlorides and sodium halides (after [10.11])

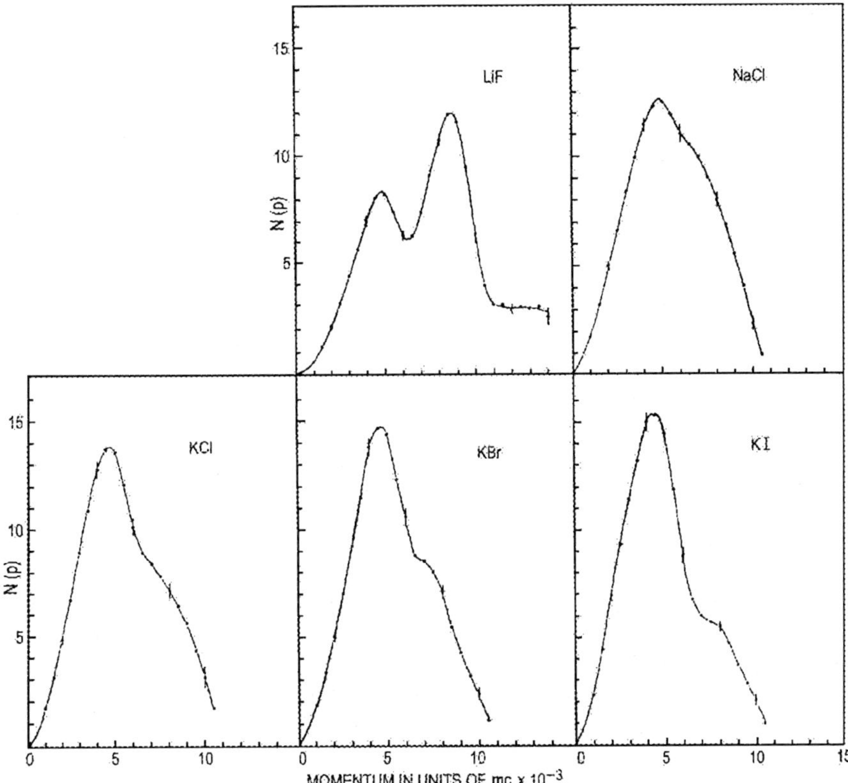

Fig. 10.2 $N(p)$ versus p plots for LiF, NaCl, KCl, KBr and KI (after[10.12])

10.6.3 Positron Annihilation Lifetimes

Table 10.7 Lifetimes τ_1, τ_2, τ_3 and intensities I_1, I_2, I_3 of annihilation spectral components; room temperature data

Crystal	$\tau_1\,[10^{-10}\,\mathrm{s}]$	$I_1\,[\%]$	$\tau_2\,[10^{-10}\,\mathrm{s}]$	$I_2\,[\%]$	$\tau_3\,[10^{-9}\,\mathrm{s}]$	$I_3\,[\%]$	Ref.
NaCl Structure							
LiF	-	-	4.02 ± 0.20	70	2.7	0.8	[10.13]
LiCl	1.8 ± 0.6	-	4.25 ± 0.25	25	-	-	[10.14]
LiBr	-	-	4.58 ± 0.20	85	3.6	1.0	[10.13]
NaF	-	-	5.05 ± 0.35	49	2.5	1.4	[10.13]
NaCl	1.6 ± 0.8	-	4.79 ± 0.29	26	-	-	[10.14]
	3.20 ± 0.22	79	6.70 ± 0.27	17	-	-	[10.15]
	-	-	4.96 ± 0.24	52	2.0	0.9	[10.13]
NaBr	-	-	5.56 ± 0.20	34	3.6	0.9	[10.13]
NaI	-	-	5.11 ± 0.18	61	3.3	1.2	[10.13]
KF	-	-	6.20 ± 0.40	33	2.7	0.4	[10.13]

Table 10.7 (Continued)

Crystal	$\tau_1\,[10^{-10}\,s]$	$I_1\,[\%]$	$\tau_2\,[10^{-10}\,s]$	$I_2\,[\%]$	$\tau_3\,[10^{-9}\,s]$	$I_3\,[\%]$	Ref.
NaCl Structure							
KCl	2.2 ± 0.9	-	6.10 ± 0.45	25	-	-	[10.14]
	-	-	6.07 ± 0.12	50	-	-	[10.13]
	2.89 ± 0.20	62	6.68 ± 0.27	41	-	-	[10.15]
KBr	-	-	6.73 ± 0.22	37	-	-	[10.13]
KI	-	-	6.40 ± 0.21	43	-	-	[10.13]
RbCl	2.2 ± 0.9	-	5.90 ± 0.19	30	-	-	[10.14]
RbI	-	-	6.36 ± 0.18	56	3.9	0.2	[10.13]
CsCl Structure							
CsCl	1.6 ± 0.5	-	6.39 ± 0.18	25	-	-	[10.14]
	-	-	6.39 ± 0.18	25	-	-	[10.13]
	1.64 ± 0.16	29	4.44 ± 0.18	62	12.3 ± 0.50	10	[10.15]
CsBr	-	-	6.84 ± 0.33	31	-	-	[10.13]
CsI	-	-	6.36 ± 0.29	25	3.9	0.8	[10.13]

Notes and Comments

1. The formation of a positronium atom (a positron-electron bound system in free space) is energetically not favoured [10.10]. The first (short lifetime) component corresponds to the annihilation of free positrons. The second component is attributed to annihilation of the positron through the formation of a Wheeler compound: $Cl\,(e^-\,e^+)$ i.e. a complex where the two particles in parenthesis are more closely bound to one another than the third [10.13]. A third component has been observed in some cases but its origin was not understood.
2. Bertolaccini and Dupasquier [10.15] studied the temperature variation of positron annihilation in NaCl, KCl and CsCl over the range 295–1200 K i.e. up to the molten state. They found a slight increase in τ_1 with temperature but did not observe any increase in τ_2. The experimental results of these workers support the A-centre model for the second component of annihilation i.e. a positron captured by a lattice defect (e^+-cation vacancy).

10.6.4 Positron Annihilation Rates

Table 10. 8 The annihilation rates λ_1 and λ_2 for the 1st and 2nd annihilation states and the transition rate k_{12} for transition from state 1 to state 2; [10.16]

Crystal	$\lambda_1\,[10^{10}\,s^{-1}]$	$\lambda_2\,[10^{10}\,s^{-1}]$	$k_{12}\,[10^{10}\,s^{-1}]$
NaCl Structure			
LiF	0.482 ± 0.024	0.337 ± 0.010	0.276 ± 0.037
LiCl	0.329 ± 0.021	0.230 ± 0.007	0.095 ± 0.016
LiBr	0.310 ± 0.025	0.180 ± 0.006	0.027 ± 0.004
LiI	0.286 ± 0.026	0.152 ± 0.005	0.020 ± 0.002

Table 10. 8 (Continued)

Crystal	$\lambda_1 \, [10^{10} \, s^{-1}]$	$\lambda_2 \, [10^{10} \, s^{-1}]$	$k_{12} \, [10^{10} \, s^{-1}]$
NaCl Structure			
NaF	0.384 ± 0.018	0.324 ± 0.009	0.134 ± 0.028
NaCl	0.269 ± 0.020	0.146 ± 0.004	0.050 ± 0.006
NaBr	0.309 ± 0.025	0.134 ± 0.004	0.033 ± 0.004
NaI	0.287 ± 0.026	0.129 ± 0.004	0.018 ± 0.002
KF	0.345 ± 0.021	-	-
KCl	0.275 ± 0.019	0.159 ± 0.005	0.089 ± 0.012
KBr	0.250 ± 0.020	0.132 ± 0.004	0.038 ± 0.005
KI	0.285 ± 0.020	-	-
RbF	0.317 ± 0.018	-	-
RbCl	0.296 ± 0.019	-	-
RbI	0.275 ± 0.017	-	-
CsF	0.351 ± 0.025	-	-
CsCl Structure			
CsCl	0.327 ± 0.024	-	-
CsBr	0.382 ± 0.033	-	-
CsI	0.341 ± 0.025	-	-

Notes and Comments

1. The annihilation rates cannot be directly obtained from the lifetime. They are obtained from data on lifetimes as well as intensities using certain relations which can be found in [10.15].
2. According to Dirac's theory of the positron, the annihilation rate λ_1 is related to other crystal properties through the relation:

$$\lambda_1 = \pi \, r_0^2 \, c \, \xi \, N \, \rho / M \tag{10.4}$$

where r_0 is the classical electron radius, c the velocity of light, ξ the effective number of electrons per ion pair, N the Avogadro number, ρ the density and M the molar mass. The values of ξ are given in Table XXV.

Table XXV Values of the effective number of electrons from Eq. (10.4); [10.15]

Crystal	ξ
LiCl	15.0 ± 0.5
NaCl	16.6 ± 0.6
KCl	21.6 ± 2.0
RbCl	29.4 ± 1.2
CsCl	27.5 ± 2.0

3. Bisi et al. [10.13], found a linear relation between the molar density (ρ / M) and λ_2 for each halide series which may be expressed as

$$\lambda_2 = \lambda_0 + \alpha \, (\rho / M) \tag{10.5}$$

Further, they found

$$\alpha = k \, r_-^{\,2} \tag{10.6}$$

where r_- is the radius of the halogen ion and k is a constant $= 1.90 \times 10^2$ cm s^{-1}.

10.7 Radiation Detection

Table 10.9 Scintillator parameters for alkali halides used as scintillation detectors [10.17]

Crystal Parameter ↓	CsI (Tl)	CsI (Na)	CsI	NaI (Tl)
Radiation length* [cm]	1.86	1.86	1.86	2.59
Decay constant [ns]	900	630	10	230
Photons / MeV	51800	38500	16800	37770
Emission wavelength [nm]	570	420	310	410
Mechanical stability	V. good	V. good	V. good	Fair; cleaves; hygroscopic

* Radiation length is the length of the material needed to attenuate the energy of the incident beam to (1/e) of its intensity. It is given by $(180 \, A / Z^2 \rho)$ where A is the mass number, Z the mean atomic number and ρ the density in g cm^{-3}.

10.8   Surface Properties

Table 10.10 Values of surface tension (τ) and surface energy (σ) for the {100} surface

Crystal	τ [dyne cm^{-1}]	σ [erg cm^{-2}]		
	Theory [10.18]	Theory [10.19]	Experimental	Ref.
NaCl Structure				
LiF	494	142	340	[10.20]
LiCl	624	107	-	-
LiBr	591	86	-	-
LiI	546	73	-	-
NaF	741	221	-	-
NaCl	438	158	276	[10.21]
			310	[10.20]
			400	[10.22]
NaBr	386	138	-	-
NaI	341	118	-	-
KF	495	188	-	-
KCl	264	145	318	[10.20]
			200 ± 50	[10.22]
			110 ± 5	[10.23]

Table 10.10 (Continued)

Crystal	τ [dyne cm^{-1}]	σ [erg cm^{-2}]		
	Theory [10.18]	Theory [10.19]	Experimental	Ref.
NaCl Structure				
KBr	229	130	306	[10.20]
			380 ± 100	[10.22]
KI	191	113	530 ± 130	[10.22]
RbF	427	176	-	-
RbCl	222	140	-	
RbBr	192	125	-	
RbI	176	110		
CsF	371	166		

Notes and Comments

1. Calculated values of τ are also given by Lennard Jones and Dent [10.24], Nicolson [10.25] and Shuttleworth [10.26].
2. Directly determined values of τ are not available. For NaCl, a value of 438 dyne cm^{-1} is estimated from the observed variation of lattice constant with particle size [10.18].
3. The theoretically calculated values of surface energy [10.19] in Table 10.10 correspond to a distorted surface and include a relaxation contribution. These calculated values differ considerably from the values theoretically calculated for an undistorted surface (not given in the table but available in [10.27]).
4. Van Zeggeren and Benson [10.27] showed that the theoretical values of $\sigma\{100\}$ for the NaCl-type alkali halides lie on a smooth curve when plotted against the interionic distance.
5. Sangwal [10.1] showed that the surface energy (σ) of alkali halides with NaCl structure correlates with the hardness (H_v) and the solubility (S) as follows:

$$\sigma = 280\, S^{-1/5} \qquad (\sigma \text{ in erg cm}^{-2} \text{ and } S \text{ in g per 100 cc}) \qquad (10.7)$$

$$\sigma^2 = 1000\, H_v \qquad (\sigma \text{ in erg cm}^{-2} \text{ and } H_v \text{ in kg mm}^{-2}) \qquad (10.8)$$

For the CsCl structure, the $\{100\}$ surface is not stable. Values of σ calculated for the $\{110\}$ surface for the Cs halides by Van Zeggeren and Benson [10.27] are given in Table XXVI.

Table XXVI Values of σ for the $\{110\}$ surface of Cs halides

Crystal	$\sigma\{110\}$ [erg cm^{-2}]
CsCl	244
CsBr	225
CsI	202

10.9   Secondary Electron Emission

Table 10.11 Values of the parameters δ_{max} and $(E_P)_{max}$; for definitions see Notes and Comments; [10.8]

Crystal	Sample form	δ_{max}	$(E_P)_{max}$ [eV]
NaCl Structure			
LiF	crystal	8.5	-
	layer	5.6	-
NaF	crystal	14	1200
NaCl	crystal	14	1200
NaBr	crystal	24	1800
	layer	6	1800
NaI	crystal	19	1300
KCl	crystal	12	1600
KBr	crystal	14	1800
KI	crystal	7.5	1200
RbCl	layer	5.8	-
CsCl Structure			
CsCl	crystal	6.5	-

Notes and Comments

1. δ is the average number of secondary electrons emitted from a bombarded material for every incident primary electron; δ is a function of the primary electron energy E_p. The E_p vs δ plot is a bell-shaped curve which tails off at high E_p. δ_{max} and $(E_p)_{max}$ are values of E_p and δ for which δ is maximum (the peak of the E_p-δ plot).
2. δ_{max} is less by a factor of $\approx$ ½ for samples in the form of layers (films) than for crystals.

10.10   Nuclear Quadrupole Relaxation Time

Table 10.12 Values of the nuclear quadrupole relaxation time (T_1)

Nucleus and Crystal	T_1 [s]		
	Experimental	Theoretical	
	[10.28]	[10.28]	[10.29]
NaCl Structure			
$Na^{23}Cl$	12	30	6.22
$NaCl^{35}$	5.2	5.8	2.43
$Na^{23}Br$	6	10	9.89
$NaBr^{79}$	0.052	0.082	0.059
$Na^{23}I$	5	2.8	8.54

Table 10.12 (Continued)

Nucleus and Crystal	T_1 [s]		
	Experimental	Theoretical	
	[10.28]	[10.28]	[10.29]
NaCl Structure			
NaI^{127}	0.012	0.104	0.056
$K^{39}Cl$	6.2	28	4.19
KCl^{35}	8.5	7.0	1.38
KBr^{79}	0.072	0.133	0.029
KI^{127}	0.019	0.165	0.034
$Rb^{87}Cl$	0.250	0.845	0.136
$Rb^{87}Br$	0.165	0.510	0.052
$RbBr^{79}$	0.065	0.098	0.013

Notes and Comments

1. An experimental value of 10.5 s is reported for T_1 for $Na^{23}I$ by Waber [10.30].
2. A 'fair overall agreement' is found between experimental values and the calculated values reported by Joshi et al. [10.29]. The difference between experimental and theoretical values in some cases is attributed to the neglect of the Kondo-Yamashita overlap mechanism.
3. Calculated values of T_1 for KBr and KI are also given in [10.31].

References

10.1   K. Sangwal, *Etching of Crystals*, North Holland, Amsterdam, 1987.

10.2   *Landolt-Bornstein, Numerical Data and Functional Relationships in Sci. and Tech., New Series*, Ed. K.H. Hellwege and A.M. Hellwege, Group III, Vol. **7**, Part a, Springer-Verlag, Heidelberg, 1973.

10.3   G. Foëx in *Tables de Constantes et Donees Numeriques*, No. **7**, Mason and Cie, Paris, 1957 (quoted in [10.4]).

10.4   A.R. Ruffa, Phys. Rev., **159**, 742, 1967.

10.5   *CRC Handbook of Chemistry and Physics*, 60th Ed. (CRC Press, Boca Raton, Florida, 1979)

10.6   R.R. Reddy, Y.N. Ahammed and M. Ravi Kumar, J. Phys. Chem. Solids, **56**, 825, 1995.

10.7   R.R. Reddy, Y.N. Ahammed, K. Rama Gopal, P. Abdul Azeem and T.V.R. Rao, J. Magn. and Magn. Mat., **192**, 516, 1999.

10.8   *CRC Handbook of Chemistry and Physics*, 76th Ed., CRC Press, Boca Raton, Florida, US, 1995–1996.

10.9   G. Lang and S. DeBenedetti, Bull. Am. Phys. Soc., Ser. II, **1**, 69, 1956.

10.10   R.A. Ferrell, Rev. Mod. Phys., **28**, 308, 1956.

10.11   A.T. Stewart and N.K. Pope, Phys. Rev., **120**, 2033, 1960.

10.12   W.E. Millet and R. Castillo-Bahena, Phys. Rev., **108**, 257, 1957.

10.13   A. Bisi, A. Fiorentini and L. Zappa, Phys. Rev., **134**, A328, 1964.

10.14   A. Bisi, A. Fiorentini and L. Zappa, Phys. Rev., **131**, 1023, 1963.

10.15   M. Bertolaccini and A. Dupasquier, Phys. Rev., **B1**, 2896, 1970.

10.16   M. Bertolaccini, A. Bisi, G. Gambarini and L. Zappa, J. Phys., C: Solid State Physics, **4**, 734, 1971.

10.17   *Optical Crystal Handbook, Optovac*, N. Brookfield, MA, USA, 1993.

10.18   G. C. Benson and K. S. Yun, J. Chem. Phys., **42**, 3085, 1965.

10.19   G. C. Benson, J. Chem. Phys., **35**, 2113, 1961.

10.20   J.J. Gilman, J. Appl. Phys., **31**, 2208, 1960.

10.21   G. C. Benson, H.P. Schreiber and Van Zeggeren, Canad. J. Chem., **34**, 1553, 1956.

10.22   S.L. Norman, E.L. Zwicker and L. I. Grossweiner, Amer. J. Phys., **30**, 51, 1962.

10.23   A.R.C. Westwood and T.T. Hitch, J. Appl. Phys., **34**, 3085, 1963.

10.24   J.E. Lennard Jones and B.M. Dent, Proc. Roy. Soc. Lond., **A121**, 247, 1928.

10.25   M.M. Nicolson, Proc. Roy. Soc. Lond., **A228**, 490, 1955.

10.26   R. Shuttleworth, Proc. Phy. Soc., **A63**, 444, 1950.

10.27   F. Van Zeggeren and G.C. Benson, J. Chem. Phys., **26**, 1077, 1957.

10.28   E.J. Wikner, W. Blumberg and E.L. Hahn, Phys. Rev., **118**, 631, 1960.

10.29   S.K. Joshi, R. Gupta and T.P. Das, Phys. Rev., **134**, A693, 1964; calculations based on Karo's general lattice distribution function for the phonon spectrum and including monopole as well as dipole mechanisms.

10.30   M.J. Waber, Phys.Rev., **130**, 1, 1963.

10.31   B.I. Kocheleev, Sov. Phys., JETP, **10**, 171, 1960.

Index

Anderson-Gruneisen parameter, 97
Absorption coefficient
    linear, 124
    three-photon, 126
    two-photon, 125
Auger-free luminescence, 256

Band gap energy, 190
    pressure derivative of band gap, 191
Band structures
    KBr, 189
    LiF, 187
    NaCl, 188
Born repulsion parameters, 176
Brillouin zones, 2
Bulk damage threshold parameters, 126
Bulk modulus, 25

Circular Dichroism, 241
Colour centre information storage, 253
Colour centre lasers, 254
Colour centres, 221, 235, 264, 265
Compressibility
    at RT and 0 K, 25
Compression data
    equation of state parameters , 34
    Murnaghan equation, 32
    up to 45 kbars (experimental), 31
Coordination numbers, 1, 174

Debye temperature
    at ~ 0 K, 82
    at room temperature, 81
    from compressibility, 81
    from high pressure data on elastic
        constants, 90
    from room temperature elastic
        constants, 81
    from specific heats, 82, 83
    from X-ray/neutron diffraction, 81
    from X-ray diffraction at high
        temperatures, 87

from Mossbauer scattering, 75, 78, 81
pressure variation of, 90
temperature variation at low
    temperature, 83

Debye-Waller factors
    at low temperatures, 77
    at RT, 76
    close to melting point, 77
    temperature variation of, 79
Deformation bleaching, 246
Dielectric polarisability, 145
Diffusion parameters, 226
Dislocations
    dislocation mobility parameters, 261
    etchants for observation of, 261
    slip (glide) systems, 260
    stacking fault energy, 260

Effective ionic charge
    temperature and volume derivatives,
        148
Effective mass, 193
Elastic constants
    second-order elastic constants, 15
    third-order elastic constants, 38
    fourth-order elastic constants, 40
Electric breakdown, 150
Electric field induced Raman spectra,
    165
Electron affinity, 180
Electron density distribution
    electron density projection, 182
    radial distribution curves, 182
Electron energy loss spectra, 217
Electronic dielectric constant, 144
Electro-optic effect, 121
Electrostriction, 149
Enthalpy of formation of a Schottky
    pair, 227
Exciton spectra
    exciton energy, 214
    pressure derivative of , 216

F aggregate centres, 247
F centres
    deformation bleaching, 246
    dissociation energy of, 241
    EPR and ENDOR parameters, 240
    F centre formation energy, 238
    mechanoluminescence, 246
F_A centres, 251
Faraday effect
    dispersion of Verdet constant, 120
    Verdet constant, 118
Faraday rotation, 241
Force constant, 183
Fourth-order elastic constants
    experimental, 40
    theoretical, 41

Gruneisen parameter
    as functions of reduced temperature,
        93
    at room temperature, 91
    from elastic properties, 91, 92
    high temperature limit, 83, 93, 94
    low temperature limit, 83, 93
    mode Gruneisen parameters, 96, 97,
        144, 160
    temperature variation of, 95
    volume derivative, 95

Hardness
    anisotropy, 46
    at room temperature, 44
    Knoop hardness, 46
    Moh hardness, 44
    pressure variation of hardness, 47
    surface hardness, 47
temperature variation of, 46
Harmonic generation
    second harmonic generation, 127
    third harmonic generation, 127
Hyper Raman spectra, 165

Interband transition energy, 193
Interionic distances, 7
Ionic conductivity
    pressure variation of, 228, 231
    temperature variation of, 223–225
Ionic radii, 8
Ionicity
    Pauling, 181, 217

Phillips, 26, 72, 147, 181
IR spectra, 155
    damping constant, 159
    pressure derivative of ν_{TO} frequency,
        159
    pressure variation of ν_{TO} and ν_{LO},
        frequencies 160
    temperature variation of TO
        frequencies, 157
    TO and LO Frequencies, 155

Laser Raman spectra, 163
Lattice constants, 6
Lattice energy
    experimental, 178
    interatomic potentials for, 173
    theoretical, 178
Linear absorption coefficient, 124
Luminescence
    Auger-free luminescence, 256
    intrinsic luminescence, 255
    thermoluminescence, 259

Madelung constant, 2, 16, 25, 174, 215
Magnetic susceptibility, 270
Mass related parameters
    mass ratio, 267
    molar mass, 267
    reduced mass, 267
Mechanoluminescence, 246
Melting temperatures
    at atmospheric pressure, 70
    Lindemann parameter, 75
    melting parameters, 71
    pressure coefficient at zero pressure,
        74
    pressure variation of melting point,
        72, 73
Metallisation and superconductivity, 220
Molar volume, 10
Molecular volume, 10

Neutron inelastic scattering, 166
Nonlinear refractive index, 126
Nuclear quadrupole relaxation time, 278

Optical transmittance, 122

Phonon dispersion relations, 166, 172
Phonon frequencies, 171

Photoelasticity
  dispersion of photoelastic constants, 113
  polycrystalline photoelastic constants, 116
  strain-optical constants, 111–113,
  stress-optical constants, 111–113
Plasma oscillation frequency, 219
Point groups, 1
Polarisability
  electronic polarisabilities, 37, 43, 72, 104, 128
  pressure variation of, 131
  quadrupole and octupole, 133
  strain polarisability constant, 130, 147
Polarons
  effective band mass, 233–235
  polaron coupling constant, 233
  polaron mass, 234
Polishing agents, 268
Polycrystalline elastic moduli
  at RT, 28
  pressure derivatives of, 30
  temperature derivatives of, 30
Polymorphic transitions, 12
Positron annihilation
  lifetimes, 273
  rates, 274
Pressure variation of M, R_2 and N frequencies, 249

Radiation detection, 276
Raman spectra
  electric field induced Raman spectra, 165
  laser Raman spectra, 163
  second order Raman spectra, 162, 164
Reciprocal lattice, 1
Reciprocal vectors, 1
Refractive index
  dispersion equation, 105, 108
  electronic polarisabilities, 37, 43, 72, 104, 128
  density derivative of, 109
  pressure derivative of, 109
  pressure variation of, 110
  temperature derivative of, 107

Schottky defects, 223
Second-order elastic constants, 15
  at room temperature, 15
  Cauchy inequality, 19
  elastic anisotropy, 18, 44
  elastic compliances, 17, 112
  high temperature adiabatic, 21
  low temperature adiabatic, 20
  pressure derivatives of, 24
  temperature derivatives at constant volume, 23
  thermoelastic constants, 22
Secondary electron emission, 278
Solubility, 268
Solution enthalpy of divalent defects, 233
Space groups, 1
Specific heat
  at high temperatures, 55
  at low temperatures, 51, 54
  pressure variation of, 55
Static dielectric constant
  at low temperatures, 138, 139
  at room temperature, 137
  first and second temperature derivatives, 140
  pressure coefficient of, 142
  temperature coefficient of, 139
Surface properties, 276
Szigeti charge, 146

Temperature variation of M band half-width, 248
Thermal conductivity, 64–67
  at room temperature, 64
  at selected temperatures, 65
  of doped alkali halides, 67
  pressure variation of, 66
  temperature variation of, 65
Thermal expansion
  as a polynomial in temperature, 61
  at high temperatures, 60
  at low temperatures, 59
  at room temperature, 56
  at very low temperatures, 58
  pressure variation of, 63
  surface linear expansion, 57
Thermoluminescence, 259, 260
Third-order elastic constants
  experimental, 38
  theoretical, 39
Three-photon absorption coefficient, 126
Two-photon absorption coefficient, 125

U centre
    localised IR bands, 250
    main UV bands, 249
U_2 centres, 251
Unit cell, 2
UV photoelectron and X-ray photon
    emission, 216

Vacancy-impurity dipoles, 232
Valence band width, 192
Van der Waal constants, 175

Velocity of sound
    longitudinal waves, 41
    mean velocity, 41
    second sound velocity, 43
    shear waves, 41

X-ray monochromator parameters, 271
X-ray powder diffraction data, 3

Z centres, 252

Springer Series in
MATERIALS SCIENCE

Editors: R. Hull   R. M. Osgood, Jr.   H. Sakaki   A. Zunger

1 **Chemical Processing with Lasers***
By D. Bäuerle

2 **Laser-Beam Interactions with Materials**
Physical Principles and Applications
By M. von Allmen and A. Blatter
2nd Edition

3 **Laser Processing of Thin Films
and Microstructures**
Oxidation, Deposition and Etching
of Insulators
By. I. W. Boyd

4 **Microclusters**
Editors: S. Sugano, Y. Nishina, and S. Ohnishi

5 **Graphite Fibers and Filaments**
By M. S. Dresselhaus, G. Dresselhaus,
K. Sugihara, I. L. Spain, and H. A. Goldberg

6 **Elemental and Molecular Clusters**
Editors: G. Benedek, T. P. Martin,
and G. Pacchioni

7 **Molecular Beam Epitaxy**
Fundamentals and Current Status
By M. A. Herman and H. Sitter 2nd Edition

8 **Physical Chemistry of, in and on Silicon**
By G. F. Cerofolini and L. Meda

9 **Tritium and Helium-3 in Metals**
By R. Lässer

10 **Computer Simulation
of Ion-Solid Interactions**
By W. Eckstein

11 **Mechanisms of High
Temperature Superconductivity**
Editors: H. Kamimura and A. Oshiyama

12 **Dislocation Dynamics and Plasticity**
By T. Suzuki, S. Takeuchi, and H. Yoshinaga

13 **Semiconductor Silicon**
Materials Science and Technology
Editors: G. Harbeke and M. J. Schulz

14 **Graphite Intercalation Compounds I**
Structure and Dynamics
Editors: H. Zabel and S. A. Solin

15 **Crystal Chemistry of
High-T_c Superconducting Copper Oxides**
By B. Raveau, C. Michel, M. Hervieu,
and D. Groult

16 **Hydrogen in Semiconductors**
By S. J. Pearton, M. Stavola,
and J. W. Corbett

17 **Ordering at Surfaces and Interfaces**
Editors: A. Yoshimori, T. Shinjo,
and H. Watanabe

18 **Graphite Intercalation Compounds II**
Editors: S. A. Solin and H. Zabel

19 **Laser-Assisted Microtechnology**
By S. M. Metev and V. P. Veiko
2nd Edition

20 **Microcluster Physics**
By S. Sugano and H. Koizumi
2nd Edition

21 **The Metal-Hydrogen System**
By Y. Fukai

22 **Ion Implantation in Diamond,
Graphite and Related Materials**
By M. S. Dresselhaus and R. Kalish

23 **The Real Structure
of High-T_c Superconductors**
Editor: V. Sh. Shekhtman

24 **Metal Impurities
in Silicon-Device Fabrication**
By K. Graff   2nd Edition

25 **Optical Properties of Metal Clusters**
By U. Kreibig and M. Vollmer

26 **Gas Source Molecular Beam Epitaxy**
Growth and Properties of Phosphorus
Containing III–V Heterostructures
By M. B. Panish and H. Temkin

27 **Physics of New Materials**
Editor: F. E. Fujita   2nd Edition

28 **Laser Ablation**
Principles and Applications
Editor: J. C. Miller

* The 2nd edition is available as a textbook with the title: *Laser Processing and Chemistry*

Printing: Saladruck, Berlin
Binding: H. Stürtz AG, Würzburg